Global Issues in Water Policy

Volume 8

More information about this series at http://www.springer.com/series/8877

Agnes C. Rola • Juan M. Pulhin
Rosalie Arcala Hall
Editors

Water Policy in the Philippines

Issues, Initiatives, and Prospects

Springer

Editors
Agnes C. Rola
Institute for Governance and Rural Development, College of Public Affairs and Development
University of the Philippines Los Baños
Los Baños, Laguna, Philippines

Juan M. Pulhin
Department of Social Forestry and Forest Governance, College of Forestry & Natural Resources
University of the Philippines Los Baños
Los Baños, Laguna, Philippines

Rosalie Arcala Hall
Division of Social Sciences, College of Arts and Sciences
University of the Philippines Visayas
Miagao, Iloilo, Philippines

ISSN 2211-0631 ISSN 2211-0658 (electronic)
Global Issues in Water Policy
ISBN 978-3-319-70968-0 ISBN 978-3-319-70969-7 (eBook)
https://doi.org/10.1007/978-3-319-70969-7

Library of Congress Control Number: 2017962567

Printed on acid-free paper

This Springer imprint is published by Springer Nature
The registered company is Springer International Publishing AG
The registered company address is: Gewerbestrasse 11, 6330 Cham, Switzerland

Foreword

Water advocates had warned of an impending water crisis in the Philippines. The projections for 2025 show that in a high-economic-growth scenario, the water balance is predicted to be negative for some regions in the Philippines. The reason behind the crisis could either be because of increased water demand arising from economic growth and population rise or because of decreased water supply associated with watershed degradation and climate change.

Notwithstanding, the Philippine freshwater ecosystems also face severe problems because of pollution. Economic activities have considerably increased the effluents being discharged to water bodies. Domestic sewage has contributed about 52% of the pollution load, while industries account for the remaining 48%, according to recent data. Other causes of water pollution include improper and inefficient operation of landfills and lack of public cooperation on the proper disposal of sewage and solid wastes.

This book is a timely reference for water policy decision-makers. The analysis and recommendations contained in this publication to ensure water security can turn the tide for both declining and degraded water resources.

As the last chapter prescribes, "there is a need to implement policies and corresponding policy instruments in an integrated manner instead of the usual piecemeal approach." As deemed by the authors, the policy implementation requires putting investments into human and institutional capacities as well as on modern technologies. These investments, from the government point of view, are necessary for the water sector's improved planning, proper and efficient implementation of water programs, and sustained management and development of water resources in the country.

Roy A. Cimatu
DENR Secretary

Preface

Rapid population growth and urbanization were observed to cause increasing competition in the use of water, the supply of which is also becoming scarce due to widespread water pollution and degradation of watersheds. According to a 2007 Asian Development Bank report, some Asian countries will face a water crisis in the future due to inadequate or inappropriate water governance mechanisms. Cases all over the world, the Philippines included, have shown that water governance is a very complex process. Decisions about policies, laws, institutional structure, incentives, and capacity development are made by a multilayer of decision-makers: national, regional, local, and even ethnic authorities. Water conflicts have also been observed especially at the local level.

To address this issue, a group of University of the Philippines faculty members conducted a research program that aimed to understand the current status of water governance. Funded by the University of the Philippines during the years 2012 to 2016, the study characterized water governance in urban, urbanizing, and rural areas of the country; conducted an actor-based assessment of water governance at the watershed level; and developed a platform for adaptive collaborative water governance in three sites differentiated by level of urbanization and ease of access to water.

The writing of this book was motivated by the aspiration to share the results of this 4-year study. The original proposal for the book chapters focused on the water policy supporting the water dynamics within the watershed and national levels. While water governance at the watershed level has been observed to be improved by the program, national-level water governance reforms will not be that easy.

This book aims to contribute to the national water governance reforms through incisive analysis of present water policies. Using perspectives from the biophysical and social sciences, this volume maps and examines the current state of the water sector in the Philippines in terms of demand, supply, and uses; legal, institutional, and policy arrangements; sector performance (domestic, industrial, agriculture and fisheries, environment, and multiple-use sectors); initiatives for water supply sustainability and improved water demand management; and prospects to achieve water security. With increasing population, urbanization, land use conversion, and the threats associated with climate change, water policy and its implementation must improve.

Los Baños, Philippines

Agnes C. Rola

Acknowledgment

This book reflects the authors' passion to contribute to the improvement of the Philippine water policy environment. All of the authors have in one way or another been engaged with the water sector as researcher, consultant, and administrator. Each has his/her own thinking, stock knowledge, and practical experience about what ails the sector, and thus, the actual writing of the manuscript was not a difficult exercise. We therefore thank the various authors who gave their time and expertise freely to come up with this volume.

The University of the Philippines (UP) through the Emerging Interdisciplinary Research Program (EIDR) funded most of the research work in this volume. We acknowledge with gratitude the unconditional support of then UP President Alfredo E. Pascual and Vice President for Academic Affairs, Dr. Gisela P. Concepcion, in making this volume a reality.

This volume also benefitted from the contribution of UP colleagues, Dr. Guillermo Q. Tabios lll and Dr. Rex Victor O. Cruz, who wrote two chapters each, and our non-UP colleagues, Dr. Arlene B. Inocencio, Dr. Rafael D. Guerrero lll, and Mr. Antonio "Tony" R. de Vera, whose chapter contributions made the story more complete.

To Professor Ariel Dinar of the School of Public Policy, University of California, Riverside, and the editorial board of the Issues in Water Resources Policy series of Springer, our heartfelt thanks for the encouragement and for the confidence that we can come through with this publication.

Finally, we thank the Springer Nature publication and its staff, particularly Mr. Joseph Daniel, for guiding the editors in the publication process.

The Editors

Contents

List of Figures

List of Tables

List of Boxes

Chapter 1
Water Resources in the Philippines: Overview and Framework of Analysis

Agnes C. Rola, Juan M. Pulhin, and Rosalie Arcala Hall

Abstract This chapter provides the context, setting and framework of water policy analysis for this volume. It gives an introduction to the state of water resources in the Philippines, supply and demand situation and the extent of pollution of water bodies. It presents the framework of analysis that guided the discussion in the book. The framework showcases the water supply sustainability issues vis a vis the growing demand and the needed policy support to make this happen. Factors identified to drive water scarcity are the increasing population and urbanization, weak institutional arrangements in the water sector, lack of policy instruments and weak implementation of environmental policies and laws. The last part outlines the organization of the book.

Keywords Water resources • Water policy framework • Policy instruments • Water institutions • Philippines

1.1 Introduction

Water scarcity in the Philippines does not seem to be a pressing concern because of the perception of too much water, the country being an archipelago. It is composed of an estimated 7107 islands, with a land area of 299,764 km^2 (115,831 mi^2). Its length measures 1850 km, starting from the point near the southern tip of Taiwan and ending close to northern Borneo. Its breadth is about 965 km. The Philippine

A.C. Rola (✉)
Institute for Governance and Rural Development (IGRD), College of Public Affairs and Development, University of the Philippines Los Baños, College, Los Baños, Laguna, Philippines
e-mail: acrola@up.edu.ph

J.M. Pulhin
Department of Social Forestry and Forest Governance, College of Forestry and Natural Re-sources (CFNR), University of the Philippines Los Baños, Los Baños, Laguna, Philippines

R.A. Hall
Division of Social Sciences, College of Arts and Sciences, University of the Philippines Visayas, Miagao, Iloilo, Philippines

A.C. Rola et al. (eds.), *Water Policy in the Philippines*, Global Issues in Water Policy 8, https://doi.org/10.1007/978-3-319-70969-7_1

coastline adds up to 17,500 km. The 11 largest islands contain 94% of the total land area. The largest of these islands is Luzon (about 105,000 km^2 or 40,541 mi^2). The next largest island is Mindanao (95,000 km^2 or 36,680 mi^2). The archipelago is around 800 km (500 mi) from the Asian mainland and is located between Taiwan and Borneo. Three prominent bodies of water surround the archipelago: the Pacific Ocean in the east, the South China Sea in the west and north, and Celebes Sea in the south (PIA 2012).

The Philippines is divided into three island groups: Luzon, Visayas, and Mindanao. The current (2016) population of the Philippines is estimated at 102,638,895, equivalent to 1.37% of the total world population. It ranks number 12 in the list of countries by population (Worldometer 2016). The economy of the Philippines is the 33rd largest in the world (IMF 2016) and is one of the largest economies in ASEAN. The Philippines is also regarded as one of the emerging markets. Based on 2009 official statistics (NSCB 2009), overall poverty incidence in the country increased slightly, reaching 26.5% of the population in 2009, up from 26.4% in 2006 and 24.9% in 2000. Importantly, income disparity between the country's urban and rural populations continued to widen over the period, as urban poverty incidence declined faster than did rural poverty. Results of the 2008 Annual Poverty Indicator Survey (PSA 2008) showed a significantly lower poverty incidence in urban areas than in rural areas (19.7% vs 42.5%). As in other developing countries in the region, urban areas in the Philippines are seen to be offering better employment opportunities and higher incomes than rural areas.

As to form of government, it is a presidential, representative, and democratic republic, meaning that the president of the Philippines is both head of state and head of government within a pluriform multiparty system. The government has three interdependent branches: legislative, executive, and judicial (Philippine Government Official Gazette 2016a). The country has a decentralized form of governance, where power and authority from a central institution have been transferred to lower or local levels. Currently, local government units number 80 provinces, 143 cities, 1491 municipalities, and 42,028 villages/barangays (Dorotan n.d.). Each of these is headed by a local executive.

Per capita water availability in the country has been declining over the years brought about, on one hand, by increased water demand arising from economic and population growth, and, on the other, by decreasing water supply associated with degradation of watersheds (Rola and Francisco 2004) and low surface runoff during dry months in some areas (see, for instance, Peñaranda et al. 2015). In terms of spatial differences across the country, projections for 2025 show that, in a high-economic growth scenario, water balance is predicted to be negative for some regions in the Philippines due to rising water demand in the metropolis such as in Metro Manila (JICA/NWRB 1998).

The Philippine freshwater ecosystem faces severe problems of pollution and rising costs of potable water supply (DENR EMB, 2008). The increase in population and economic activities has considerably increased the effluents being discharged to these water bodies. Other sources of water pollution are inefficient and improper operation of landfills or incinerators and inadequate public cooperation on the proper disposal of sewage and solid wastes. Toxic red tide outbreaks occur regularly, and their frequency is also increasing through time.

The World Resources Institute (WRI) predicted that the Philippines will experience a "high" degree of water shortage in the year 2040 (WRI 2015 as cited in Almaden 2014). It ranked 57th out of 167 countries that are most likely to be water stressed by 2040. The sector that will bear the brunt of water shortage by that year is agriculture, a major component of the country's economy. The study likewise predicted the degree of water shortage for three specific sectors: industrial, domestic, and agricultural. For the Philippines, the predicted degree of water stress for agriculture is the highest among the three. Water stress for the industrial sector and domestic use was graded "medium to high," whereas overall water stress projection for the country is "high," specifically for regions such as Mindanao, which could experience more extreme cases of water shortage than the national average.

The impending water shortage is an outcome of weak governance. This weakness is further revealed through destruction of tropical rainforests and catchment areas, water pollution, excessive extraction of groundwater, general water misuse, and poor resource management. It must be known that the Philippines has a multitude of laws and policies as governance instruments, but structural constraints in water policy in the Philippines persist.

Water as a policy area in the Philippines exhibits the kind of multilayered complexity and fragmentation (Hall et al. 2015). Multiple institutions with hierarchical areas of coverage, varying mandates (regulatory and customary), and sectoral representations inhabit its universe. Participants include state agencies and non-state actors (NGOs and civil society), including the private sector, but their ability to influence outcomes is highly uneven (Malayang 2004). The power of each actor is, in turn, determined by its mandate, resources, and public recognition of its legitimacy. There are more than 30 agencies (national, catchment-based, private, and local) managing the water resources of the Philippines.

Coordination among these agencies is not observed (Rola et al. 2015, 2016). Some national agencies have no field presence in many localities. There are some catchment-based formations, but these are mostly loose and informal. There is no nationally legislated funding scheme for water resource management. As such, there is a national policy, but plural governance schemes at the local level exist, many of which have customary bases. In addition, local water managers have a poor understanding of the formal laws (Hall et al. 2015).

1.2 State of Water Resources in the Philippines

The Philippines is an archipelagic country with an average annual rainfall of 2500 mm. It is endowed with vast water resources (Fig. 1.1) consisting of marine (bays), inland waters (rivers and lakes), and groundwater with a total area of 2,257,499 km^2. These resources are vital for the welfare of its people and the country's economic development. Groundwater reservoirs have an estimated storage

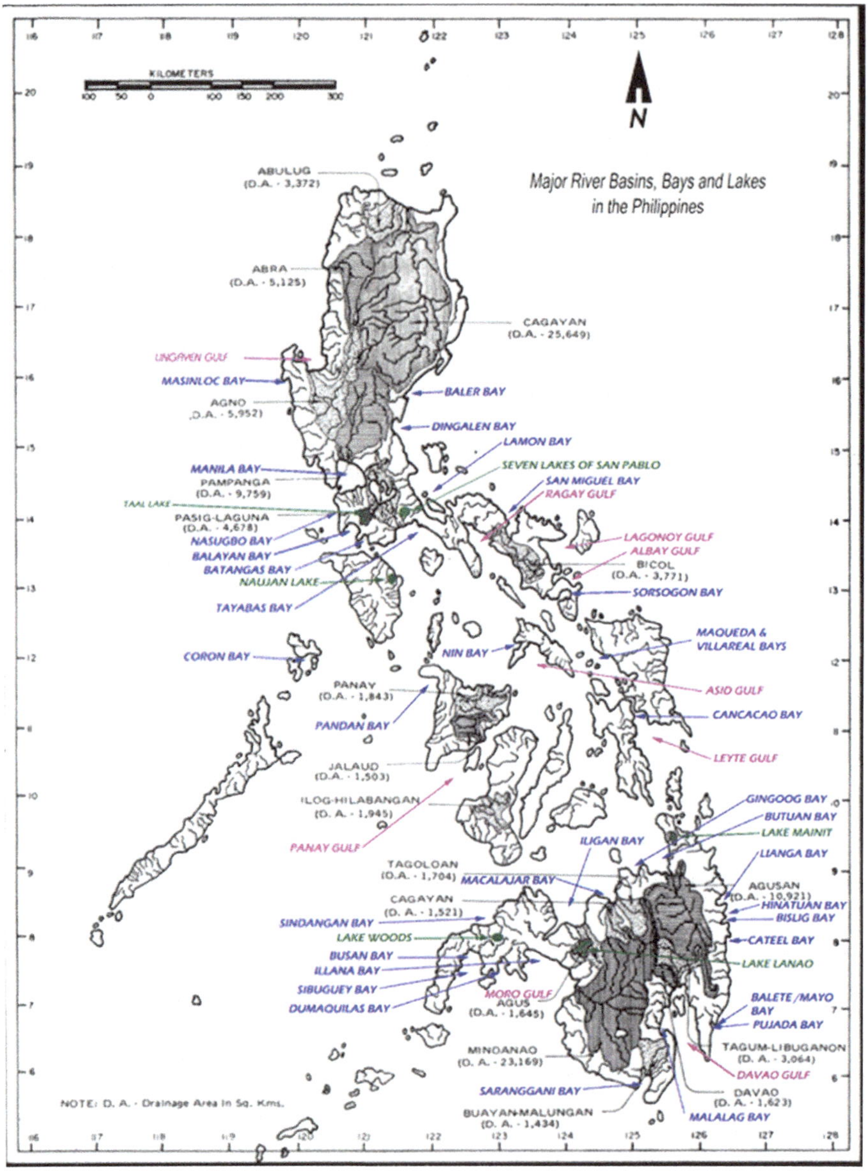

Fig. 1.1 Major rivers, dams, bays, and lakes in the Philippines (Source: WEPA n.d.)

capacity of 251,100 million m^3 and a dependable supply of 126,000 million m^3 per year (DENR 2016).

The country's surface water use is largely devoted to agriculture, with irrigation, livestock, and fisheries representing 86% of total water use. Industry and domestic sectors both comprise the rest of the total. On the other hand, groundwater is distributed according to these uses: 63% for domestic consumption, 17% for industry, 13% for agriculture, 1% for power generation, and 6% for other sectors (Gamolo 2008).

1.2.1 Water Supply and Demand

Global projection states that, for 2025, only 10% of total renewable water shall have been withdrawn. Yet, there are spatial variations in water supply conditions. In the Asia-Pacific region, only a small portion of renewable water sources can be tapped, even if, statistically, per capita annual use of 400 m^3 is only 12% of the 3360 m^3 per capita renewable water resources in the area (Webster and Le-Huu 2003). This pattern was also noted in the Philippines, where annual water use accounts for only 12% of available supply (FAO 2002). Viewed in isolation, this figure tends to suggest that the need to manage water use and conserve water resources in the region and in the Philippines, in particular, is not a pressing concern. Several facts, however, quickly dispel this notion (Rola and Francisco 2004).

First, per capita water availability has been declining over the years (Webster and Le-Huu 2003). This situation is brought about, on one hand, by increased water demand arising from economic and population growth, and, on the other, by decreased water supply associated with degradation of forest watersheds in the country. Second, the data on aggregate availability are illusory in that they indicate *average* supply per capita per year, without regard to the distribution of available supply. True availability is contingent on time, place, quality, and cost. The Philippines, like all other Asian developing countries, has regions and times of year in which water for specific uses is scarce (Fig. 1.2).

1.2.2 Trends in Water Supply and Demand

Available data also show that water demand is increasing and water supply is dwindling. This can be seen by analyzing trends of surface and groundwater supply and demand. Water demand, while considered stable, is actually increasing in terms of per capita use (NSCB 2004; Bagares et al. 2012).

1.2.3 Trends in Water Use, by Sector

As a proxy for water use, data on water permits issued and the corresponding volume of water allocated were examined for the periods 2006–2010 and 2011–2014 (see Chap. 2). There was an increase in total water permits issued and in the volume of water allocated. In terms of sectoral water use, irrigation remained to be a major

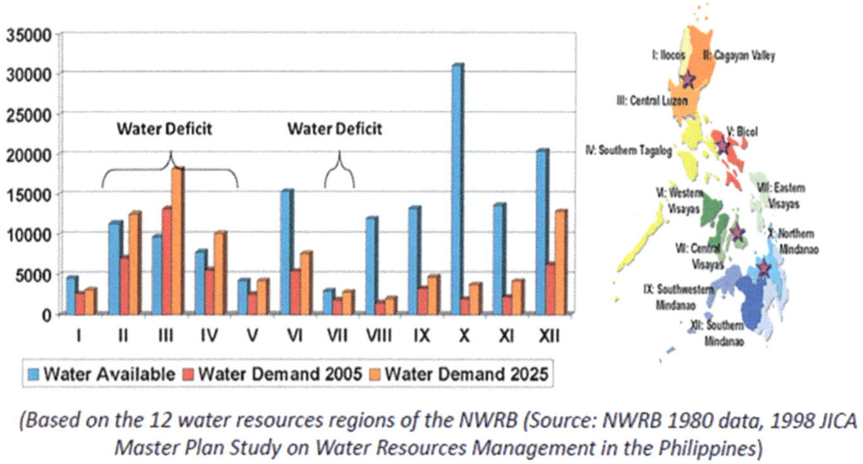

Fig. 1.2 Projected supply and demand of water (‘000 m^3) in various regions (2005–2025) (Source: JICA/NWRB 1998)

water use, followed by municipal or domestic water use. Industrial water use was also becoming prominent. Future conflicts are expected, brought about by competition between domestic water use and irrigation, the latter being tied to the major goal of food security.

1.2.4 Extent of Water Pollution

Of the 688 classified water bodies in the country, only about 27% have potable water. Many of the major rivers and lakes are heavily polluted. Of the 40 water bodies monitored as sources of drinking water supply, only 28% satisfied the criterion established for total suspended solids, signifying the effects of sand and gravel-quarrying activities and runoff sediments from denuded forest and agricultural land (DENR-EMB 2014). Domestic sewage has contributed about 52% of the pollution load, while industries account for the remaining 48% (NSCB 2006). On the other hand, pollution of groundwater occurs when contaminants coming from domestic wastewater, agricultural runoffs, and industrial effluents reach the aquifer or water table in the form of leachates (Ancheta et al. 2003). Of these, domestic wastewater is the main cause of bacterial contamination to groundwater supply. Water-borne diseases such as diarrhea, cholera, dysentery, hepatitis A, and others can be caused by the presence of coliform bacteria in drinking water.

Saline water intrusion has likewise emerged as a problem in some areas, reducing the availability of groundwater supply. This is caused by overexploitation or excessive withdrawal of groundwater. As salt water enters the water table, water availability for domestic usage, including drinking and agricultural usage, is reduced. This poor water quality (see also Chap. 5) plus the dwindling water supply due to environmental degradation (Chap. 9) and climate change can result in water scarcity looming in the horizon.

1.3 Factors Affecting Water Scarcity: A Framework of Analysis

The Philippines faces water shortage due to a variety of factors: population increase and urbanization constitute the direct drivers of change, and then there is weak water governance characterized by fragmented and multiple institutional arrangements within the water sector, lack of effective policy instruments, and weak enforcement of environmental protection policies.

1.3.1 Population Growth and Rapid Urbanization as Drivers of Water Demand

The population of the Philippines when it was first surveyed in 1903 was 7.6 million people. In merely two decades, the national population grew from 60 million in 1990 to 92 million by 2010.

In addition, as early as 1980, urbanization (proportion of urban population to total population) reached 37.2%; by 1990, it was 48.6%. In 2000, urban areas were home to 48.1% of the country's 76.5 million inhabitants. By 2010, those areas accounted for 45.3% of a total population of 92 million. While annual population growth rate has declined over a 25-year period, urban growth rate remains higher than national average due to high birth rates, in-migration, and, to some extent, the reclassification of local government units (LGUs).

Rapid increase in population and creeping urbanization gave rise to conflicting demands for water for domestic, agricultural, and industrial purposes (see also Chap. 2). Trends show that overall access of the population to water supply services and sanitation facilities has declined, along with the ability of water service providers to expand their services. Both the World Health Organization (WHO) and the United Nations Children's Fund (UNICEF) had reported a decline in overall access to improved water services in the Philippines, from 87% in 1990 to 85% by 2002 (Almaden 2014).

For households without formal access to safe water, the alternatives are self-provision (e.g., private wells, fetching from river/spring systems) or availing of services of informal providers such as small-scale independent providers, entrepreneurs with water tankers, or neighborhood water vendors (Hall et al.2015).

1.3.2 Weak Institutional Arrangements Within the Water Sector

Water decisions in the country are made and implemented under multilayered and fragmented institutional arrangements and policy complements (see Chap. 3). The Water Code of the Philippines (PD 1067) is the overarching law that governs water access, allocation, and use. However, there are at least eight legal frameworks that govern the water sector in the Philippines; in some instances, there are conflicting provisions (Philippine Government Official Gazette 2016a, b, c) that make matters worse.

Water governance is also a mandate of multiple water institutions that were a result of the creation of offices corresponding to new laws but without the old systems being abolished. There are 30 agencies (national and LGU-based) managing the water resources of the Philippines (Tabios and Villaluna 2012). Their regulatory mandates cover water quality and quantity, water resource and water services. Institutional concerns as expected are also varied: water sanitation and quality, watershed management, integrated area development, data collection, flood management, irrigation, hydropower, water supply, research and cloud seeding. While not all are present at every locality, the sheer number of potential actors and the assumed plurality of mandates (no mandate is deemed a priority over the others) make for serious political inertia in terms of getting the job done. As a consequence of fragmented and overlapping functions, there is no central planning body for water; no available integrated water data bases for decision support; and there are known conflicts in the enforcement of the various water-related laws (Hall et al. 2015).

In terms of planning for water use, there are also no observed vertical and horizontal linkages among the water-related institutions (Rola et al. 2016). The weak institutional arrangements result in high costs of coordination (Elazegui 2004), politically contingent, and selective application of formal rules at the local level, and inherent tensions between formalization and traditional/customary norms in water provisioning.

Within particular water sectors, institutional concerns are also an issue. For the domestic water sector, the water districts and municipal-government-administered water works play a key role in supply, distribution, and management at the local level (see Chap. 4). Historically, these water institutions have paid less attention to sanitation and focused more on water provisioning. The patterns of water use in the domestic household sector also follow the country's demographic distribution where urban and town centers are favored infrastructure-wise over rural and urbanizing areas. Urbanizing areas suffer from competing uses between households and small-scale industries and the attendant pollution arising from unregulated wastewater-dumping activities. Especially in the urban centers, private companies are now also becoming popular service providers.

With regard to industrial water use, the Philippine Environmental Management Bureau (EMB) regulates the toxic emissions into water bodies. The Laguna Lake

Development Authority (LLDA) regulates the water quality of the Laguna Lake, where industrial agencies abound within its perimeter. Water is a driver for industry both from the perspective of securing stable supply sources as well as from the standpoint of enhancing operating efficiency (see Chap. 5). It is also increasingly recognized by businesses as fundamental to the health of its customers, workers, and communities. In 2010, based on the NWRB reports, industrial users represented the second largest user of water, with agriculture still the sector with the highest water use. Fifteen percent of wastewater generated is attributable to industries.

In irrigation, currently, there are three government entities concerned with irrigation governance: the National Irrigation Administration (NIA), the Bureau of Soils and Water Management (BSWM) under the Department of Agriculture (DA), and the local governments. NIA constructs and manages the large reservoirs/dams. The DA-BSWM is charged with the promotion of small-scale irrigation projects. It provides supplemental irrigation, incidental functions such as flood control structure, and other economic uses like those for fishery and livestock production. With devolution of powers to local governments, LGUs were also given the mandate to construct communal irrigation systems and to build inter-barangay irrigation infrastructure. This multiplicity of institutions hinders the preparation of an integrated irrigation plan (see also Chap. 6).

For the aquaculture sector, the Bureau of Fisheries and Aquatic Resources (BFAR) under the DA is responsible for the development, conservation, management, protection, and utilization of fisheries resources, with regional offices throughout the country. The Local Government Code of 1991 and the Fisheries Code transferred government supervision and licensing of all types of aquaculture to the LGUs. The only licensing function left with BFAR as far as aquaculture is concerned is the granting of fishpond lease agreements for public land (see Chap. 7).

Another government body, the Fisheries and Aquatic Resources Management Councils (FARMC) were created at national level and coastal municipalities to advise the national government and LGUs on fisheries policy and planning. In addition, the government has been increasingly engaging non-government organizations and people's organizations on fisheries co-management. The participation of the business sector in the development and management of the aquaculture industry has not yet been institutionalized or strengthened.

1.3.3 Lack of Effective Policy Instruments

The government has attempted to establish market-based mechanisms founded on a socially conscious system of water tariffs. Practices in water allocation across multiuse are concentrated through public and user-based allocation systems and equity still is the dominant driving force in drawing the country's water policy. This has been justified on account of the public good characteristics of water resources, the large investment required in water infrastructure, and the need to provide for the basic requirements of people in water-scarce regions. While efforts to price water more appropriately to reflect its marginal opportunity cost are already being considered, initiative on its implementation has not been achieved (Almaden 2014).

For domestic water tariff, there are as many forms of pricing as the typology of domestic water suppliers (Rola et al. 2016). This ranges from full-cost pricing by the water district to free water by the community-based systems. The latter is the more dominant form of water service delivery in the country. Self-provisioning is very common, where again water is free. This non-uniformity of water tariffs raises issues of equity in water access and inefficiency in the system.

For industrial users, payment of user fees by virtue of the 'polluters pay' principle is also being implemented in some water bodies; however, water pollution still exists (see also Chap. 5). The utilization of these fees by the regulatory agencies has to be established.

For irrigation, only the farmers served by the NIA pay for the irrigation facility maintenance, called the irrigation service fees (ISF) to cover operation and maintenance expenditures. This fee is used primarily to finance the continuous operation of the irrigation systems (IS). Rates are based on the IS development scheme (run-off-the-river, reservoir, and pump), crops planted, and season. Farmers whose irrigation facility is constructed by the BSWM do not pay any fee. Farmers whose irrigation system was established communally have to pay for the capital outlay and for maintenance of the systems.

As discussed, water-pricing policy is vague in the Philippines. Seemingly, water is not treated as a commodity. Several subsidy schemes reflect the Philippine policy that water is a social good and should be accessible to all as a human right. The policy is payment for the delivery and other maintenance services. Until this day, there is no definite bulk water price as it is deemed that water is abundant and free at the level of source. Increasingly, water transfers, competition for multiple uses (see also Chap. 8), and payments for environmental services are emerging issues as water scarcity becomes real.

1.3.4 Weak Enforcement of Environmental Protection Policies

In the natural resource and environment sector, two earlier laws issued during the Martial Law years provided a sound legal basis for the conservation of water resources. In forestry, the Revised Forestry Code of the Philippines (1975) adopted policies on multiple uses of forest lands and their protection, development, and rehabilitation to ensure the continuity of their productive condition, including water resources (Philippine Government Official Gazette 2016a, b, c). Another presidential decree provided the legal definitions of a watershed and watershed reservation, with the latter "established to protect or improve the conditions of the water yield thereof or reduce sedimentation." Likewise, it provided for the definition of a "critical watershed" and accorded protected status to these areas by prohibiting commercial logging and grazing operations therein. To date, there are 146 proclaimed watershed forest reserves covering a total area of 2,675,687 ha, which constitute about 16.94% of the country's total classified forest land of 15,805,325 ha (FMB 2014).

In spite of these two landmark laws, the state of the Philippine watersheds and water resources has continued to deteriorate through time (see also Chap. 9). Forest cover loss in many watersheds in the country has been severe (Cruz et al. 2010). The ratio of forest cover to irrigated and irrigable lands is generally quite low, and this could have serious implications on soil erosion and the availability and quality of water for irrigation. More recent environmental policies and programs impinge on the water quantity and quality. However, these policies and programs do not explicitly link these environmental initiatives with the water outcomes. The institutional arrangements and mechanisms to promote synergy among the various water-related agencies and overall efficiency in the water sector remains a challenge (see Chap. 10).

1.4 Organization of the Book

Using perspectives from the biophysical and social sciences, this book maps and examines the current state of the water sector in the Philippines in terms of demand, supply and uses; legal, institutional, and policy arrangements; and sector performance (domestic, industrial, agriculture and fisheries, environment and multiple-use sectors). It also analyzes the initiatives for water supply sustainability and improved water demand management as well as the ways forward to achieve water security.

The book is divided into three parts with a total of 12 chapters. The first part (Chaps. 1, 2 and 3) will set the context of the Philippine water sector. The second part (Chaps. 4, 5, 6, 7, 8 and 9) is an assessment of water sector performance. Sectors include domestic, agriculture, aquaculture, industry/commercial, environmental protection, and the multiple uses of water. In this part of the book, each chapter will discuss the state of the sector and the current policy environment, then highlight the problems; assess the sector from the policy/governance lens in terms of different indicators such as efficiency, effectiveness, equity, sufficiency, social profitability, transparency, accountability, and participation, whenever appropriate; and finally ends with a set of policy recommendations for sector improvement. The last part of the book (Chaps. 10 and 11) discusses current and future initiatives for water supply security and demand management, respectively. The last chapter (Chap. 12) maps out the ways forward for a more responsive and integrative water policy in the country.

References

Almaden, C. R. C. (2014). Protecting the water supply: The Philippine experience. *Journal of Social, Political and Economic Studies, 39*(4), 467–493.

Ancheta, C., Hiroyama, M., Illangovan, P., Lorenz, J., Mingoa, Y., Porras, A., Shah, J., Sy, J., Tavares, L. C., Tuyor, J., Verzola, E., Verzosa, D. G., & Villaluz, M. G. (2003). *Philippines environment monitor: Water quality* (p. 42). Manila: The World Bank Group.

Bagares, I. B., Rola, A. C., Bello, R. T., & Elazegui, D. D. (2012). Water rights in the Philippine uplands: Policy and practice. In D. C. Catacutan, A. R. Mercado Jr., M. E. Chiong-Javier, V. B. Ella, M. V. O. Espaldon, A. C. Rola, M. C. Palada, C. Duque-Pinon, J. A. Saludades, A. M. Penaso, M. R. Nguyen, C. T. Pailagao, I. B. Bagares, N. R. Alibuyog, D. Midmore, M. R. Reyes, R. Cajilig, W. Suthumchai, K. Kunta, & S. Sombatpanit (Eds.), *Vegetable-agroforestry systems in the Philippines, Special Publication No. 6b*. Beijing: World Association of Soil and Water Conservation and Nairobi, Kenya: World Agroforestry Center. 457 p.

Cruz, R. V. O., Pulhin, J. M., & Mendoza, M. D.. (2010). *Reinventing CFNR: Leading the way in integrated tropical forest and natural resource management education, research and governance (2011–2025).* UPLB Centennial Professorial Lecture delivered on 14 Dec 2010, College of Forestry and Natural Resources, UP Los Baños, College, Laguna, Philippines.

DENR-EMB (Department of Environment and Natural Resources-Environmental Management Bureau). (2008). *EMB accomplishment report, CY 2008*. Quezon City: DENR-EMB.

DENR-EMB (Department of Environment and Natural Resources-Environmental Management Bureau). (2014). *National water quality status report 2006–2013*. Quezon City: DENR-EMB.

Department of Environment and Natural Resources (DENR). (2016). *Water resources of the Philippines*. Quezon City: Department of Environment and Natural Resources.

Dorotan, E. (n.d.). Decentralization in the Philippines. https://wpqr4.adb.org/lotusquickr/cop-mfdr/PageLibrary482571AE005630C2.nsf/0/7E0AEAB811AEB05348257BA30024A265/$file/Session%202b_PHILIPPINES%20Eddie%20Dorotan.pdf. Accessed 24 Oct 2016.

Elazegui, D. D. (2004). Water resource governance: Realities and challenges in the Philippines. In A. C. Rola, H. A. Francisco, & J. P. T. Liguton (Eds.), *Winning the water war: Watersheds, water policies and water institutions* (pp. 85–104). Makati City: Philippine Institute of Development Studies and Los Baños, Laguna, Philippines: Philippine Council for Agriculture, Aquatic and Natural Resources Research and Development.

FAO (Food and Agriculture Organization of the United Nations). (2002). *Aquastat database*. http://www.fao.org/ag/agl/aglw/aquastat. Accessed 1 Oct 2014.

FMB (Forest Management Bureau). (2014). *Philippine forestry statistics* (pp. 2, 21–24). Quezon City: FMB, Department of Environment and Natural Resources.

Gamolo, N.O. (2008). Philippines 'water resources explained. *The Manila Times*. https://news.google.com/newspapers?nid=2518&dat=20080803&id=p01aAAAAIBAJ&sjid=KSgMAAAAIBAJ&pg=1980,15905370&hl=en. Accessed 16 Jan 2017.

Hall, R., Lizada, J. C., Dayo, M. H. F., Abansi, C. L., David, M. E., & Rola, A. C. (2015). To the last drop: The political economy of the Philippine water policy. *Water Policy, 17*(5), 946–962.

IMF (International Monetary Fund). (2016). World economic outlook (database). http://www.imf.org/external/pubs/ft/weo/2016/01/weodata/weorept.aspx?pr.x=58&pr.y=10&sy=2014&ey=2021&scsm=1&ssd=1&sort=country&ds=.&br=1&c=566&s=NGDPD%2CNGDPDPC%2CPPPGDP%2CPPPPC&grp=0&a=. Accessed 16 Nov 2016.

JICA/NWRB (Japan International Cooperation Agency)/(National Water Resources Board). (1998). *Master plan study on water resources management in the Republic of the Philippines*. Tokyo: Nippon Koei Ltd./Nippon Jogesuido Sekkei, Ltd. 527p.

Malayang, B. S., III. (2004). A model of water governance in the Philippines. In A. C. Rola, H. A. Francisco, & J. P. T. Liguton (Eds.), *Winning the water wars: Watersheds, water policies and water institutions* (pp. 59–84). Makati City: Philippine Institute for Development Studies.

NSCB (National Statistical Coordination Board). (2004). *Compendium of Philippine environment statistics*. http://www.nscb.gov.ph/peenra/Publications/Compendium/CPES%202004.pdf. Accessed 6 June 2016.

NSCB (National Statistical Coordination Board). (2006). *Compendium of Philippine environment statistics*. Makati City: NSCB. 447p.

NSCB (National Statistical Coordination Board). (2009). *PressCon on the 2009 official poverty statistics*. http://www.focusonpoverty.org/download/data/2009%20Poverty%20Statistics%20(NSCB).pdf. Accessed 16 Nov 2016.

Peñaranda, E., Pulhin, J., & Combalicer, E.. (2015). Water balance and land cover change assessments of the Maasin Watershed: implications to local governance. *Philippine Journal of Social Sciences and Humanities, 20*(1).

Philippine Government Official Gazette. (2016a). *Philippine government*. www.gov.ph/about/gov/. Accessed 8 Oct 2016.

Philippine Government Official Gazette. (2016b). Presidential decree No. 1067, s. 1976. http://www.gov.ph/1976/12/31/presidential-decree-no-1067-s-1976/. Accessed 17 Nov 2016.

Philippine Government Official Gazette. (2016c). Presidential decree No. 705, s. 1975. http://www.gov.ph/1975/05/19/presidential-decree-no-705-s-1975/. Accessed 17 Nov 2016.

PIA (Philippine Information Agency). (2012). *General profile of the Philippines*. Quezon City: Presidential Communications Operations Office.

PSA (Philippine Statistics Authority). (2008). *Annual poverty indicators survey (APIS)*. Quezon City: PSA.

Rola, A. C., & Francisco, H. A. (2004). Toward a win-win water management approach in the Philippines. In A. C. Rola, H. A. Francisco, & J. P. T. Liguton (Eds.), *Winning the water war: Watersheds, water policies and water institutions* (pp. 1–26). Makati City/Los Baños: Philippine Institute of Development Studies, Philippine Council for Agriculture, Aquatic and Natural Resources Research and Development.

Rola, A. C., Pulhin, J. M., Tabios, G. Q., III, Lizada, J. C., & Dayo, M. H. F. (2015). Challenges of water governance in the Philippines. *Philippine Journal of Science, 144*(2), 197–208.

Rola, A. C., Abansi, C. L., Arcala-Hall, R., Lizada, J. C., Siason, I. M. L., & Araral, E. K., Jr. (2016). Drivers of water governance reforms in the Philippines. *International Journal of Water Resources Development, 32*(1), 135–152.

Tabios, G. Q., III, & Villaluna, R. A. L. (2012). *Development of the implementation and operational plan for the National Water Resources Management Office* (p. 96). Quezon City: NEDA.

Webster, D. & Le-Huu, T. (2003). *Guidelines on strategic planning and management of water resources*. United Nations Economic and Social Commission for Asia and the Pacific. http://www.asia-water.org/WaterCOntrol%20Backup/library/guideline_spm.pdf. Accessed 1 Oct 2014.

WEPA (Water Environment Partnership in Asia). (n.d.). *State of water environmental issues and policies*. http://www.wepa-db.net/policies/state/philippines/overview.htm. Accessed 16 Jan 2017.

Worldometer. (2016). *Philippine population*. www.worldometers.info/world-population/philippines-population/. Accessed 8 Oct 2016.

Dr. Agnes C. Rola is full Professor at the University of the Philippines Los Baños (UPLB), former Dean of the College of Public Affairs and Development, UPLB, and member of the National Academy of Science and Technology- Philippines. She has degrees in Statistics (BS) and Agricultural Economics (MS) from the UP; and PhD in Agricultural Economics (Major in Natural Resource Economics) from the University of Wisconsin Madison, USA. She attended the Summer Certificate on Environmental Leadership Program at the University of California-Berkeley and has more than 20 years' research experience in sustainable agriculture at the watershed level with a research focus on water governance. With colleagues, she has written and edited an award winning book, "Winning the water wars: watersheds, water policies and water institutions" (2004), whose recommendations were adopted in the Philippines' Clean Water Act. For the past 4 years, she led two major research programs on water in the Philippines, namely, water governance for development and water security under climate risks.

Dr. Juan M. Pulhin is full Professor and former Dean of the College of Forestry and Natural Resources, University of the Philippines Los Baños (UPLB). He earned his Bachelor of Science and Master of Science in Forestry degrees in UPLB and Ph.D. degree in Geographical Sciences from The Australian National University. He was a Visiting Professor at The University of Tokyo for four times and has more than 30 years of experience in natural resources education, research and development. He has authored more than 100 technical publications on various aspects of natural resources management and climate change. He was a Coordinating Lead Author of the Intergovernmental Panel on Climate Change (IPCC) Fifth Assessment Report and a Lead Author of the Fourth Assessment Report. He has been involved in an interdisciplinary research project on water governance for development of the University of the Philippines and in numerous watershed development planning projects since 1998.

Dr. Rosalie Arcala Hall is a Full Professor at the University of the Philippines Visayas (UPV). She completed her Master's degree in Political Science and PhD in International and Public Affairs (2002) at Northeastern University, Boston, Massachusetts. She completed research projects with The Asia Foundation, The Nippon Foundation and Toyota Foundation concerning policy frameworks, their localization, resistance and competing narratives behind contestations in the areas of hard and soft security. She has contributed chapters to the books Rosalina Palanca-Tan ed. Nature and Culture: Environmental Issues in Asia (Ateneo de Manila Press, 2014), and Justine Vaz and Narumol Aphineves eds. Living Landscapes Connected Communities: Culture, Environment and Change Across Asia (Areca Books, 2014). She is co-author to two articles in Water Policy and International Journal of Water Resources Development. Currently, she is a member of the Philippine Commission on Higher Education Technical Committee on Political Science and the Philippine Political Science Association Board.

Chapter 2
Water Supply and Demand and the Drivers of Change

Juan M. Pulhin, Rhodella A. Ibabao, Agnes C. Rola, and Rex Victor O. Cruz

Abstract This chapter synthesizes the existing information and knowledge on the state of water resources in the Philippines by providing a general overview on water supply, demand and uses at the national level. The major sources of water, namely surface and ground water, will be examined in terms of its adequacy considering present and future supply based available studies and projections. Similarly, the demand side will be analyzed considering the sectoral needs and priorities in relation to the present and projected future water supply. Major drivers of change that are likely to shape the water supply and demand scenarios such as demographic shifts, urbanization, and climate change will be highlighted in the analysis. The chapter concludes with the analysis of the gap in the water supply and demand and its implications on the water governance of the country.

Keywords Water supply and demand • Water sustainability • Demographic processes • Urbanization • Climate change

---

J.M. Pulhin (✉)
Department of Social Forestry and Forest Governance, College of Forestry and Natural Re-sources (CFNR), University of the Philippines Los Baños, Los Baños, Laguna, Philippines
e-mail: jmpulhin@uplb.edu.ph

R.A. Ibabao
Department of Management, College of Management, University of the Philippines Visayas, Iloilo City, Philippines

A.C. Rola
Institute for Governance and Rural Development (IGRD), College of Public Affairs and Development, University of the Philippines Los Baños, College, Los Baños, Laguna, Philippines

R.V.O. Cruz
Environmental Forestry Programme, College of Forestry and Natural Resources, University of the Philippines Los Baños, College, Los Baños, Laguna, Philippines

A.C. Rola et al. (eds.), *Water Policy in the Philippines*, Global Issues in Water Policy 8, https://doi.org/10.1007/978-3-319-70969-7_2

2.1 Introduction

The Philippines is well-endowed with abundant water resources, both surface and underground, to meet its water requirements. However, as in many parts of the world, water resources in the Philippines are facing increasing pressure from a combination of naturally occurring conditions and people's actions (UNESCO 2006). Considering water availability against increasing demand, as well as the worsening quality of both surface and underground water, such precious resources are now precariously approaching the critical limit. Even within a low economic development scenario, projections on water availability indicate that water stress will worsen in the future.

More recent literature analyzed the various drivers of change that influence the supply, demand, and long-term sustainability of water resources (Gleick 2013; Chang et al. 2013; Schnoor 2015). Among the major drivers that directly affect water stress and sustainability (these also apply in the Philippines) are demographic factors, which include population growth and rural-urban migration; urbanization and increasing economic activities; land use change; and climate change (Gleick 2013; Chang et al. 2013; Schnoor 2015). Their impacts have a significant bearing on the balance between water demand and supply–usually in uncertain ways–thereby creating new risks for water managers and users (UNESCO 2006, 2012). They also pose a greater challenge for achieving water sustainability.

This chapter synthesizes the current state of knowledge in terms of present and future demand and supply of water resources in the Philippines in the context of the changing times. More than reviewing available literature and project documents, it also attempts to provide updated data and information from various government sources, which are not yet currently available as published materials, in order to have a clearer picture of the water resource situation in the country. Based on water demand and supply assessment, the chapter analyzes the major factors that drive the demand-supply dynamics, which, in turn, shapes the achievement (or non-achievement) of the country's water sustainability goal. It concludes with a brief statement on how to ensure that water supply will meet long-term demand by focusing on the need to better align scientific understanding with management of water resources, specifically in addressing the environmental and socioeconomic stressors that determine water availability and sustainability.

2.2 State of Water Supply and Demand in the Philippines

At the outset, the Philippines' water resources are abundant, considering the current volume of actual renewable water resources, both surface and underground. However, looking at water availability in the face of increasing demand, as well as noting the deteriorating quality of both surface and underground water resources, one may surmise that such resources are persistently approaching their critical limit.

Table 2.1 Groundwater and surface water potential of water-resource regions in the Philippines (million m^3)

Region	Groundwater potential	Surface water potential	Total water resource potential	Percent groundwater to total potential
X Northern Mindanao	2116	29,000	31,116	6.8
VI Western Visayas	1144	14,200	15,344	7.45
IX Western Mindanao	1082	12,100	13,182	8.21
XII Southern Mindanao	1758	18,700	20,458	8.59
XI Southeastern Mindanao	2375	11,300	13,675	17.37
III Central Luzon	1721	7890	9611	17.91
IV Southern Tagalog	1410	6370	7780	18.12
VIII Eastern Visayas	2557	9350	11,907	21.47
II Cagayan Valley	2825	8510	11,335	24.92
V Bicol	1085	3060	4145	26.18
I Ilocos	1248	3250	4498	27.75
VII Central Visayas	879	2060	2939	29.91
Total	20,200	125,790	145,990	13.84

Sources: Ancheta et al. (2003), Data from NWRB (2003)

2.2.1 *Water Sources and Supply*

The major sources of water in the Philippines include rainfall, surface water resources (rivers, lakes, and reservoirs), and groundwater resources. Annual rainfall in the country ranges from 1000 mm to 4000 mm, of which 1000–2000 mm are captured as runoff by natural topography consisting of rivers, lakes, and swamps (NWRB 2003). Tropical cyclones contribute 38% of the annual rainfall in the country, while monsoon rains from the southwest and northeast account for the remainder. Average rainfall is 2348 mm/year. (AQUASTAT 2012) with huge variations: from about 960 mm in General Santos City in southeast Mindanao to more than 4050 mm in the municipality of Infanta in Central Luzon.

The Philippines' actual renewable water resource is estimated to be about 145,990 million m^3 (Table 2.1). Of these, 125,790 million m^3 are surface water, which constitutes around 86% of the country's total water resource potential. Its reliable surface water supply is estimated at 833 million m^3 per day, with only about 28% being consumed (AQUASTAT 2012). In terms of volume, Northern Mindanao has the highest potential source of surface water, whereas Central Visayas has the least.

On the other hand, total groundwater potential for all water resource regions in the country is around 20,200 million m^3 per year, which accounts for about 14% of water resource potential. Groundwater is replenished or recharged by rain and seepage from rivers, making it a renewable resource. The values in Table 2.1 are arranged from lowest to highest percentage of groundwater potential to total potential.

The country's reliable surface water supply is estimated at 833 million m^3 per day, of which only 28% is consumed (AQUASTAT 2012). Surface water supply comes mainly from major watersheds or river basins and lakes. At least 70% of the country's land area is considered as watershed areas. There are 421 principal river basins, 146 of which are proclaimed watershed forest reserves covering a total area of 2,675,687 ha (FMB 2014). From among the principal river basins, 18 are regarded as major river basins with drainage areas of at least 1400 km^2 (Kho and Agsaoay-Saño 2006). They have an aggregate area of around 11.62 million ha (about 36% of the country's total land area).

In spite of this seeming abundance, both surface and groundwater resources of the Philippines are under threat. The 2010 land use and land cover map of the Philippines reflects the critical condition of the major river basins of the country, which poses a great danger to its surface water potential.[1] Barely 25% of these basins are covered with forest vegetation. Six river basins, particularly Bicol, Buayan-Malungan, Ilog-Hilabangan, Jalaur, Panay, and Pasig-Laguna, have less than 10% forest cover. Similarly, significant portions of the total area of the basins are under cultivation (33.45%), which implies that soil and water conservation is a challenge, particularly those in sloping areas and high elevation. The Bicol, Jalaur, Panay, Pampanga, and Mindanao river basins have the largest cultivated land, more than 40% of the basin area. Moreover, about 20% of the total basin area is covered with shrubs, while another 7% is either barren or covered with grass and hence are very prone to forest fires, especially during summer months.

On the other hand, supply of groundwater has been declining through time (NSCB 2004). This may be attributed to unregulated groundwater extraction in many parts of the country. The 2003 Philippine Environmental Monitor published by the World Bank reported the absence of water-right permits in about 60% of groundwater extraction, resulting in indiscriminate withdrawal (Ancheta et al. 2003). This endangers the future supply of a high percentage (86%) of piped water systems that use groundwater as a source. More importantly, it threatens the very well-being of about half the country's population who depend on groundwater for drinking.

[1] Please see Chap. 9, Table 9.3 for details of land use and land cover in river basins in 2010.

Table 2.2 Water demand (million m^3), by sector, 1988–2016

	Demand						
Year	Domestic	% of Total demand	Agricultural	% of Total demand	Industrial	% of Total demand	Total demand
1988	5199.62	12.00	35,736.63	82.45	2404.94	5.55	43,341.19
1989	5318.80	12.01	36,453.76	82.34	2501.96	5.65	44,274.52
1990	5560.06	12.63	36,031.91	81.84	2435.45	5.53	44,027.42
1991	5811.88	13.09	36,030.53	81.16	2549.98	5.74	44,392.39
1992	5949.12	14.59	32,210.34	78.97	2626.15	6.44	40,785.61
1993	6112.07	13.47	36,453.26	80.35	2800.63	6.17	45,365.96
1994	6235.91	13.71	36,314.14	79.83	2941.16	6.47	45,491.21
1995	6411.86	16.04	30,030.82	75.12	3535.37	8.84	39,978.05
1996	6560.80	16.43	29,853.00	74.78	3506.82	8.78	39,920.62
1997	6709.34	16.57	30,228.52	74.64	3559.73	8.79	40,497.59
1998	6857.87	16.31	31,483.35	74.87	3711.81	8.83	42,053.03
1999	7006.41	16.41	31,974.20	74.87	3722.98	8.72	42,703.59
2000	6936.96	16.15	32,236.79	75.03	3789.34	8.82	42,963.09
2016	7073.27	5.3	92,266.59	69.1	34,214.04	25.61	133,581.96

Sources: NSCB (2004), DENR-Compendium of Philippine Environment Statistics (2014), NWRB (2016)

2.2.2 *Water Uses and Demand*

Based on the *2004 Compendium of Philippine Environmental Statistics* (NSCB 2004), the agricultural sector has the highest demand for water resources in the country, ranging from around 31,974 to 36,453 million m^3 annually or about 75–82%, respectively, of the country's total water demand within the 1988 to 2000 period (Table 2.2). This is followed by domestic demand, which constitutes 12–17%. The industrial sector has the lowest demand, only about 6–9% within the same period. In terms of trend, however, a comparative analysis across the three sectors indicates that both domestic and industrial sectors have increasing water demand, while that of agriculture is declining through time. The latest available figures (2016) show that the proportion of agricultural demand has declined and that of industrial demand has increased tremendously, which reveals the high rate of industrialization in the country. The latter may be attributed to continuous conversion of agricultural areas into settlements and diversion to other land uses.

Future projections, however, indicate that water demand in all three sectors will significantly increase by the year 2025 under scenarios of both low and high economic growth (Table 2.3). Even assuming a low–economic-development scenario, only 32% of the anticipated demand by 2025 will be met by the groundwater recharge.

On the other hand, gauged from the volume of water allocated by the National Water Resources Board (NWRB) in terms of water-right permits issued to different users, sectoral water demand has been continuously increasing through time.

Table 2.3 Water demand in the Philippines (million m³/year)

Water demand	1996	2025		% of Total (1996)
		Low	High	
Municipalities	2178	7430	8573	7.27
Industrial	2233	3310	4997	7.46
Agriculture	25,533	51,920	72,973	85.27
Irrigation	18,527	38,769	53,546	61.87
Livestock	107	224	309	0.36
Fishery	6899	14,437	19,939	23.04
Total demand	**29,944**	**62,660**	**86,543**	**100.00**
Groundwater (GW)				
Recharge	20,200	20,200	20,200	
%GW Potential/				
Total Demand	67.46	32.24	23.34	

Sources: NWRB (2003) and JICA, Master Plan Study on Water Resources Management in the Republic of the Philippines (1998)

From a total of 135,313 million m³ allocated to various users (19,190 permits issued in 2006), the allocation increased to 199,706 million m³ issued to 21,459 permit holders in 2014. This represents an 8% increase in 8 years (Fig. 2.1a).

Recent records from the Department of Environment and Natural Resources (DENR) indicate that the power sector has the highest volume of water allocation, representing 57.72% of the total allocation in 2014, which is equivalent to 115,275 million m³ (Fig. 2.1b).[2] This is followed by the irrigation sector, which receives 33.55% (67,005 million m³). The industrial (4.55%) and municipal (3.34%) sectors are the two other sectors with relatively small percentage allocation. The remaining 0.84% is shared by other sectors that include recreational, fisheries, and livestock.

Similarly, viewed from the perspective of demand and supply, total demand for groundwater has increased from 1998 to 2001 by 3197 million m³, while supply has decreased by 76,573 million m³ in the same period (NSCB 2004). This dual pressure from both demand and supply is important since groundwater, as mentioned earlier, is used for drinking by about 50% of the country's population. On the other hand, the demand for surface water declined in the same period, but it was more than compensated for by a decline in its potential supply. Overall, the national annual water demand is catching up with potential supply as gleaned from the increasing percentage of demand over supply, which implies that more and more water resources of desired quality are needed through time.

[2] Although the power sector has the highest volume of water allocation, it is generally non-consumptive and recycled to irrigation and other uses. The data for the power sector therefore overlap with those of irrigation and other water uses.

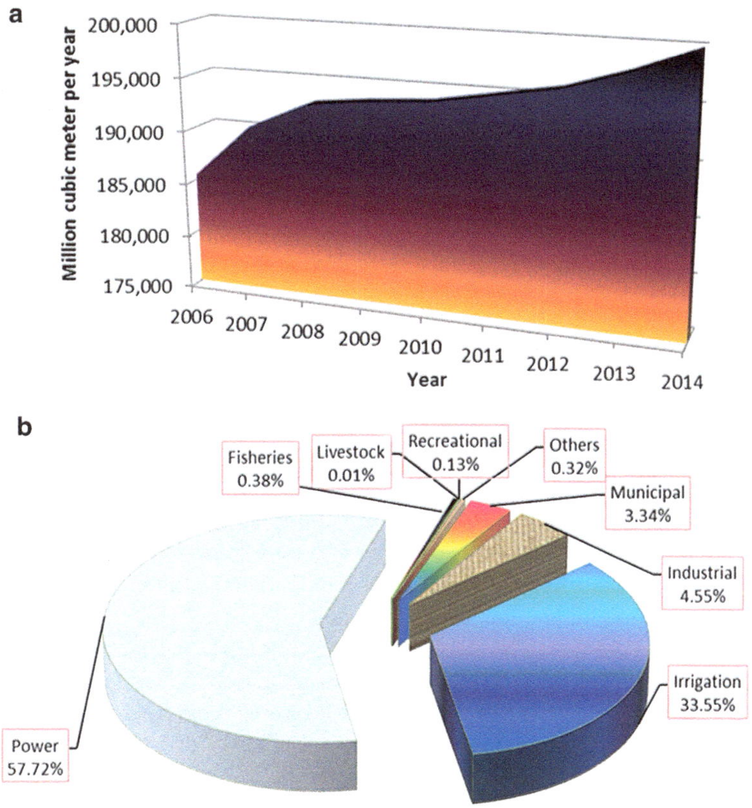

Fig. 2.1 (**a**) Volume of water allocated: 2006–2014. (**b**) Volume of water allocated, by water use, 2014 (DENR 2014)

2.2.3 Water Availability

The country has an estimated annual average runoff of 444 km^3. In 9 out of 10 years, annual runoff exceeds 257 km^3. With an annual average rainfall of more than 2400 mm, there are sufficient surface runoff and groundwater resources (FAO 2012; Kho and Agsaoay-Saño 2006). Theoretically, the high rate of precipitation assures the country of adequate water supply for agricultural, industrial, and domestic uses (Greenpeace 2007). However, due to climate variability and geography, rainfall in the country is highly unevenly distributed across time and space, often resulting in water shortages in densely populated areas, especially during the dry season (Greenpeace 2009).

The Philippines ranks second to Thailand in terms of lowest per capita water availability per year among Southeast Asian countries. It has only 1907 m^3, which is roughly half of the Southeast Asian per capita of 3668 m^3 or close to one-third of the 7045 m^3 global average for the year 2000 (Ancheta et al. 2003). The Lao People's

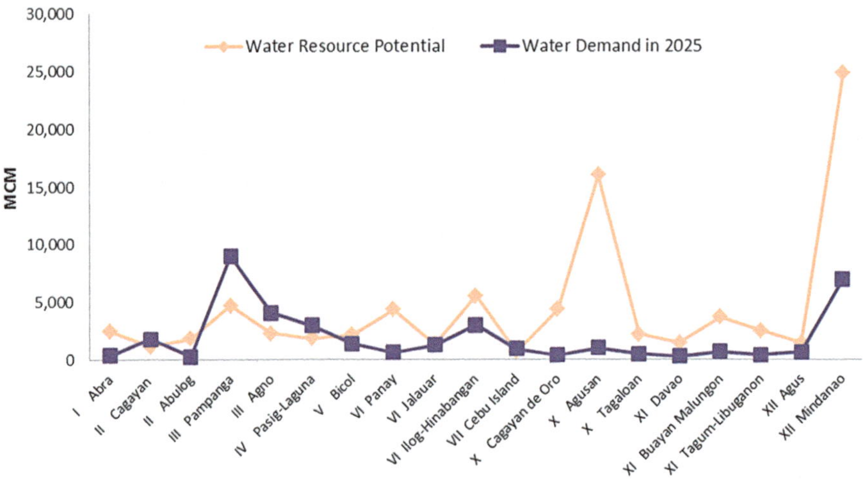

Fig. 2.2 Water potential and demand, by river basin (Ancheta et al. 2003)

Democratic Republic has the highest per capita water availability at 35,049 m^3, more than 18 times higher than the Philippines'.

According to the World Resources Institute (2000–2001), water stress is experienced in areas where per capita water supply drops below 1700 m^3/year, while water scarcity is experienced in areas with per capita water supply below 1000 m^3/year. This means that the Philippines, as a whole, is already close to experiencing water stress. In fact, the 2003 Philippine Environmental Monitor categorizes four river basins in the country as already water-scarce: Pampanga, Agno, Pasig-Laguna, and the island of Cebu (Ancheta et al. 2003). Indeed, the stability and integrity of most river basins are under increasing stress associated with a growing population, poverty, and inefficient governance. These stresses manifest as loss of forest cover, expansion of agricultural areas, prevalence of erosive and pollutive farming practices, unregulated land use conversion, poor solid waste and wastewater management, and excessive surface soil erosion (Cruz 2014).

Future projections pose a major challenge in terms of making water available to meet the increasing demands of the different sectors. In the 1998 Master Plan Study on Water Resource Management in the Philippines conducted by the Japan International Cooperation Agency (JICA), it is projected that, by 2025, even under a low-economic-growth scenario, there will be a deficit in water availability in major river basins in at least six water-resource regions (WRR). These are (1) Pasig-Laguna (WRR IV); (2) Pampanga and Agno (WRR III); (3) Bicol (WRR V); (4) Cagayan (WRR II); (5) Jalaur and Ilog-Hilabangan (WRR VI); and (6) the island of Cebu (WRR VII) in the Visayas (Fig. 2.2). Moreover, seven (7) of nine (9) major cities of the Philippines will experience groundwater supply deficit in 2025—Metro Manila, Metro Cebu, Davao, Baguio, Bacolod, Cagayan de Oro, and Zamboanga. Only Angeles and Iloilo City are not expected to experience such a deficit, although seasonality of sustained supply, particularly in the latter, may be a problem.

2.2.4 Water Demand and Supply

The latest available data show that the amount of groundwater granted to the various sectors is about 19% of total groundwater potential (Table 2.4). Almost all regions still have high groundwater potential. The surface water source, however, is a different story. The total amount of surface water available has reached the negative mark, as demand has surpassed supply, nationally. Five of the 12 WRRs already have negative supply (Table 2.5). Most water for irrigation, industry, and power rely on surface water source.

2.2.5 Water Quality Assessment[3]

The Philippines classifies its water bodies according to uses for easy monitoring. This classification aims to maintain safe quality and satisfactory condition according to best usage. Most existing and future beneficial use of said bodies of water and the land bordering them includes residential, agricultural, aquaculture, commercial, industrial, navigational, recreational, wildlife conservation, and aesthetic purposes.

Water bodies under fresh surface waters (rivers, lakes, reservoir, etc.) are classified into five types based on their best usage (DENR – EMB 2008): (1) Class AA or waters that require only approved disinfection in order to meet the Philippine national standards for drinking water (PNSDW); (2) Class A or waters that require complete treatment (coagulation, sedimentation, filtration, and disinfection) in order to meet the PNSDW; (3) Class B or waters that can be used for primary recreation such as bathing and swimming, skin diving, etc.; (4) Class C or fishery water for propagation and growth of fish and other aquatic resources; and (5) Class D or waters allowed for use in agriculture, irrigation, livestock watering and cooling in industrial facilities.

The coastal and marine water group (coastal, offshore, and estuarine) is classified into: (1) Class SA or waters suitable for propagation, survival, and harvesting of shellfish for commercial purposes and designated as marine parks and reserves; (2) Class SB or waters suitable for bathing, swimming, and skin diving; (3) Class SC described as Recreational Water Class II suited for boating and commercial sustenance fishing; and (4) Class SD waters rated as Industrial Water Supply Class II for cooling purposes in industrial facilities.

Among those classified in 2008 were 283 principal rivers (67.22% of the total 421 principal rivers nationwide) and 340 lakes/small rivers/bays. Class AA, which is the most potable, is found only in very small percentage in six regions. Of the 623 water bodies classified, 36% were in Class C for fisheries and 33% were in Class A

[3] Data drawn from DENR (2000–2008), Compendium of Basic Environment and Natural Resources (ENR) Statistics for Operations and Management (Second Edition).

Table 2.4 Groundwater assessment by region, 2016

Water-resource region	Groundwater potential (million m^3)	Amount of water granted (million m^3)				Total amount of water granted		Amount of water available (million m^3)
		Domestic	Irrigation	Industrial	Power	(million m^3)	(%)	
I	1248	215.167	73.292	6.780	0.073	295.312	24	952.688
II	2825	33.269	30.213	4.347		67.829	2	2757.171
III	1721	362.795	220.875	94.642		678.312	39	1042.688
IV	1410	456.260	102.432	189.035	4.415	752.141	53	657.859
V	1085	122.131	39.504	3.842		165.477	15	919.523
VI	1141	121.119	91.697	42.586		255.402	22	885.598
VII	879	344.220	87.239	36.196		467.654	53	411.346
VIII	2557	369.143	16.122	14.318		399.583	16	2157.417
IX	1082	27.274	2.656	3.952	3.564	37.445	3	1044.555
X	2116	125.986	44.184	54.675		224.845	11	1891.155
XI	2375	184.538	151.236	9.712		345.486	15	2029.514
XII	1758	108.023	69.316	14.891		192.231	11	1565.769
Total	**20197**	**2469.925**	**928.766**	**474.976**	**8.051**	**3881.718**	**19**	**16315.282**

Source: NWRB (2016)

Table 2.5 Surface water assessment by region, 2016

Water-resource region	Surface water (80% dependable flow) in million m^3	Amount of water granted (million m3)				Total amount of water granted (with Power)		Total amount of water granted (w/o power)		Amount of water available (million m^3)
		Domestic	Irrigation	Industrial	Power (Non-consumptive)	(million m^3)	(%)	(million m^3)	(%)	
I	3250	172.828	3447.254	78.102	6123.385	9821.569	302	3698.184	114	−448.184
II	8510	7.090	9361.782	24.642	34150.148	43543.662	512	9393.514	110	−883.514
III	7890	1121.158	16781.985	3816.842	14240.166	35960.151	456	21719.985	275	−13829.985
IV	6370	3068.142	7463.938	3643.232	18974.894	33150.206	520	14175.312	223	−7805.312
V	3060	35.099	2925.161	37.494	805.266	3803.021	124	2997.754	98	62.246
VI	14,200	105.274	5845.925	606.090	3662.461	10219.750	72	6557.289	46	7642.711
VII	2060	9.717	26288.046	24989.543	1854.406	53141.712	2580	51287.306	2490	−49227.306
VIII	9350	23.732	2517.598	195.028	107.794	2844.151	30	2736.358	29	6613.642
IX	12100	25.584	1179.561	4.707	340.072	1549.924	13	1209.852	10	10890.148
X	29000	4.745	4683.802	54.675	21288.420	26031.642	90	4743.222	16	24256.778
XI	11300	34.830	4126.486	199.776	2273.745	6634.837	59	4361.092	39	6938.908
XII	18700	15.151	6716.288	88.936	12315.375	19135.750	102	6820.375	36	11879.625
Total	**125790**	**4623.349**	**91337.827**	**33739.068**	**116136.131**	**245836.375**	**195**	**129700.244**	**103**	**−3910.244**

Source: NWRB (2016)

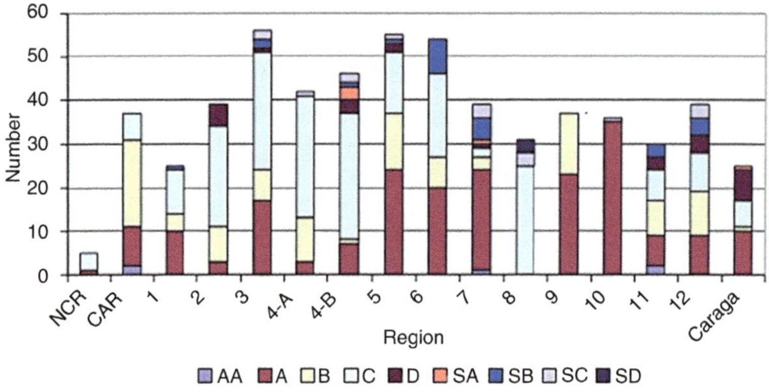

Fig. 2.3 Number of classified water bodies (Including principal and small rivers), CY 2007

(need complete water treatment to be potable) (Fig. 2.3). These figures are a far cry from the situation 50 years ago where river water everywhere is potable and where fishes in these rivers abound.

Biochemical oxygen demand (BOD) and dissolved oxygen (DO) are the main parameters used in assessing water quality. BOD is a measure of the amount of oxygen used by microorganisms to decompose organic waste. The BOD criterion standard is 5.0 mg/L (maximum) for classes A and B, 7.0 mg/L (maximum) for Class C, and 10.0 mg/L (maximum) for Class D. DO is an indicator of how well the water can support aquatic life. The DO criterion standard is 5.0 mg/L (minimum) for Classes AA to C and 2.0 mg/L (minimum) for Class D per DENR Administrative Order 2016–08.

For BOD and DO, regular water quality monitoring is mainly focused on the 19 priority rivers identified under the *Sagip- Ilog* (Save the River) Program. Assessment results revealed that all 19 priority rivers have improved significantly from 2003 to 2008 in terms of DO level. Two rivers seen to have exceeded the BOD standards are found in urbanizing areas, implying that the higher the population density, the lower the quality of river water. This is mainly caused by solid wastes being poured into the body of water and the failure of government to regulate these actions.

Moreover, water quality monitoring of the Pasig River that crosses Metro Manila was also done. For the 2004–2008 period, general water quality was poor, consistently failing to meet BOD and DO criteria for Class C waters. The most polluted parts of this river system are located downstream of the San Juan and Marikina rivers (DENR-EMB 2008; Naz 2012/2013), which are also highly populated areas.

Two other major bodies of water near the metropolis are the Manila Bay and the Laguna Lake. For the Manila Bay area, regular monitoring of the quality of the coastal waters is based on fecal coliform (FC) count. Available data show that the FC counts exceeded the maximum limit; high values are consistently registered, among others, near Luneta Park, a popular tourist attraction in the Philippines. On the other hand, BOD and DO levels of Laguna Lake showed that the lake water is still appropriate for propagation and growth of fish and other aquatic resources (DENR-EMB 2008).

Pollution of groundwater occurs when contaminants coming from domestic wastewater, agricultural runoffs, and industrial effluents reach the aquifer or water table in the form of leachates (Ancheta et al. 2003). Of these, domestic wastewater is the main contributor of bacterial contamination to groundwater supply. Water-borne diseases such as diarrhea, cholera, dysentery, hepatitis A, and others can be caused by the presence of coliform bacteria in drinking water. Based on limited data compiled from various feasibility studies of water districts, LWUA (1990–1997), and the NWRB-NWIN Project, up to 58% of groundwater intended for drinking water supplies (75 out of 129 wells) are contaminated with FC and would need treatment (Ancheta et al. 2003).

Saline water intrusion has likewise emerged as a problem in some areas, thereby reducing the availability of groundwater supply. This is caused by overexploitation or excessive withdrawal of groundwater in coastal areas. As salt water enters into the water table, water availability for domestic (including drinking) and agricultural usage, is reduced. Metro Manila, Cebu, Bulacan, Pampanga, Capiz, and Sorsogon are considered representative areas where a range of problems related to deterioration of water quality in wells yielding saline water is present (NEPC 1987 as cited in PWR 2000, p. 182).

2.3 Drivers of Change in Water Demand and Supply

A study conducted by the World Resources Institute predicts that the Philippines will experience a "high" degree of water shortage in 2040 (WRI 2015). By this time, it would rank 57th (out of 167) in a list of the most water stressed countries in the world. The study defined water stress as "the ratio between total water withdrawals and available renewable surface water at a sub-catchment level"(WRI 2015, p. 3). The sector that will have the worst impact of water shortage by that year is agriculture, a major component of the Philippine economy. Water stress will also be experienced in the industrial and domestic sectors. The study, however, does not reflect future water scarcity for smaller localities. Thus, although overall water stress projection for the country is "high," specific regions such as Mindanao could experience more extreme cases of water shortage than the national average.

Water sustainability is one of the crucial challenges that confront many societies, including the Philippines. It has been defined as *the continual supply of clean water for human uses and for the use of all other living organisms* (Schnoor 2015). Such a definition refers to a sufficient quantity of quality water for the foreseeable future for humans and all biota. Water sustainability is therefore mainly concerned with ensuring that water supply meets changing demand through time. The combined effects of demographic factors, urbanization, land use conversion, and climate change drive many regions of the world to face issues of water scarcity and water pollution, which threaten the long-term sustainability of water resources (Gleick 2003). In the Philippines, four major factors drive water demand and supply that impact on the long-term sustainability of water.

2.3.1 Demographic Factors

Demographic processes such as population growth and migration create some of the greatest pressures on water resource quantity and quality (UNESCO 2009). These processes directly affect water availability and quality through increased water demand and consumption and through pollution resulting from water use. In addition, demographic processes affect water resources indirectly through changes in land use and water use patterns, with significant implications at the local, regional, and global levels.

The demographics of the global population are changing, with important implications for water resources (UNESCO 2009). Asia-Pacific countries now have more working-age people and fewer dependents than at any point in history, providing a springboard for growth (UNDP 2016). Region-wide, 68% of the people are of working age and only 32% are dependents. The Philippine population has been steadily growing for many years (World Population Prospects 2016). It is listed as the 12th most populated country in the world, between Mexico and Ethiopia, at a population of 100,981,437 (World Population Prospects 2016). This is higher by 8.64 million compared with the population of 92.34 million in 2010 and by 24.47 million compared with the population of 76.51 million in 2000. Philippine population increased by an annual average of 1.72% from 2010 to 2015. By comparison, the rate at which the country's population grew during the 2000 to 2010 period was higher at 1.90%.

The country's fertility rate, which pertains to the number of children that a woman wants to have in her lifetime, had historically been going down (Crisostomo 2016). From 1.9 in 2010, it went down to 1.7 in 2015, meaning people are choosing to have fewer children. The increase in the population is driven primarily by 23 million women (ages 15 to 49) who are of reproductive age. In 5–10 years, the country is presumed to have the biggest number of women of reproductive age in history at 25–30 million. This is because there are many 5- to 10-year-old girls who are going to reach reproductive stage in the next 5 years.

A relationship exists between age structure and consumption and production patterns with increasing longevity. This means that with people living longer, there is greater provision for medicine, medical facilities and health care providers (UNESCO 2009). The interaction, however, between age structure and demand and supply on water resources has been inconclusive since there are other factors, such as population density, land degradation or improvement, that can influence the production and consumption of water (Sherbinin et al. 2007).

In terms of mortality, the country ranked 72nd in under-five mortality when the estimated global figure of UNICEF shows that 2000 children under the age of 5 die every day from diarrheal diseases. Of these 2000, some 1800 deaths are linked to water, sanitation, and hygiene (UNICEF-WHO 2012). Moreover, access to water and sanitation systems is a key health issue. Those who have less access tend to have higher rates of disease due to poor drinking water quality and reduced availability of water for hand washing (Giles and Brown 1997). Water-borne diseases are spread when drinking water is contaminated with pathogens from waste matters from infected humans or animals and then ingested by humans. Safe drinking water and good sanitation are effective tools to fight water-borne diseases (Boberg 2005, p. 58).

The same UNICEF-WHO report mentions that the Philippines has an almost 90% of child deaths that are due to diarrheal diseases caused by contaminated water, lack of sanitation, or inadequate hygiene (UNICEF-WHO 2012). An estimated 26% of Filipinos do not have improved sanitation, translating into more than 24 million people. Almost 8 million Filipinos are openly defecating, which is the third highest total in the Asia-Pacific region. Over the last 20 years, the poorest 20% of the rural population went from 36% to 48% open defecation (Matilla 2013). Those in urban areas have better access to water services and toilets than those in rural areas, with the rural poor four times more likely to practice open defecation than those in urban areas. The poor provinces of Masbate and Maguindanao have sanitation coverage that is as low as 38% and 30%, respectively (FIES 2009 as cited in Matilla 2013).

2.3.2 *Urbanization*

Urbanization is an important population trend that affects water resources (Boberg 2005). The process of urbanization has been used in several ways. These include migration from rural areas to urban areas, absolute growth in the urban population (urban growth), and urban growth that is faster than rural growth (UN-DESA 2015).

The level of urbanization affects the level of water use within a country (Boberg 2005). It can influence levels of per-capita use, overtax water resources by concentrating demand in a small area, and overwhelm existing infrastructure. The redistribution of population by migration can also shift pressures on water resources, primarily as a major contributor to urbanization.

In 2014, the urban population accounted for 54% of the total global population, up from 34% in 1960; it is expected to increase to 66% by 2050 (WHO 2016). The urban population growth is concentrated in the developing regions of the world and it is estimated that, by 2017, a majority of people will be living in urban areas. Global urban population is thus expected to grow approximately 1.84% per year between 2015 and 2020, 1.63% per year between 2020 and 2025, and 1.44% per year between 2025 and 2030. This rapid urbanization is caused by an interplay of natural increase, high levels of rural-urban migration and the transformation of rural settlements into cities (Boberg 2005).

Table 2.6 Urban population and level of urbanization, by region, 2007 and 2010

	Urban population (million)		Level of urbanization (%)	
Region	2007	2010	2007	2010
Philippines	35.580	41.856	42.4	45.3
NCR	11.566	11.856	100	100
CAR	0.298	0.425	19.6	26.3
Region 1-Ilocos Region	0.520	0.601	11.4	12.7
Region 2-Cagayan Valley	0.268	0.373	8.8	11.6
Region 3-Central Luzon	4.685	5.233	48.3	51.6
Region 4A-CALABARZON	6.404	7.527	54.5	59.7
Region 4B-MIMAROPA	0.465	0.613	18.2	22.3
Region 5-Bicol Region	0.648	0.831	12.7	15.3
Region 6-Western Visayas	2.206	2.246	32.2	34.7
Region 7-Central Visayas	2.556	2.969	39.9	43.7
Region 8-Eastern Visayas	0.223	0.358	5.7	8.7
Region 9-Zamboanga Peninsula	1.026	1.157	31.8	33.9
Region 10-Northern Mindanao	1.512	1.773	38.3	41.3
Region 11-Davao Region	2.255	2.649	54.2	59.3
Region 12-SOCKSARGEN	1.645	1.911	43	46.5
ARMM	0.731	0.466	17.7	13.7
CARAGA	0.569	0.667	24.8	27.5

Source: PSA (2013)

Urbanization and overall growth of the world's population could add another 2.5 billion people to urban populations by 2050, with about 90% of the increase concentrated in Asia and Africa (UN-DESA 2015). The largest urban growth will take place in India, China, and Nigeria which will account for 37% of the projected growth of the world's urban population between 2014 and 2050.

In 2014, 17 countries in Asia were more than 75% urban, including several of the region's most populous countries, such as Japan (93% urban), Republic of Korea (82% urban), and Saudi Arabia (83% urban) (UN-DESA 2015). China's population has grown to 1.4 billion people in 2014, 54% of whom resided in urban settlements. Indonesia has surpassed the 50% mark where 53% of its 253 million inhabitants resided in urban settlements in 2014.

In the Philippines, the level of urbanization or proportion of urban population to total population was 45.3% in 2010 (PSA 2013). This means that, of the 92.3 million people in the Philippines in 2010, 41.856 million lived in areas classified as urban (Table 2.6). The rural population or those who live in areas classified as rural numbered 50.5 million, accounting for 54.7% of the total population.

Secondary cities, which are growing fast, are putting pressure on urban infrastructure (Navarro 2014), increasing the need to establish more water system facilities to meet the rising water demand of industries and urban population. Of the 17 regions in the Philippines, the CALABARZON corridor (comprising the urban areas of Cavite, Laguna, Batangas, Rizal, and Quezon that are adjacent to Metro Manila) is the most populated area in the country with 12.61 million inhabitants. The population of CALABARZON has surpassed that of the National Capital Region (NCR), which is composed of Metro Manila (11.86 million), and Central Luzon (10.14 million). Further, 33 cities, including all 16 cities in the NCR, are now classified as "highly urbanized cities," four of these being home to more than 1 million inhabitants. These four HUCs include three located in the NCR (Caloocan City [1.49 million], the City of Manila [1.65 million], and Quezon City [2.76 million]), as well as Davao City (1.45 million) in the southern island of Mindanao. Cities in the Philippines are contending with urban problems such as congestion, overcrowding, poor quality of life, and rapidly growing urban poor communities.

While the annual population growth rate has declined over a period of 25 years, the urban growth rate remains higher than the national growth rate due to a high birth rate, in-migration, and, to some degree, the income reclassification of local government units (Navarro 2014). Compared with other countries, the Philippines ranks 11th among countries or areas with declining percentage of urban residents between 1990 and 2014 (UN-DESA 2015).

Rapid urban growth can also affect water quality when formerly vegetation-covered land is changed into pavements and buildings (Boberg 2005). These infrastructure developments can increase the volume of runoff and pollution levels, degrading or eliminating the ability of the land to absorb rainwater and possibly infecting water systems with human wastes.

A report on the Philippines indicated worsening water quality caused by rapid urbanization (ADB 2007). Only about 33% of its river systems are classified as suitable public water supply sources and up to 58% of groundwater is contaminated. Further, water availability will be unsatisfactory in eight of the 19 major river basins and in most major cities before 2025. The depletion of groundwater resources is an increasing problem in Metro Manila and Cebu, and the ability of groundwater to meet future demand is also very limited, amounting to only 20% of the total water requirement of the country's nine main urban centers by 2025. Water quality is poorest in urban areas, the main sources of pollution being untreated discharges of industrial and municipal wastewater.

Although groundwater resources are generally abundant, downstream water courses and aquifers have been polluted by over abstraction and poor environmental management of extractive resource industries such as mining and forestry (ADB 2007). This has caused siltation and lowered water tables. In addition, water pollution, wasteful and inefficient use of water, saltwater intrusion, high non-revenue water levels due to leaks and illegal connections, and denudation of forest cover are placing major strains on water resources, making it more difficult to provide basic water services.

Domestic waste is responsible for 48% of pollutants (ADB 2007). Thirty-seven percent comes from agricultural waste, while 15% comes from industrial waste. Metro Manila is estimated to generate 5345 tons of solid waste per day and only 65–75% is collected and a measly 13% is recycled. Meanwhile, 700 industrial establishments in the country generate about 273,000 tons of hazardous waste annually. The Philippines, however, has no integrated treatment facility to deal with it, although there are around 95 small to medium-scale facilities. Due to lack of proper treatment and landfill facilities, about 50,000 tons of hazardous waste is stored on- or off-site.

Urbanization also affects the level of water use within a country (Boberg 2005). This is particularly true for the domestic and municipal sector, where urbanization—and the infrastructure that often accompanies it—can make a significant difference in per capita use. While domestic water use in many countries is a relatively small part of the freshwater demand burden, densities of urban areas can mean that local demand can be extremely high, outstripping the resources available locally, and localized water shortages.

Rural to urban migration is a major component of urban population growth in developing nations (Population Division of the Department of Economic and Social Affairs of the United Nations Secretariat 2001). Natural disasters, insurgency, and perceived lack of economic opportunities are just some of the negative factors that make people leave an area. In developed nations, the underlying factor for the process of urbanization is industrialization. Migration from rural to urban areas poses a major challenge for city planners. The provision of basic drinking water and sanitation services to peri-urban and slum areas to reach the poorest people is important to prevent outbreaks of cholera and water-related diseases in these often overcrowded places (WHO-UNICEF 2006).

This rural-to-urban migration pattern remains the main migration stream in the Philippines (Tirona 2013). The 2000 census of population and housing showed that about 48% of the Filipinos live in urban areas compared with 37% over two decades ago. The urban population grew rapidly at an annual rate of about 5% from 1960 to 1995, albeit showing a decline of about 3% annually from 1995 to 2000. It is estimated that, by 2030, about eight out of 10 Filipinos will be living in cities and urban areas. Rural-to-rural and urban-to-rural migration flows and counterflows are also evident, giving rise to urban sprawls. Peripheral rural areas receive the spillovers of population from the highly urbanized cities. This is observable in Central Luzon and the CALABARZON regions where rural-urban or "rurban" communities accommodate relocating residents from Metro Manila.

A two-way relationship exists between water and migration. Water stressors influence migration and migration contributes to water stress (UNESCO 2009). Water stressors, such as water scarcity and flooding, can trigger migration decisions. The social, economic, and political contexts in which water stresses occur will influence the migration response. That is, if the natural environment becomes inhospitable, people may move to areas where their locally specific knowledge may no longer apply.

The arrival of more people requires that their places of destination must provide them with water resources (UNESCO 2009). Migration can strain the capacity of the urban infrastructure and aggravate water-related conflicts. It can also upset the fragile balance of human populations and water resources. Climate change, which is predicted to lead to greater frequency and intensity of extreme weather events, is likely to result in an overall increase in the displacement of people in the future.

2.3.3 Land Use Change

Land use changes contribute to water resource conditions (FAO 2000). Between 1988 and 2010, changes in land use and land cover have been largely unregulated as the use of land in major river basins in the Philippines shifted from one major use to other major uses—i.e., forest land, cultivated land, shrub and grassland (Table 2.7). For instance, from a total area of 3,262,407 ha in 1988, forest cover was reduced to 2,913,627 ha in 2010. This represents a total loss of 348,780 ha of forest cover or an annual loss of 15,854 ha in 22 years. On the other hand, barren and grassland more than doubled in the same period from 1,057,856 ha in 1988 to 2,623,210 ha in 2010. Continuous land use conversion to non-forest uses and the degradation of the watersheds will have adverse impacts on the quantity, quality, and water regime, which will, in turn, threaten the sustainability of water supply.

Changes in land use patterns (e.g., conversion of watersheds, rapid urbanization) and increasing discharges of untreated wastes and various pollutants affect the availability of water for human consumption. For example, some rivers in Metro Manila are already heavily polluted and are fit only for navigation. Their potential as sources of water supply is lost. As a result, the Metropolitan Waterworks and Sewerage System (MWSS) has to get its water supply from Angat River, which is located in another river basin.

Land use changes, apart from pollution and increasing siltation, also influence the change in the quality of lakes, rivers, and reservoirs (PWR 2000, p. 196). On a regional basis, the critical problem in the NCR is the poor quality of its surface water, which further widens the gap between increasing demand and declining supply.

2.3.4 Climate Change

Climate change has adverse impacts on the water sector. The Asian chapter of the Intergovernmental Panel on Climate Change (IPCC) Fifth Assessment Report states that water scarcity will be a major challenge for most of the region, including the Philippines, due to the combined impacts of increased water demand and lack of

Table 2.7 Land use and land cover change between 1988 and 2010 in the major river basins in the Philippines (in ha)

Land Cover	1988	2010	Percent change	Remarks
Forest	3,262,407	2,913,627	−10.69	Decline in forest cover mainly attributed to increase in barren areas and grassland (10.63%), shrubs (9.32%), and cultivated areas (4.94%)
Plantation	150,964	38,260	−74.66	Decrease in plantation areas mainly due to increase or conversion to cultivated areas (71.49%), shrubs (11.45%), barren areas and grassland (8.55%), forests (4.95%), and mangrove (1.24)
Cultivated	6,753,565	4,494,350	−33.45	Reduction in cultivated areas mainly due to increase in barren areas (23.6%), shrubs (9.38%), and forests (6.8%)
Shrubs	0	1,024,386		Shrubs mostly came from cultivated and forested areas
Barren and grassland	1,057,856	2,623,210	147.97	Increase in barren and grassland areas (61.04%) mainly from cultivated areas (25.9%), shrubs (6.2%), and forests (5%)
Mangrove	10,452	13,843	32.44	Increase in mangrove areas (47.13%) mainly from inland water (26.12%), cultivated (18%), barren areas and grassland (6.14%), forests (1.3%), and shrubs (1.15%)
Marshland	99,825	126,930	27.15	Increase in marshland (73.49%) mainly from cultivated (19.4%), inland water (5.83%), and shrubs (1.08%)
Inland water	283,971	384,433	35.38	Increase in inland water (79.46%) mainly from cultivated (9%), barren areas and grassland (8.05), and marshland (1.46%)
Total	**11,619,040.00**	**11,619,039**		

good management (Hijioka et al. 2014). The anticipated drier summer season and wetter rainy season brought about by climate change would have profound effects on stream flow, dam operation and water allocation, domestic water supply, irrigation, hydropower generation, depth and recharge of aquifers, water quality (e.g., salt-water intrusion), and even on water infrastructure and management systems (Pulhin and Tapia 2015). There is a grave threat of water scarcity during summer and too much precipitation during the rainy season, which can trigger more floods and landslides.

The vulnerability of the country's water sector to natural hazards associated with climate change is already evident in the recent extreme weather events like typhoons and droughts (Pulhin and Tapia 2015). For instance, Tropical Storm Ketsana (Ondoy) exposed the deficiencies in water infrastructure and management systems in the country, catching the sector off-guard to extreme climate variability. Pumping facilities to ease floodwaters in Metro Manila were unable to handle beyond 100 mm of rainfall per hour, leaving the greater part of the area and adjacent municipalities submerged in floodwaters. More than US$18.7 M (PhP 820 M) worth of irrigation facilities, including dikes and canals that serviced 53,000 ha of farmland in Central Luzon, were also destroyed by the same extreme event. Moreover, water supply in the city was affected and halted, affecting more than 100,000 households without piped-in water (Climate Change Commission n.d.). Further, the situation submerged more than 500 barangays or communities in Region III because the National Irrigation Administration (NIA) was forced to open the gates of some water reservoirs, such as La Mesa Dam, Ipo Dam, Ambuklao Dam, and Binga Dam, as water levels already reached critical status (NDCC 2009).

On the other hand, droughts usually caused by a strong El Niño produced significant dips in water inflows of major water reservoirs, which led to shortages in domestic water and irrigation supply (Jose 2002). The 1997–1998 El Niño in the country significantly reduced the water level of Angat Dam (from 37 to 22 m^3/s), which supplies more than 90% of domestic water in Metro Manila. As a result, the MWSS had to augment water supply through rationing; water supply was limited only to 4 h a day to contain the shortage. To address agricultural needs, the Bureau of Soils and Water Management had to resort to cloud seeding, spending an additional US$0.83 M (PhP 36.7 M) (Pulhin and Tapia 2015). Moreover, many people had to use water wells indiscriminately, contributing to groundwater depletion and salt-water intrusion (Juanillo 2011).

The many threats to water resources brought about by climate change interact with other factors in a complex manner. To better focus and prioritize regional action, **United Nations Economic and Social Commission** for Asia and the Pacific (ESCAP) has identified hotspots of multiple challenges. Hotspots are countries, areas, or ecosystems with overlapping challenges of poor access to water and sanitation, deteriorating water quality, limited water availability, and increased exposure to climate change and water-related disasters. Many of these challenges directly relate to the Philippines. The Philippines is among five countries in the Asia-Pacific region with compound hotspots in six categories (Fig. 2.4). These are (1) water utilization level, (2) water quality, (3) frequency of floods, (4) frequency of cyclones, (5) frequency of droughts, and (6) climate change pattern. These challenges need to be addressed if the country is to move forward with its quest of reducing poverty and achieving sustainable development.

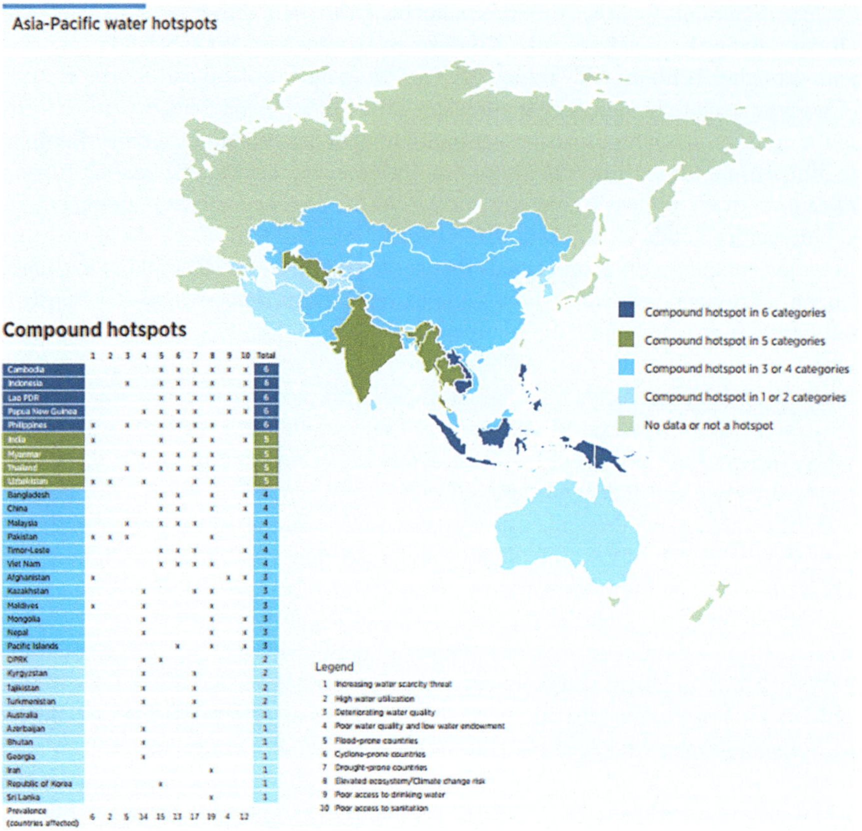

Fig. 2.4 Asia-Pacific hotspots (UNESCO 2012, p. 683)

2.4 Conclusion: Ensuring That Water Supply Meets Long-Term Demand

Water sustainability is one of the great challenges facing society in the twenty-first century (Falkenmark 2008). With ongoing land use conversion driven by population growth and urbanization and the anticipated impacts of climate change, many countries of the world, including the Philippines, face issues of water scarcity and pollution, which threaten the long-term sustainability of water resources (Gleick 2003). Considering the Philippines' water availability as against increasing demand, as well as the degenerating quality of both surface and underground water resources, such precious resources are now approaching critical limits. Projections on water availability indicate that water stress will worsen in the future, brought about by limited supply amidst higher demand and worsening water quality.

The paucity of more updated data and reliable information on water supply and demand is one of the major constraints that limit the effectiveness and efficiency of water management in the Philippines. Such information is crucial for policy and decision-making both at the national and local levels to develop more appropriate and strategic policies and programs to advance water resource management and sustainability. Water governance needs to better align scientific understanding with the management of water resources, specifically in addressing the environmental and socioeconomic stressors that shape water availability and sustainability (Falloon and Betts 2010). Unless the country can overcome or adapt to these driving forces, future generations will inherit a legacy of declining and degraded water resources that threaten their livelihoods and well-being, particularly those of the poorer sector of society. In the next chapter, we discuss water governance, which is one of the important pillars for pursuing water sustainability.

References

ADB (Asian Development Bank). (2007). *Water quality worsening*. Asian water development outlook. http://pcij.org/blog/2007/12/05/water-quality-worsening-adb. Accessed 8 Aug 2016.

Ancheta, C., Hiroyama, M., Illangovan, P., Lorenz, J., Mingoa, Y., Porras, A., Shah, J., Sy, J., Tavares, L. C., Tuyor, J., Verzola, E., Verzosa, D. G., & Villaluz, M. G. (2003). *Philippine Environment Monitor 2003: Water quality* .(42 p. Manila: The World Bank Group.

Boberg, J. (2005). *Liquid assets: How demographic changes and water management policies affect freshwater resources*. Santa Monica: RAND Corporation. http://www.rand.org/content/dam/rand/pubs/monographs/2005/RAND_MG358.pdf. Accessed 1 Aug 2016.

Chang, H., Il-Won, J., Strecker, A., Wise, D., Lafrenz, M., Shandas, V., Moradkhani, H., Yeakley, A., Pan, Y., Bean, R., Johnson, G., & Psaris, M. (2013). Water supply, demand, and quality indicators for assessing the spatial distribution of water resource vulnerability in the Columbia River Basin. *Atmosphere-Ocean, 51*(4), 339–356.

Climate Change Commission. (n.d.). *National climate change action plan 2011–2028*. Manila: Climate Change Commission.

Crisostomo, S. C. (2016). Philippine population seen to hit 104 M. *The Philippine Star*. http://www.philstar.com/headlines/2016/01/03/1538653/philippines-population-seen-hit-104m. Accessed 1 Aug 2016.

Cruz, R. V. O. (2014). *Watersheds in a changing climate: Issues and challenges*. SEARCA Policy Brief Series 2014–2. ISSN 1656–5818. SEARCA, College, Laguna.

DENR (Department of Environment and Natural Resources). (2000–2008). *Compendium of basic ENR statistics for operations and management* (2nd ed.). http://www.focusonpoverty.org/download/data/Compendium%20of%20Basic%20ENR%20Statistics%20(DENR).pdf. Accessed 6 June 2016.

DENR (Department of Environment and Natural Resources). (2014). *Compendium of ENR statistics*. http://www.denr.gov.ph/e-library/compendium-enr-statistic-2014.html. Accessed 6 June 2016.

DENR-EMB (Department of Environment and Natural Resources-Environmental Management Bureau). (2008). *EMB accomplishment report CY 2008*. Manila: DENR-EMB.

Falkenmark, M. (2008). Water and sustainability: A reappraisal. *Environment, 50*, 4–16.

Falloon, P., & Betts, R. (2010). Climate impacts on European agriculture and water management in the context of adaptation and mitigation—the importance of an integrated approach. *Science Total Environment, 408*, 5667–5687.

FAO (Food and Agriculture Organization). (2000). Land-water linkages in rural watersheds. Discussion paper 1. Land use impacts on water resources: A literature review. *FAO Land and Water Bulletin*. ftp://ftp.fao.org/agl/aglw/docs/lw9e.pdf. Accessed 9 Aug 2016.

FAO (Food and Agriculture Organization). (2012). *AQUASTAT, FAO's Global Information System on water and agriculture*. http://www.fao.org/nr/aquastat. Accessed 9 Aug 2016.

FMB (Forest Management Bureau). (2014). *The Philippine forestry statistics* (pp. 2, 21–24). Quezon City: FMB.

Giles, H., & Brown, B. (1997). And not a drop to drink: Water and sanitation services to the urban poor in the developing world. *Geography, 82*(2), 97–109.

Gleick, P. H. (2003). Global freshwater resources: Soft-path solutions for the 21st century. *Science, 302*(5650), 1524–1528.

Gleick, P. H. (2013, September). A drop in the bucket. *Finance and Development, 50*(3). http://www.imf.org/external/pubs/ft/fandd/2013/09/gleick.htm. Accessed 15 Aug 2016.

Greenpeace Philippines. (2009). *Climate change and water: Impacts and vulnerabilities in the Philippines*. http://www.greenpeace.org/seasia/ph/Global/seasia/report/2009/12/climate-change-water-impacts-philippines.pdf. Accessed 1 Aug 2016.

Greenpeace Southeast Asia. (2007). *The state of water resources in the Philippines*. http://www.greenpeace.org/seasia/ph/Global/seasia/report/2007/10/the-state-of-water-in-the-phil.pdf. Accessed 1 Aug 2016.

Hijioka, Y., Lin, E., Pereira, J. J., Corlett, R. T., Cui, X., Insarov, G. E., Lasco, R. D., Lindgren, E., & Surjan, A. (2014). Asia. In V. R. Barros, C. B. Field, D. J. Dokken, M. D. Mastrandrea, K. J. Mach, T. E. Bilir, M. Chatterjee, K. L. Ebi, Y. O. Estrada, R. C. Genova, B. Girma, E. S. Kissel, A. N. Levy, S. MacCracken, P. R. Mastrandrea, & L. L. White (Eds.), *Climate change 2014: Impacts, adaptation, and vulnerability, Part B: Regional aspects. Contribution of Working Group II to the Fifth Assessment Report of the Intergovernmental Panel on Climate Change* (pp. 1327–1370). Cambridge: Cambridge University Press.

JICA/NWRB (Japan International Cooperation Agency/National Water Resources Board). (1998). Master plan study on water resources management in the Republic of the Philippines (257 p). Tokyo: Nippon Koei Ltd./Nippon Jogesuido Sekkei, Ltd.

Jose, A.M. (2002, July–September). El Niño-related droughts in the Philippines: Impacts and mitigation efforts. *Asian Diasters Management News, 8*(3).

Juanillo, E. (2011). *El Niño impacts on Philippine water resources: Focus on Angat Dam*. Abstract of paper presented at the World Climate Research Programme (WCRP) Climate Research in Service to Society Conference, Denver. http://www.wcrp-climate.org/conference2011/abstracts/C1/Juanillo_C1_M222B.pdf. Accessed 22 Nov 2016.

Kho, J., & Agsaoay-Saño, E. (2006). *Country study on customary water laws and practices*. Rome: Food and Agriculture Organization of the United Nations.

Matilla, D. R. (2013). Majority of global child deaths caused by contaminated water. *The Philippine Star* http://www.philstar.com/health-and-family/2013/03/22/922799/unicef-majority-global-child-deaths-caused-contaminated-water. Accessed 8 Aug 2016.

National Disaster Coordinating Council (NDCC). (2009). *NDCC update: Final report on tropical storm "Ondoy" (Ketsana) and Typhoon Pepeng (Parma)*. http://www.ndrrmc.gov.ph/attachments/article/1543/Update_Final_Report_TS_Ondoy_and_Pepeng_24-27SEP2009and30SEP-20OCT2009.pdf. Accessed 8 Aug 2016.

Navarro, A. (2014). *Scrutinizing urbanization challenges in the Philippines through the infrastructure lens* (PIDS Discussion Paper Series No. 2014–37). http://dirp3.pids.gov.ph/webportal/CDN/PUBLICATIONS/pidsdps1437.pdf. Accessed 1 Aug 2016.

Naz, A. C. (2012/2013). The State of the Philippine environment: An update on Chapter 4 of the 1994 Philippine Human Development Report (HDN Discussion Paper Series, PHDR No. 10, 50 p). Quezon City: Philippine Human Development Network: School of Economics, University of the Philippines.

NSCB (National Statistical Coordination Board). (2004). *Compendium of Philippine environment statistics*. http://nap.psa.gov.ph/peenra/Publications/Compendium/CPES%202004.pdf. Accessed 18 Nov 2016.

NWRB (National Water Resources Board). (2003). *National Water Resources Board strategic lanning and management of integrated water resources in the Philippines*. http://www.fao.org/DOCREP/004/AB776E/ab776e03.htm. Accessed 18 Nov 2016.

NWRB (National Water Resources Board). (2016). *Data on surface and ground water assessments by region, 2016*. Quezon City: National Water Resources Board.

Population Division of the Department of Economic and Social Affairs of the United Nations Secretariat. (2001). *The components of urban growth in developing countries*. Preliminary unedited version. https://esa.un.org/unpd/wup/Archive/Files/studies/United%20Nations%20(2001)%20-%20The%20Components%20of%20Urban%20Growth%20in%20Developing%20Countries.pdf. Accessed on 24 Nov 2017.

PSA (Philippine Statistics Authority). (2013). *Urban barangays in the Philippines (based on 2010 CPH)*. https://psa.gov.ph/content/urban-barangays-philippines-based-2010-cph. Accessed 8 Aug 2016.

Pulhin, J. M., & Tapia, M. A. (2015). Vulnerability and sustainable development: Issues and challenges from the Philippines' agricultural and water sectors. In *Sustainable development and disaster risk reduction* (pp. 189–206). Tokyo: Springer.

PWR (Philippine Water Resources). (2000). *Environmental and natural resources accounting*. http://hopkins.addu.edu.ph/moda/wp-content/uploads/2016/03/water.pdf. Accessed 9 Aug 2016.

Schnoor, J. L. (2015). Water unsustainability. *Dædalus, the Journal of the American Academy of Arts & Sciences, 144*(3), 1–11.

Sherbinin, A., Carr, D., Cassels, S., & Jiang, L. (2007). Population and environment. *Annual Review of Environment and Resources, 32*, 345–373. https://doi.org/10.1146/annurev.energy.32.041306.100243. Accessed 19 June 2017.

Tirona, M. G. A. (2013, April 22–26). *Philippine statement at the 46th Session of the Commission on Population and Development*, New York. http://www.un.org/en/development/desa/population/pdf/commission/2013/country/Agenda%20item%204/Philippines_Item4.pdf. Accessed 8 Aug 2016.

UN-DESA (United Nations Department of Economic and Social Affairs/Population Division). (2015). *World urbanization prospects: The 2014 revision*. https://esa.un.org/unpd/wup/Publications/Files/WUP2014-Report.pdf. Accessed 8 Aug 2016.

UNDP (United Nations Development Programme). (2016). *How changing demographics can power human development*. Asia-Pacific Human Development Report—Shaping the Future. New York: UNDP.

UNESCO (United Nations Educational, Scientific and Cultural Organization). (2006). *Water: A shared responsibility*. United Nations World Water Development Report 2. UNESCO:World Water Assessment Program. http://unesdoc.unesco.org/images/0014/001444/144409E.pdf. Accessed 8 Aug 2016.

UNESCO (United Nations Educational, Scientific and Cultural Organization). (2009). *Water in a changing world*. United Nations World Water Development Report 3. http://webworld.unesco.org/water/wwap/wwdr/wwdr3/pdf/12_WWDR3_ch_2.pdf. Accessed 21 Nov 2016.

UNESCO (United Nations Educational, Scientific and Cultural Organization). (2012). *Managing water under uncertainty and risk*. The United Nations World Water Development Report 4 (Vol. 1). Paris: UNESCO.

UNICEF-WHO (United Nations Children's Fund-World Health Organization). (2012). *Progress on drinking water, 2012 update*. https://www.unicef.org/media/files/JMPreport2012.pdf. Accessed 8 Aug 2016.

WHO (World Health Organization). (2016). *Urban population growth*. http://www.who.int/gho/urban_health/situation_trends/urban_population_growth_text/en/. Accessed 8 Aug 2016.

WHO-UNICEF (World Health Organization- United Nations Children's Fund). (2006). *Meeting the MDG drinking water and sanitation target: The urban and rural challenge of the decade*. Joint Monitoring Programme for Water Supply and Sanitation. http://www.who.int/water_sanitation_health/monitoring/jmpfinal.pdf. Accessed 8 Aug 2016.

World Population Prospects. (2016). http://worldpopulationreview.com/countries/philippines-population/. Accessed 1 Aug 2016.

WRI (World Resources Institute). (2015). *Aqueduct projected water stress country ranking*. http://www.wri.org/sites/default/files/aqueduct-water-stress-country-rankings-technical-note.pdf. Accessed 8 Aug 2016.

Dr. Juan M. Pulhin is full Professor and former Dean of the College of Forestry and Natural Re-sources, University of the Philippines Los Baños (UPLB). He earned his Bachelor of Science and Master of Science in Forestry degrees in UPLB and Ph.D. degree in Geographical Sciences from The Australian National University. He was a Visiting Professor at The University of Tokyo for four times and has more than 30 years of experience in natural resources education, research and development. He has authored more than 100 technical publications on various aspects of natural resources management and climate change. He was a Coordinating Lead Author of the Intergovernmental Panel on Climate Change (IPCC) Fifth Assessment Report and a Lead Author of the Fourth Assessment Report. He has been involved in an interdisciplinary research project on water governance for development of the University of the Philippines and in numerous watershed development planning projects since 1998.

Dr. Rhodella A. Ibabao holds a BA degree in Sociology-Psychology, a MA in Urban and Regional Planning, and a doctorate degree in Environmental Planning and Management. She is a faculty member of the University of the Philippines Visayas (UPV) teaching courses in urban and regional planning. She is a licensed environmental planner with areas of specialization in social development and land use planning. She has completed research projects on land use planning, solid waste management, urban river management and resettlement planning. She has special interest in community-based planning with vulnerable groups such as waste pickers, indigenous people, and women.

Dr. Agnes C. Rola is full Professor at the University of the Philippines Los Baños (UPLB), former Dean of the College of Public Affairs and Development, UPLB, and member of the National Academy of Science and Technology- Philippines. She has degrees in Statistics (BS) and Agricultural Economics (MS) from the UP; and PhD in Agricultural Economics (Major in Natural Resource Economics) from the University of Wisconsin Madison, USA. She attended the Summer Certificate on Environmental Leadership Program at the University of California-Berkeley and has more than 20 years' research experience in sustainable agriculture at the watershed level with a research focus on water governance. With colleagues, she has written and edited an award winning book, "Winning the water wars: watersheds, water policies and water institutions" (2004), whose recommendations were adopted in the Philippines' Clean Water Act. For the past four years, she led two major research programs on water in the Philippines, namely, water governance for development and water security under climate risks.

Dr. Rex Victor O. Cruz is a Full Professor at the University of the Philippines Los Baños (UPLB). He obtained his bachelor and Master's degree in forestry at UPLB and his doctoral degree at the University of Arizona. His major fields of expertise are forestry, watershed management, environment and natural resources management, upland development, land use planning and climate change. The author is a member of the UN Intergovernmental Panel on Climate Change (IPCC) in 1992–1995; 1997–2000; and 2004–2007. Currently he leads three national programs, the National Re-search and Development Program for Watershed Management, National Conservation Farming Village Program, and the Monitoring and Detection of Changes in Ecosystems for Resiliency and Adaptation. He is a member of the Asia Pacific Forestry Network Board of Directors, the People Survival Fund Board, and the National Pool of Experts of Climate Change Commission.

Chapter 3
Laws, Institutional Arrangements, and Policy Instruments

Rosalie Arcala Hall, Corazon L. Abansi, and Joy C. Lizada

Abstract The chapter maps the array of formal arrangements between water apex bodies, national government agencies, local government units and water organizations with mandates on water supply and demand management. Institutional arrangements are complex, multilayered and fragmented, with duplication of tasks such as planning and monitoring, whilst few are involved in operations and financing. Local government units directly operate facilities and replicate the work of water apex bodies, which have no local presence. The role of the state is central as owner of property rights to water, regulator and subsidy provider. Despite increasing private sector participation in water provisioning mainly in urban centers, many rural and village-level waterworks continue to rely on grants from the government for crucial infrastructure, operations and management funds. The government uses water permits, subsidy, legal requirements for inter-sector transfer, and penalties for pollution as policy instruments. However, serious gaps remain as mechanisms are not fully articulated in order to meet social equity and resource sustainability goals. The outcomes are highly localized, politically contingent, and diverse water governance schemes. Select cases illustrate how the absence of clear property rights and rules for orderly contracting lead to political mobilization for preferred outcomes, local social arrangements, and water pricing innovations.

Keywords Water laws • Water rights • Water permits • Decentralization • Water apex bodies

R.A. Hall (✉)
Division of Social Sciences, College of Arts and Sciences, University of the Philippines Visayas, Miagao, Iloilo, Philippines
e-mail: rbarcalahall@up.edu.ph

C.L. Abansi
Institute of Management, University of the Philippines Baguio, Baguio City, Philippines

J.C. Lizada
Department of Management, College of Management, University of the Philippines Visayas, Iloilo City, Philippines

A.C. Rola et al. (eds.), *Water Policy in the Philippines*, Global Issues in Water Policy 8, https://doi.org/10.1007/978-3-319-70969-7_3

3.1 Introduction

The Philippine water policy is an amalgam of laws, derived institutional arrangements, and instruments that were built up over time by various administrations. Together, these formal water rules reflect shifts in government mindset and political context. From highly decentralized domestic water systems with minor irrigation projects, the Marcos dictatorship formalized the water sector by centralizing control over permit issuance, extending regulation toward water districts (reconstituted from municipal waterworks) and pursuing large-scale, gravity-based irrigation projects with foreign international funding. This trend was supported by an international public financing environment that favors state investments for water supply provisioning (Hall et al. 2015; Rola et al. 2015b; David 2004). After the democratic transition in 1986, decentralization efforts gave local governments mandate in providing water supply and sanitation, even while regulation and permit issuance remain centralized. New legislations, in turn, added new environmental concerns within the government such as watershed protection to ensure water supply, anti-pollution, health safety, and water crisis.

Correspondingly, new mandates and functions were given to various government agencies. The end result is a multilayered, complex, and fragmented system with different platforms and diverse actors (Hall et al. 2015).

3.2 Locating the State in the Formal Policy Waterscape

Nine national legislations together form the core of the legal framework: Presidential Decree 1067, Water Code (1976); Presidential Decree 198, Provincial Water Utilities Act (1973); Presidential Degree 522, Prescribing Sanitation Requirements for the Traveling Public (1974); Republic Act 7160, Local Government Code (1991); Republic Act 7586, National Integrated Protected Area System Act (1992); Republic Act 8041, National Water Crisis Act (1995); Republic Act 8371, Indigenous People's Rights Act (1997); Republic Act 9275, Clean Water Act (2004), and Republic Act 8435, Agriculture and Fisheries Modernization Act (1997). The legal framework addresses the following: (1) property rights (permits/licenses/franchises for collection and distribution; private rights granted to individuals versus a collective; whether rights can be leased, transferred, or recalled by the granting authority); (2) distinction between types (surface or ground) and sectors/uses; (3) intersectoral prioritization and re-prioritization in terms of crisis or emergency; (4) regulation of water service providers; (5) integrated treatment of the water law with other laws on land, forest, and the environment, and for water planning and development; (6) state, private sector, and civil society participation in water planning and development; and (7) openness to market solutions in providing water services.

Property rights to all water (surface, ground, air) belong exclusively to the state, although other rights (e.g., individual household, common or pooled) are also present. The state awards individual or municipal/collective rights through water permits, which are assumed to carry the weight of the state's enforcement apparatus behind, notably that of the National Water Resources Board (NWRB). However, as borne by survey data from Rola et al. (2016), local water managers overall have a weak understanding of water rights and managers of community-based organizations (CBOs), and irrigation associations are less grounded on the legal basis of water rights than are water district managers. In the same survey, some managers erroneously claimed that water permits are issued by government agencies other than the NWRB.

As water is an economic and social good, the state plays a key role in the management of this resource. Water is not free; payment is collected for its production and its distribution, although under varying cost-recovery schemes across sectors (Rola et al. 2016).[1] Water, to the extent that it is provided by the state, is heavily subsidized; its price, a consequence of regulation, does not adequately reflect actual production cost and environmental service values. The set cost of water permits is antiquated and treats the bulk water resource almost as a free good (Tabios and David 2004). The Water Code also specifies schemes for provisioning that meet some social equity goals (e.g., public water faucets for poor households).

The state exercises regulatory power over water organizations. Water districts, concessionaires, private operators, and irrigation associations are subject to regulation by the NWRB, the Local Water Utilities Administration (LWUA), and the Metropolitan Water and Sewerage System (MWSS) in terms of tariff, financing, personnel, accounting, investment decisions, and planning. Presumably, the more numerous rural formations—rural and barangay water works associations (RWSA and BWSA, respectively) and cooperatives—are also subject to regulation by LWUA. To the extent that these water organizations are required to register with the state, they also fall under the ambit of the Securities and Exchange Commission (SEC) and the Cooperative Development Authority (CDA). It is also assumed that local government unit (LGU)-run water organizations are supervised by the Department of Interior and Local Government (DILG). Most formations rely on local government funds for their organization and maintenance (Greenpeace 2007; The Water Dialogues n.d.). The Water Code also has provisions for resolving conflicts on competing water rights claims, in addition to regular courts as intergovernment agency adjudication court. For competing uses, the principle followed in the Water Code is "first in time, priority in right."

The Water Code assigns the provision of both water and sanitation and sewerage to MWSS and water districts. Parallel responsibilities are given to local governments: water provisioning for village/barangay governments, and sanitation and sewerage system for municipal governments. In actuality, very little investments are put into sanitation and sewerage because of the low priority given by these formal water organizations (Ebarvia 1997; Singh 2006). In his analysis of foreign donor-funded water system projects to CBOs, Singh (2006) noted that sanitation was also neglected. Only about three public sewerage systems still operate (Baguio City,

[1]The water manager survey by Rola et al. (2016, 9–11) showed a remarkable diversity in pricing structures of water organizations. Only 74% of irrigation association managers claimed that their water pricing is based on rules set up by the National Irrigation Administration (NIA) and 55% of water districts said their price is based on LWUA or set by their board. Meanwhile, some LGU-managed organizations and CBOs charge for water, but, in most instances, it is free.

Zamboanga City, and Vigan City); outside of Metro Manila, public sanitation services are almost non-existent (ADB 2013; World Bank 2005; Robinson 2003).

Metro Manila, Baguio City, Boracay Island, and the Clark and Subic economic zones are the only urban centers with sewage treatment facilities (World Bank 2013). Some private residential areas and commercial establishments in Manila run their own water treatment system (e.g., Ayala), but, for the rest, sewage goes straight into Manila Bay.

Of the many uses of water (domestic, industrial, agriculture, energy production, etc.), the Water Code allows for transfers on the basis of contract or interagency agreement among permit holders. Under the Water Crisis Act, water can be reallocated from agriculture to domestic use in case of severe shortage. Under the Water Code, the physical transfer of surface or groundwater from one territorial jurisdiction to another through a purchase contract involving the permit holder is theoretically allowed. However, no mechanism for orderly contracting of this kind has been developed. So called local water transfers involving the tacit agreement of LGUs have been the subject of several court filings by concerned citizens or groups.[2]

Private sector participation in water provisioning is allowed under the Water Code. In fact, there has been a trend toward more private sector participation in water provisioning since the time of President Ramos (1992–1998) (Hall et al. 2015). Under the National Water Crisis Act (1995), the state shifted toward inviting more private sector involvement through build-operate-and-transfer schemes for funding, construction, repair, rehabilitation, and improvement in water utilities and supply. The Clean Water Act (2004) enumerates government incentives and market-based instruments, and promotes the role of private individual enterprises in the water sector. In 1997, water provisioning for Metro Manila was privatized (i.e., awarded to two private concessionaires). Since then, a wide variety of private water enterprises has emerged, ranging in size from the two concessionaires in Manila to bulk suppliers and small water vendors.[3] Since the Philippine Supreme Court ruling

[2] In July 2014, the Philippine Court of Appeals issued a Temporary Environmental Protection Order against PTKO H2O Corporation's plan to extract 50,000 m^3 per day from four rivers in Cavite province to supply the growing needs of Tagaytay City. The NWRB issued a water permit to PTKO H2O, but the court ruled in favor of the petitioners' (a citizen advocacy group from the affected Cavite towns) argument that such an extraction will compromise the health of the watershed and result in much lower surface water flow for use by the other riparian Cavite towns. In another case, in Majayjay, Laguna, concerned citizens filed a graft case against local government officials in 2011 for irregularities in the 50-year bulk water contract (with 50-year automatic renewal) granted to a private water enterprise. The mayor, vice mayor, and town councilors were found guilty by the Office of the Ombudsman for extending unwarranted benefits, preference, and advantage to the private company despite its lack of experience in water system development. CA stops water extraction in four rivers in Cavite Town. (July 17, 2014) http://www.journal.com.ph/news/provincial/ca-stops-water-extraction-in-4-rivers-in-cavite-town; Ex-Laguna town mayor, nine others indicted for graft (May 23, 2014) http://www.abs-cbnnews.com/nation/regions/05/23/14/ex-laguna-town-mayor-9-others-indicted-graft

[3] A documentation for an initiative to support policy rationalization among various water organizations in the Philippines cited the formation of the National Water and Sanitation Association (NaWaSA) in 2007, a grouping of small scale water providers. The same report cited that most small scale water providers were unaware of policies or national government agencies responsible for regulating them or water policy in general (The Water Dialogues n.d., 14).

in the Adala case (Supreme Court of the Philippines. G.R. No. 168914 Metropolitan Cebu Water District vs Margarita Adala. 04 July 2007), NWRB can now issue more permits in the form of certificate of compliance to competing water service providers in the same service area of water districts (Hall et al. 2015) (Box 3.1).

Box 3.1: Inter-local Government Informal Water Sharing ("Protecting Lon-oy Waters: Mayor Velasco's Continuing Crusade" http://launion.gov.ph/news.php?extend.196. Posted 19 May 2011. Accessed 12 April 2016)

The Case of San Fernando and San Gabriel in La Union Province

Large disparities in water availability exist within a country's borders. The rising water demand, particularly in urbanizing cities, has led municipalities and cities to rely on cooperation to manage shared resources like rivers and at the same time address water scarcity problems. The highly urbanized city of San Fernando and the rural municipality of San Gabriel, both in La Union province, have a long-standing water sharing agreement. Upstream San Gabriel does not consume the entire volume available from the river and its tributaries and instead passes some of it to downstream San Fernando by means of an informal agreement cited by both parties to have been made a long time ago. The surface water provides for both irrigation and domestic water requirements of both towns. This informal/tacit arrangement was sustained over time and has become a social norm. According to key informants, this cooperation generated benefits beyond the river, reflecting a wide range of political, economic, and cultural transactions. These are regarded by many as having created a climate of goodwill and mutual confidence and offered a window of opportunity for other collaborative development endeavors between the two localities.

However, as the population of San Gabriel increases, anthropogenic activities become more complex and, consequently, an increase in water demand requirements. At the same time, the towns' water sources and tributaries dry up during summer, producing scarcity. Increasingly, San Gabriel residents feel an asymmetry in terms of access to water between the two localities, with much of San Gabriel's water being transferred to San Fernando City's water district. Tensions arose as evidenced by some protests raised by residents due to asymmetric benefits from the use of water resources.

This water cooperation agreement depicts an arrangement outside the legal bounds of water rights, presumably stemming from water permits. The LGUs are not water permit holders and, as such, have no formal authority to make decisions to withdraw or exclude access to the river. Yet, upstream San Gabriel LGU's assertion of "ownership" of the water is accepted by San Fernando LGU, even though either claim has no legal basis. The case illustrates the disconnect between water law as understood at the national level and by local governments, and how consensus-based decisions on water sharing can in fact be made and sustained, albeit in fragile terms despite the lack of clarity on water rights by parties.

3.3 Formal Institutional Arrangements: Norming and Gaps

The Philippine water policy is complex, fragmented, and multilayered. There are multiple institutions, state and non-state, with hierarchical areas of coverage, varying mandates and sectoral representations involved in water (Malayang 2004). Despite presumed linkages between water, land and forest resources in the laws, structures to support interconnectivity between these diverse actors and policy areas are weak or non-existent (Hall et al. 2015). There are 30 agencies (national and LGU-based) managing water resources (Paragas 2012). Among these agencies are water apex bodies with regulatory functions—NWRB, LWUA, and MWSS—and line agencies with legal mandates over water quality (DENR, DOH), claims on the water resource (Department of Energy[DOE]-NAPOCOR, DA, NIA), water-risk mitigation (i.e., flood control by the DPWH), and investment (NEDA). Local government units also have legal mandates over water and sanitation services under the Local Government Code and are presumably deputized by said national line agencies as implementer of various program-based interventions. Local governments are also tasked to provide water supply and supervise water utilities within their jurisdiction.

In addition, there are other national agencies with anti-poverty programs that include water systems, but whose main mandate is not on water. Of note is DILG's 6-year (2010–2016) *Sagana at Ligtas na Tubig Para sa Lahat* (SALINTUBIG) program, for which grants to waterless, poor, and high water-borne-disease-incidence communities were made for levels 1 and 2 domestic water system projects (LWUA n.d.). The program was funded through DILG's annual government appropriation and required counterpart funding from the LGU recipient and contribution from community beneficiaries. Another is the Department of Social Welfare and Development's (DSWD) *Kapit Bisig Laban sa Kahirapan*-Comprehensive Integrated Delivery of Social Services (Kalahi CIDSS), a community-based small infrastructure program that also includes domestic water systems. It is similarly structured as grants to village governments requiring municipal government counterpart funding and community contributions with extensive participation in planning, preparation, execution, monitoring, and evaluation of projects (ADB 2012). The Kalahi-CIDSS has been in place since 2003, funded initially through a World Bank Loan and the DSWD's share of the general appropriations. These program-based water projects enable local governments to fill-in gaps in water access by tapping national agency funding.

The key agencies with remit on water are:

National Economic and Development Authority (NEDA): formulates the national development plans, sector policies and strategies, and reviews implementation

Metropolitan Waterworks and Sewerage System (MWSS): responsible for providing water supply and sewerage services to Metro Manila through their two private concessionaires, Manila Water (MWCI) and Maynilad (MWSI)

Local Water Utilities Administration (LWUA): specialized lending institution that promotes and oversees the development of provincial water districts. It also establishes design and operating standards for utilities and acts as the economic regulator of water districts

National Water Resources Board (NWRB): coordinating and regulating agency for all water resource management development activities. It is tasked with the formulation and development of policies on water utilization and appropriation and economic regulation of private water utilities and franchises

Department of Interior and Local Government (DILG): provides capacity building training programs to local government units (LGUs) and provides information on available sector programs and financing

Local government units (LGUs): bear multiple mandates in the sector such as water supply and sanitation provision and economic regulation of their utilities

Department of Public Works and Highways (DPWH): in charge of major infrastructure projects of the government. Both MWSS and LWUA are currently under its policy supervision

Department of Environment and Natural Resources (DENR): promulgates rules and regulations for the control of water, air, and land pollution and ambient and effluent standards for water and air quality

Department of Health (DOH): sets drinking water standards and monitors compliance

Department of Finance (DOF): oversees the performance of government financing institutions that provide commercial funds to the sector

National Irrigation Administration (NIA): tasked with the development and management of irrigation systems all over the country; its management functions cover operations, repair and maintenance, fund management for organization and management, enforcement of procedures and resolution of water-related conflicts, and agricultural support services

National Power Corporation (NAPOCOR): in charge with hydroelectric generation of power, including small power utilities group for electrification of areas outside of the main grid

Philippine Atmospheric, Geophysical and Astronomical Services Administration (PAGASA), key agency providing scientific observation and forecasting of weather, floods, and other climatological conditions

Department of Energy (DOE): tasked to prepare and supervise the implementation of plans and programs relative to hydroelectric energy exploration, development, utilization, distribution, and conservation

Metropolitan Manila Development Authority (MMDA): tasked with formulation and implementation of policies, standards, programs, and projects for an integrated flood control, drainage and sewerage system for Metro Manila

Laguna Lake Development Authority (LLDA): a quasi-government agency with regulatory functions covering environmental protection and jurisdiction over the Laguna Lake basin's surface water; currently under the policy supervision of DENR

Table 3.1 Key government agencies and their water-related functions

Government Agencies Functional Areas	NWRB	LWUA	DENR	LGUS	DPWH	DOH	NIA	NAPOCOR	PAGASA	DOF	MWSS	DILG	DOE	MMDA	DOT	LLDA	NEDA
Policy planning	•	•	•	•	•	•	•	•	•	•	•	•	•	•		•	•
Data monitoring	•	•	•	•	•	•	•	•	•		•	•	•	•		•	
Scientific modeling									•							•	
Infrastructure and program development	•	•	•	•	•	•	•	•	•		•	•	•	•	•	•	
Operation of water facilities				•	•		•	•			•			•			
Regulatory functions	•	•	•	•		•					•		•	•		•	
Financing		•	•	•						•							•
Public relations, capacity development and IEC	•	•	•	•	•	•	•	•	•	•	•	•	•	•	•	•	
Local RBO development			•														

Source: Rola et al. (2012)

NWRB National Water Resources Board, *LWUA* Local Water Utilities Administration, *DENR* Department of Environment and Natural Resources, *LGUs* Local Government Units, *DOH* Department of Health, *NA* National Irrigation Administration, *NAPOCOR* National Power Corporation, *PAGASA* Philippine Atmospheric, Geophysical and Astronomical Services Administration, *DOF* Department of Finance, *MWSS* Metropolitan Waterworks and Sewerage System, *DILG* Department of Interior and Local Government, *DOE* Department of Energy, *MMDA* Metropolitan Manila Development Authority, *DOT* Department of Tourism, *LLDA* Laguna Lake Development Authority, *NEDA* National Economic and Development Authority

Table 3.1 shows key government agencies and their water-related functions (Rola et al. 2012). Certain trends are noticeable: (1) all government agencies, except Departments of Tourism and Finance, do policy planning and data monitoring, respectively; (2) only a handful of agencies are involved in the operation of water facilities and financing; (3) a single agency, PAGASA, does scientific modeling; and (4) all agencies do public relations, capacity development, and information education campaign (IEC). LGUs replicate activities by water apex bodies and, in addition, directly operate water facilities. Of all line agencies with functional remit on water, DENR has the most, having recently added local river basin organization (RBO) development and water quality management areas (WQMA). Planning and policy formulation are tasks shared by NEDA, NWRB, LWUA, and the LGU (Greenpeace 2007, 40). Nearly all agencies have infrastructure and program development on water funded under the annual General Appropriations Act (Box 3.2).

Box 3.2: Inter-sector Transfers
The Angat Dam Case

Angat Dam is a concrete multi-purpose and multi-use reservoir built within Angat Watershed and River Basin (AWRB) located in Bulacan province. It supplies water for hydropower generation for about 5% of Luzon island's power demand, and for domestic use of almost 65% of the population of Metro Manila (Tabios and David 2004). It also supplies irrigation water through the National Irrigation Administration (NIA), covering over 28,000 hectares of farmlands in Bulacan and Pampanga provinces. Presently, the Korean Water Company owns and manages the Angat hydroelectric plant. The rights to Angat dam waters are: MWSS has 31 m^3/s while NIA has 36 m^3/s. These allocations are met through water releases from the dam, which in turn are based on agreed rules by NIA, MWSS, the National Power Company (NPC) and the NWRB.

The releases of water from the Angat Dam depend on prevailing or current reservoir levels called rule curve policy. When the water level is above the rule curve (>212 m above sea level), all domestic, irrigation and power generation requirements are met. When the water level is below the lower rule curve (<180 m), the domestic water supply is satisfied first and irrigation water may be curtailed and is satisfied only by whatever water remains the reservoir. This is consistent with the Water Code proviso for transfers of water right to domestic from irrigation during drought or emergency, with compensation for those adversely affected by the transfer. In between the upper and lower rule curve, both domestic and irrigation needs are met but power generation is limited (Tabios and David 2004).

During the 1997–1998 El Nino, MWSS was granted 10 additional cubic meters per second from NIA's water allocation. NIA suspended operations of the Angat-Maasim River Irrigation System (AMRIS) for the whole of dry cropping season, presumably based on a compensation agreement with the MWSS, which covers both Irrigation Service Fee (ISF) due to NIA and unrealized harvest/income for farmer beneficiaries (Tabios and David 2004). MWSS' denial of this NIA compensation package claim was affirmed by the NWRB in 2001. Regardless, AMRIS irrigation associations and the Bulacan provincial government continued to lobby their claims (Rola and Elazegui 2008). The case was turned over to a body handling cases of government owned and controlled corporations, which has yet to issue a final verdict.

During the 2004 El Nino, the NWRB similarly gave zero allocation to NIA but the AMRIS irrigators associations opposed the ruling and went on to lobby the NWRB, politicians and the media for a change in allocation. Shortly after the NWRB finally approved the release of 10 m^3/s for irrigation, a typhoon increased the dam level prompting a release of 32 m^3/s. The farmers blamed the

(continued)

Box 3.2 (continued)
dam operators for the subsequent flooding (Rola and Elazegui 2008). The current El Niño has the water level in Angat Dam dropping to 198.57 m (Philippine Daily Inquirer, 3 April 2016). NIA advised Bulacan and Pampanga farmers to prepare for a total cut in irrigation supply or a sharp decline in water volume.

The Angat Dam case depicts the serious gaps in water policy because no guidelines exist on compensation, computation bases or the recipients of such monetary compensation. The farmers are at the short end of this policy gap as they stand to lose income either way from water scarcity or flooding. Policy decisions become subject to political mobilization. The farmers with their local government lobby to bring the NWRB and national politicians to their side, albeit with little success. From the perspective of farmers, theirs was a case of social injustice. They feel aggrieved that the burden of adjustment to climate risks disproportionately falls on them, but not to water-wasteful Metro Manila domestic users.

3.4 Policy Instruments

As the legal owner of all water resources inside the territory, the state, through NWRB issues permit or franchise for parties to access, withdraw, manage, exclude, or transfer the resource.[4] Household-based water extraction for domestic use is exempted from this permit requirement. The water permit specifies allowed extraction from surface or groundwater source expressed in liters per second (LPS). Permits can be modified, reduced, suspended, revoked, or cancelled by NWRB. The permits can also be leased or transferred between water permit holders but such is regulated by NWRB. Permits are issued to a diverse set of domestic, industrial, and commercial water providers—water districts, RWSAs/BWSAs, cooperatives, and private enterprises—whereas rights to agricultural water are given primarily to NIA.

The state subsidizes the cost of water development and production. First, the set cost of water permits is antiquated and treats the bulk water resource almost as a free good (Tabios and David 2004). The state provides subsidies to farmers and consumers by extending outright grants (with no repayment for water infrastructure such as dams and gravity-based irrigation systems), below-market-interest financing schemes for water projects, or through favorable import policies that make technologies accessible (e.g., water pumps) (Araral and Yu 2013).

[4] The NWRB is composed of the heads of the DENR, DPWH,NEDA, Department of Justice, DOH, DOF, National Hydraulic Research Center, University of the Philippines, and the executive director of NWRB (as per Executive Order 123 [2002]. As reconstituted, NWRB is headed by the Secretary of DENR.

At times, the national government also gives in to populist demands for reduced irrigation fees or waiver of dues for farmers, as was the case in 2000 when President Joseph Estrada waived the irrigation fees due for the year. At the local level, LGU-run water systems are also subsidized by the Internal Revenue Allotment (IRA) funds set aside for their maintenance and operations.

To achieve social equity in water provisioning, various government administrations fund water system projects targeting poor communities on a grant basis, such as DILG's Salintubig and DSWD's Kalahi-CIDSS project. Loan windows guaranteed by the national government have also been extended to LGUs for investment in municipal waterworks system.

The state's primary regulatory instrument is the water permit, which contains expressed amount (in LPS) of ground- or surface water the permit holder is entitled to. There are corresponding penalties leveled for exceeding these expressed limits (although amounts are outdated). To maintain water quality, the state imposes steep penalties to polluting industries. Local authorities like the LLDA have increasingly used or threatened to use judicial instruments against industries for untreated discharges to waterways, as well as injunctions from the Office of the Ombudsman to compel riparian local governments to address solid waste disposal in their localities. State agencies are also able to prescribe tariff and other management schemes, as tie-ins to financing instruments; LWUA sets performance benchmarks, for instance, to water districts to determine the latter's access to government financing.

For transfer of water between sectors, most reservoirs with multiple uses have water allocation schemes between agencies with mechanisms for reallocation (i.e., transfers from agriculture to domestic use) in the event of a water crisis. For the Angat reservoir, power generation by NAPOCOR is given priority, with recycled water up to 22 m^3/s of available inflow for MWSS (domestic, commercial, industrial) and 36 m^3/s for NIA (irrigation). The water rights also provides for MWSS to extract an extra 15 m^3/s in the event NIA does not fully use up its regular allocation. In the case of Angat Dam during the 1996–1997 water reallocation from irrigation to domestic water supply in Metro Manila, the mechanisms pertaining to who will be compensated and the bases for compensation and funding were not fully threshed out. Meanwhile, water allocation in Lake Buhi is illustrative of social arrangements beyond what are legally articulated. While a 1989 Memorandum of Agreement exists between NIA and NAPOCOR setting an operation rule curve to govern allocation for irrigation and power demand, what was followed instead is the imposed minimum limit set by the local government of Buhi to guarantee the viability of fish cage operations in the lake (Elazegui et al. 2016). The NIA staff opens the gates based on the final decision of the Buhi LGU, not on the agreed 3-day protocol. The two cases illustrate that even agreement-based transfers do not come with full mechanism complement and are rendered irrelevant by the socio-political dynamics of actors on the ground.

The policy instruments allowing for private sector participation in the water sector are water permits, certificates of compliance, private concession awards (covering a segment or the entire water system, for example, Metro Manila) or build-operate-transfer schemes for infrastructure projects. Private water service providers in the Philippines comprise roughly 10% of the total, ranging in size from small water vendors to bulk water suppliers. NWRB's *Listahang Tubig* (Water Registry) counts 354 homeowner associations, 78 real estate developers, 41 industrial locators, 193 peddlers, 4 ship changers, and 1657 other private operators.[5] Many shopping malls, private subdivisions, and industries have certificates of compliance, allowing them to put up their own groundwater-sourced systems. Manila Water and Maynilad were awarded private concessions in Manila, with Manila Water expanding its business further to get concession contracts through subsidiaries in Laguna, Boracay, Clark, and Cebu.[6]

3.5 Challenges

3.5.1 Limits of State Power on Water Policy

The extent of state involvement in water policy follows the march to formalization. Presumably, the state's reach and assertion of property rights do not extend to individual households, which can still access and withdraw the resource for their own consumption without a permit from the state. However, as urban populations grow and as water services are extended to peripheral areas, more and more households are brought into the ambit of formal water organizations, including water districts and other rural formations that fall under the state's regulatory ambit. Where sections of the population access water informally, it also follows that they are outside the state's reach. Alternative ways for collective management of water as a common pool resource predating state formation are also in place in remote areas. Among indigenous communities, a dual system is observed where customary laws coexist with state laws governing water use (Rola et al. 2015b). Conflicts between indigenous claims as customary rights over water resources , which is presumed under the provisions of the Indigenous People's Rights Acts (IPRA) and those by state agents (local governments in particular) asserting statutory rights under the Water Code, have been noted in some cases. In the Manupali watershed of Bukidnon province, the indigenous Talaandig refused to pay their bills to the local government waterworks on claims that, under customary rules, said water is free. The conflict was settled through an agreement whereby the local government gave indigenous residents funds to develop a spring, with 10% of the project's income going toward watershed protection (Duque-Piñon et al. 2010). The requirement for free and prior informed consent under IPRA has brought potential developers into contact with

[5] National Water Resources Board *Listahang Tubig*: A National Water Survey. http://listahangtubig.cloudapp.net

[6] Manila Water subsidiaries to expand water coverage in Boracay and Laguna. 15 April 2013. http://www.gmanetwork.com/news/story/303980/money/companies/manila-water-subsidiaries-to-expand-water-coverage-in-boracay-and-laguna#sthash.8QzbCOye.dpufhttp://www.gmanetwork.com/news/story/303980/money/companies/manila-water-subsidiaries-to-expand-water-coverage-in-boracay-and-laguna

indigenous communities where the water resource is located, where otherwise such local consultations are not required for water permit applications.

In Mount Banahaw, Quezon province, tensions are also rife between religious communities with customary rights over springs based on religious/sacred notions of the water resource and the water district that claims subsidiary rights from the state (Rola et al. 2015a). In these remote areas, conflicts over water rights are only resolved through agreement-making by consensus, not by statist-legal assertion. The principle in the Water Code of "first in time, priority in right" also is put to question, given its inequitable outcome. For instance, Benguet Mining Corporation, which is now venturing into the water business, holds 65 water appropriation permits issued by NWRB. The permits cover major creeks, springs, and rivers in Itogon, Benguet province, that communities use for their domestic and agricultural needs. Indigenous people from the area mobilized and articulated opposition to the mining firms' claims.[7]

The Philippine water sector profile strongly mirrors state involvement and state weakness. Considering that most water is consumed by agriculture, the state is more heavily invested in infrastructure toward providing irrigation water compared with domestic water (Greenpeace 2007). By comparison, domestic water is predominantly sourced from groundwater, extraction of which often occurs without a state permit (Greenpeace 2007). Water districts formed as government-owned and -controlled corporations depict the state's lead role, but their coverage is limited to urban areas and fringes. The social justice component of domestic water provisioning is often neglected, except for sporadic and project-based investments into water systems for urban poor communities. Water districts are supposed to provide public faucets to these communities but supply dearth meant least prioritization. Most community water systems in areas outside of water district coverage are often rudimentary (domestic levels 1 and 2) and are either external-donor-funded or local government-maintained (World Bank 2005). Meanwhile, water "elites" or those with financial capability to put up their own groundwater-sourced systems (e.g., malls, private subdivisions, industries) can easily have access.

3.5.2 Norming and Gaps in Institutional Arrangements

While water apex bodies such as NWRB, LWUA, and MWSS are present, they are institutionally weak (Rola et al. 2012). The NWRB has the exclusive task of issuing permits as well as regulation, but its limited staff and Manila-only presence pose constraints to job efficiency. NWRB and LWUA are also not Cabinet-level offices; as such, they are routinely transferred as subunits to line agencies, depending on the administration's policy rationale. Under Executive Order 213 (by President Gloria Macapagal-Arroyo), LWUA was stripped of its regulatory function over water

[7] Who owns our water? Bulatlat. Volume III, Number 48 (January 11–17, 2004). Retrieved at http://bulatlat.com/news/3-48/3-48-waterowner.html

districts and the task transferred to NWRB (NEDA 2010). Instead, LWUA's mandate as a financing institution to water districts was strengthened and as a unit transferred to the Office of the President, then to DPWH (along with MWSS). Under Executive Order 806 (by President Benigno Aquino III), LWUA's regulatory task was restored and extended with rural water and sanitation associations (RWSA) placed under its umbrella. At the same time, the Secretary of DPWH was appointed water czar, but the lack of clarity in his coordinative role was equally debilitating for reforms to take place. There had also been parallel attempts to reorganize the NWRB's regulatory remit and limit its task to permit and licensing. From being a subunit of DPWH, NWRB was transferred to DENR in 2010 under Executive Order 806 (Almaden 2014).

NIA is another agency which has seen intermittent transfers across various line agencies. Following the centralization of all government irrigation activities under the Marcos dictatorship, NIA was placed under the Department of Public Works, Transportation, and Communication. NIA's authority and capitalization expanded greatly throughout the 1970s and in 1987 remained attached to DPWH. Under Administrative Order No. 17 (1992), it was transferred to DA. NIA's poor performance in terms of collections and investment outlay has put the office under a negative light, prompting former President Benigno Aquino III to publicly flag the agency. Under the present administration of President Rodrigo Duterte, NIA was transferred from DA and came under the supervision of the Secretary to the Cabinet, which recommended to the President its closure.[8]

Figures 3.1, 3.2, and 3.3 show the evolution of these key agencies in terms of their placement. The end effect of these line-agency transfers is to treat water not as a singular policy concern but as an appendage of whatever focus that mother line agency has (e.g., infrastructure for DPWH; forest for DENR). There is neither a comprehensive picture of the country's water resources and future requirements nor a premium attached to rationalizing allocation through a robust water rights-based system. To the DA, water is simply an input to the goal of agricultural development; for DENR, water yield is a product of a healthy watershed environment. The DPWH treats water as a force to be controlled through infrastructure. These different agency lenses color the regulatory and other tasks of NWRB, LWUA, and NIA.

Gaps are also evident in the way the mandates are carried out. While LWUA oversees water districts (sets design standards and tariffs in line with financing extended to them), it does not exert the same effort toward RWSAs/BWSAs, which are often not registered and are located in remote places. The NWRB presumably has regulatory remit over private water service providers with permits, but its insufficient staff and exclusive Manila presence do not allow extensive monitoring. The LGUs are mainly recipients of national government agency water system projects or grants from Congressmen (under the latter's Priority Development Assistance Fund), but the distribution of such projects and grants are highly uneven. While there are funding windows available for local government

[8]Philippine Star, 26 August 2016, https://sg.news.yahoo.com/agriculture-budget-cut-p3-4-000000473.html?nhp=1

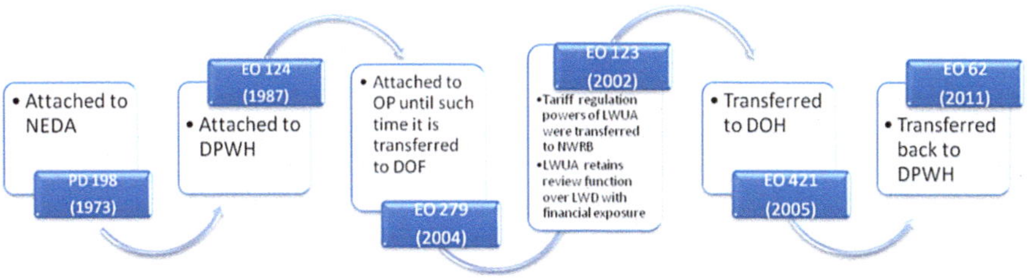

Fig. 3.1 Shifts in agency supervision over Local Water Utilities Administration (LWUA). *NEDA* National Economic and Development Authority, *DPWH* Department of Public Works and Highways, *OP* Office of the President, *DOF* Department of Finance, *LWUA* Local Water Utilities Administration, *DOH* Department of Health

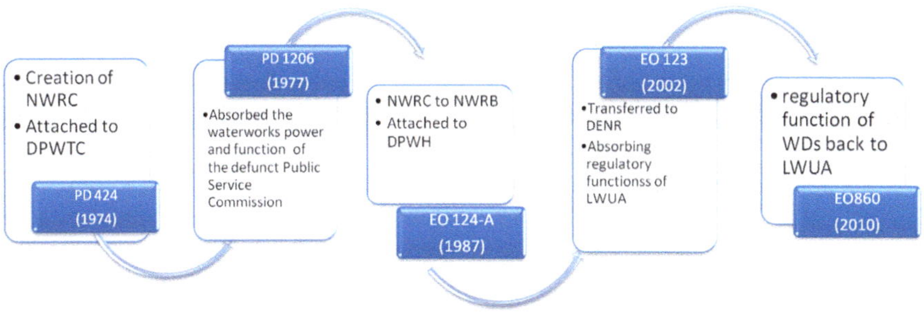

Fig. 3.2 Shifts in agency supervision over the National Water Regulatory Board (NWRB). *NWRC* National Water Resources Council, *DPWTC* Department of Public Works, Transportation and Communication, *DPWH* Department of Public Works and Highways, *DENR* Department of Environment and Natural Resources, *LWUA* Local Water Utilities Administration

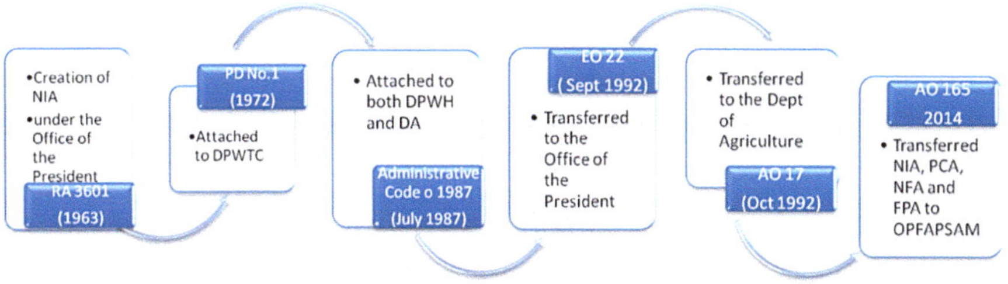

Fig. 3.3 Shifts in agency supervision over the National Irrigation Administration (NIA). *DPWTC* Department of Public Works, Transportation and Communication; *DPWH* Department of Public Works and Highways, *DA* Department of Agriculture, *OP* Office of the President, *NIA* National Irrigation Administration, *PCA* Philippine Coconut Authority, *NFA* National Food Authority, *FPA* Fertilizer and Pesticide Authority, *OPFAPSAM* Office of the Presidential Assistant for Food Security and Agricultural Modernization

waterworks' program of investment (e.g., the Local Government Units Urban Water and Sanitation Project (LGUUWSP) of the World Bank), very few LGUs actually avail of these opportunities.

There are also clear functional overlaps. Small-scale irrigation systems are done by two agencies: NIA and the Bureau of Soils and Water Management (BSWM) under the DA. While NIA has been primarily focused on large-scale gravity irrigation systems, its functions actually include planning, design, construction, and/or improvement of all types of irrigation projects and appurtenant structures. Recently, BSWM was given the mandate to coordinate the implementation of and extension of technical assistance to small-scale irrigation projects (SSIS) as a climate-change mitigation strategy. The small water impounding project (SWIP) is a structure constructed across a narrow depression or valley to hold back water and develop a reservoir that will store rainfall and run-off during rainy season for immediate or future use. They are earthen dams with structural heights of not more than 30 m and a volume storage not exceeding 50 million m^3. A similar broken picture exists for domestic water between LWUA (covering water districts and RWSA) and DILG (for LGU-operated water systems). Different roles and functions are also assigned to different layers of LGU with respect to water: barangays/villages are tasked to provide water supply and enforce pollution control laws; municipalities and cities are to put up water and sanitation facilities, while the province enforces anti-pollution laws and undertakes intermunicipal infrastructure.

Data collection is similarly fragmented among many line agencies, thus inhibiting strategic-level planning for the resource (ADB 2013). Each agency collects its own data as support for its mandate (e.g., surface water quality testing in WQMA by the DENR-Environmental Management Bureau or surface water volume and flow by Department of Science and Technology-PAGASA for its forecasting task), but these data are never collected in one place and collectively assessed. Planning is left to each line agency where water as a residual concern is lodged, not exclusively at the NWRB.

The sheer number of government actors and the plurality of mandates (where no single mandate is deemed a priority over others) make coordination a real challenge (Hall et al. 2015). Elazegui (2004) highlights the high cost of coordination given this setup. National agencies have no field presence in many localities; the Environment Management Officer is an optional position for LGUs; and there is no nationally legislated funding scheme for water resource management. In recent years, the Philippine government created multi-agency management committees to deal with particular issues. Examples of these are the Inter-Agency National Flood Management Committee and the Joint-Executive-Legislative Water Crisis Commission. But as coordinative platforms, they are found to be lacking (Greenpeace 2007). There is also a separate Water Supply Subsector composed of NEDA, DOH, and DPWH for water sector planning and investment purposes.

Local replicates of coordination schemes for water resource management exist (e.g., LLDA, watershed councils, and surface water WQMA), but they are not present in all areas. The legal framework layers responsibility over the resource between national and local government agencies, but mechanisms to effect subsidiarity and decentralization are not fully articulated (Rola et al. 2015a). Provincial governments can create watershed management councils (citing provisions of the Local Government Code) on an elective basis. Water quality management areas are identified by the DENR. There are other coordinative-management schemes that are based on social arrangements between actors (e.g., Lake Buhi between NIA, NAPOCOR, and the local government of Buhi; Manupali river between NIA, plantations and the local government representing indigenous communities). Often, any meaningful decision requires a Memorandum of Agreement or parallel contracting between the national agency, the LGU, and the water service provider for a project or initiative. As such, there is no "singular" water policy to speak of but highly localized and politically contingent water governance schemes that vary from one area to the next. The Philippines has different water governance frameworks coexisting (state, common pool, private) and different levels (national, local) and spatial configurations (within or across politico-administrative boundaries, river basin or lakewide). Rola et al. (2015a) describe these mechanisms as "nested and interlocking" and they do not make local water planning any easier. The formation of the River Basin Office (RBO) under the DENR presumably point toward a more integrated approach to water resource management (Araral and Yu 2013).

3.5.3 The Limits of Policy Instruments

The weakness of the state is also indicated by poor compliance with state laws over permit-based access or withdrawal as well as pollution. For instance, Ebarvia (1997) notes that 81% of industries in Metro Manila extract groundwater through deep wells, many of them without permit. The absence of a permit is a structural constraint, owing to the fact that while NWRB deputizes the screening of water permit applications to local government units, approval of such is still in the hands of NWRB. In Metro Manila, combined groundwater extraction for industrial and domestic use (an estimated 40% of domestic water supply) is occurring at a rate faster than natural recharging processes and is showing signs of saltwater contamination. NWRB has no robust monitoring system in place for permits once issued; it only investigates once a complaint has been filed. While shallow wells for domestic use need no permit, NWRB has no existing inventory of shallow tubewells or data on actual pump age production. Another case illustrating state weakness relates to surface water permits, for which it is assumed that those secured earlier (senior) takes precedence over those issued later (junior) (Elazegui et al. 2016). In Manupali, even though the irrigation system has an older water permit, in practice, banana plantations located upstream were diverting more water for their need, leaving little for farmers downstream. This type of conflict is too remote to be brought to NWRB attention, leading finally to an agreement between parties in which the plantations pay irrigation service fees to the irrigation system and make "side payments" to the downstream communities in the form of road maintenance and tree planting (Duque-Piñon et al. 2010). The Manupali case illustrates the inaccessibility of the formal apparatus (NWRB) to resolve local issues, prompting parties to arrive at consensus-based agreements on water allocation outside of the remit of legal rights.

State spending for water-related infrastructure and water district financing also carry drawbacks. David (2004) claims that the state's subsidy of investment cost and some operation and maintenance expenditures for large-scale gravity irrigation over alternate irrigation technologies (e.g., pumps) has had detrimental effects on patterns of water use. Singh (2006) also argues that the incentive structure for water districts in terms of financing make them more responsive to the national government rather than to equity and sustainable outcomes of investment. Thus, water districts are not keen to develop level 3 (private water point) systems in low-income areas, even though such service is direly needed, leaving these poor areas to make do with lower level systems or self-provisioning.

The state's regulatory reach is also limited. The DILG barely imposes reporting and accountability requirements to project/ grant-based or IRA-financed water projects of LGUs. There are neither benchmarks nor common standards in the way rural domestic and irrigation water are provided by CBOs (Elazegui 2004; Ebarvia 1997). Many water stakeholders operate in this "informal sector" with little or no link to formal government structures (Almaden 2014). Even foreign-donor supported water system projects run by CBOs, although registered with CDA, look to the municipal government for oversight, but there is no clear mechanism on how this is to be done (Singh 2006). The local government can set standards, corresponding penalties and enforcement mechanisms, but it does not have taxation powers over water utilities operating within its area (Hall et al. 2015). Local government authorities could also process water permit applications but the NWRB has the sole authority to decide on applications. In addition, self-installed, permit-based private water systems for domestic supply (typically in residential and industrial parks) self-regulate in practice (Dayrit 2000). The NWRB is unable to monitor whether private water organizations exceed their extraction limits.

The push toward privatization in the water sector came with gains and gaps. While both private water companies improved service and expanded coverage (especially in urban poor communities) relative to the situation under the government-owned MWSS, such came with tariff increases and below-expected target performance (Wu and Malaluan 2008; Chia et al. 2007). The differential performance of the two Manila water concessionaires and the disconnect between complex financing schemes behind the privatization and the skeptical public did not provide a fertile ground for parallel schemes to be tried in other cities. With smaller water service providers entering the market, concerns have shifted toward financing access for these small players and regulation, a task assigned yet again to personnel-challenged NWRB (Box 3.3).

Box 3.3: Communities, Local Government, and Water District Working Together Toward Sustainable Watershed (Lorenzo Lambatin. "Bago City Water District is Civil Service Awardee" The Freeman. 12 July 2012. http://www.philstar.com/region/2012/07/12/827002/bago-city-water-district-civil-service-awardee)

The Case of Bago City

Bago City is a second-class component city in the province of Negros Occidental. It has a land area of 389 km^2 and a population of 170,981 (2015 Census). Domestic water is provided by the Bago Water District (BACIWAD), which supplies water to 4348 households in barangays Poblacion, Lang-asan, Balingasag, Napoles, Ma-ao, Sampinit, and Calumangan in Bago City and in the town of Pulupandan, Negros Occidental. In 2012, BACIWAD received the Citizen's Satisfaction Center Seal of Excellence Award from the Civil Service Commission in recognition of the good quality service BACIWAD has provided its customers.

Aside from directing efforts toward economic efficiency and social equity in the provision of service, BACIWAD ensures the environmental sustainability of water provision by protecting and conserving its water source through a mechanism called Payment for Watershed Services (PWS). PWS creates an incentive system for conservation by supporting positive externalities through the transfer of financial resources from beneficiaries of certain environmental services to those who provide these services or are fiduciaries of environmental resources (Silvertown 2015).

The PWS mechanism in Bago City works as follows:

1. Each concessionaire or household receives two bills from BACIWAD, the first is the bill for water consumption and the second is a payment for the protection of the watershed (PWS). The second payment is supported by a resolution passed by Bago City in 2015 under a different city executive/mayor and implemented in 2016 under a new city executive. It is worthwhile to note that such scheme was fully supported under the two political leaderships.
2. The watershed fee is equivalent to 1% of the total amount of the water bill and is therefore progressive.
3. The fund collected from the fee is used for reforestation projects to ensure a healthy watershed. Fees are collected by the water district and turned over to the LGU who acts as the provider of the watershed service through reforestation. The first reforestation activity was done in June 2016. Contractual workers are hired by the LGU to serve as guardians of the watershed. These workers are local residents of the areas where reforestation is being done and where the water district extracts water for distribution.

(continued)

Box 3.3 (continued)

Under this scheme, the buyer of the service is the household, the provider is the LGU, and the intermediary is the water district. This is an example of a mixed scheme where business, community members, and government are all involved (Leo Espada, BACIWAD manager, Personal communications, 25 August 2016).

There are parallel cases of corporations or agencies that voluntarily pay or collect an "environmental fee," proceeds from which are used exclusively for rehabilitation and management of the watershed (e.g., NAPOCOR administers this fund through the Buhi-Bahit Watershed Area Team in Lake Buhi, Camarines Sur province; The LLDA also administers the Environmental Users Fee for rehabilitation and conservation measures). The BACIWAD PWS scheme is unique in that the LGU where the watershed is located becomes the contracted party for the task, while the funding arm is the water district. This agreement bridges the policy gap where the water permit holder does not enjoy management rights over water supply. It provides a platform where social justice goals are met because the watershed residents and their LGU feel that their livelihood needs are not imperiled in securing the supply for domestic users of the water district.

3.6 Conclusion

Philippine water policy draws from multiple legislations that map the dominant and diverse roles played by the state in this area. As owner of water resources in the country, the state alone extends rights to access, utilize, and develop the resource. In addition, the state has the following roles: (1) directly engages in provisioning by establishing public enterprises that produce and/or distribute water, (2) provides investments for water infrastructure, (3) regulates activities of the water sector, (4) decides on sectoral prioritization in a crisis event or scarcity scenario, and (5) resolves conflicts. From the legislations, the formal institutional arrangements are laid out between national level agencies through which the state carries out the tasks of service provision, regulation, and sectoral prioritization, among others. The formal institutions along these lines are the water apex bodies (NWRB, LWUA, and NIA), line agencies with subsidiary mandates on water (DA, DENR, DPWH, DOH), and other agencies with water program concerns (DSWD and DILG). Under the principle of decentralization, the state downloads some of the tasks (e.g., provisioning, enforcement of standards and penalties) to LGUs. Given that water is an economic and social good, the state, through its agents, submit to social equity goals by providing subsidies to consumers (farmers and households) and funding projects that seek to improve water access among the poor. In recent years, the state has allowed for greater private sector participation in the water sector as direct providers or concessionaires.

The country's water policy is fragmented and layered. Formal authority over water matters is dispersed among many government agencies and between the national and local levels of government. The national government agencies have overlapping or competitive mandates—e.g., NIA and BSWM under DA on small-scale irrigation. Regulatory power is given to three different agencies (NWRB, LWUA, and DILG), each with its own capability and personnel deficits. Agencies collect their own data in keeping with their mandates, be it water quality, resource management, or supply. Interagency coordination platforms exist for flood control, water crisis, and sector planning but these are weak and focused on one-dimensional concern. Water apex bodies are institutionally weak; subline agency bodies are intermittently transferred, depending on the political mood of the executive branch. The decentralization of decisionmaking on water is limited, with local governments not having the ability to supervise water organizations operating within their administrative jurisdiction, and being highly dependent upon financial transfers from the national government for water development projects.

Formalization of the water sector through the state apparatus is incomplete. Formal institutional arrangements are more prevalent in urban areas and town centers serviced by water districts. In remote locations, informal water governance schemes based on indigenous and customary rules prevail rather than water laws set by the government. In these places, the state's exclusive claim on water (legal rights) is either contested or not observed. The state's regulatory reach is weak; numerous private water providers, rural and barangay (village) water and sanitation associations, and LGU-run water utilities operate without permit and do not observe standards on tariff setting. There is poor compliance on water rules, including permit-based access and withdrawal, as well as pollution. Sanitation is neglected by the government in favor of public investments toward water provisioning (irrigation and domestic).

There remains a serious gap in mechanisms for transfers between sectors and across administrative jurisdiction. The lack of clarity regarding these mechanisms generates conflicts among actors, and subsequently, political mobilization by losing groups for favorable outcomes. As exemplified in the Angat dam case, Bulacan farmers with their local government champion resorted to lobbying in order to collect compensation which they feel they deserve from an earlier water reallocation decision toward domestic users in Metro Manila. Poor legal grounding has also produced conflicts involving LGUs over water transfers reaching the courts. Social agreements among multiple users of a common water source (e.g., waters of Lake Buhi, Manupali River) provide a temporary fix to the tensions arising from incoherent water rights claims. The tenuousness of such informal arrangement in the face of changing water supply and demand contexts is illustrated in the case of the towns of San Fernando and San Gabriel.

Given these shortcomings, reforms in water policy should be directed in these areas: (1) strengthening NWRB in the form of a Water Department with robust regulatory, coordinative, oversight, conflict resolution, and overall planning capabilities for all watershed-based use and allocation decisions to meet broad national goals; (2) polycentric governance arrangements that admit informal actors, indigenous and customary rules, and conflict-resolution mechanisms, alongside those prescribed by law; (3) clarity and local grounding of state property rights to water, but with accommodation of communal and private rights claims; (4) institutional arrangements for water governance based on the physical connectivity of surface water resource (upstream, midstream, downstream) and between water, land, and forest resources; (5) substantive decentralization that puts meaningful decisionmaking on water concerns at the local level, through inclusive and participatory platforms with government, private sector, and civil society; and (6) scientifically and empirically based decisionmaking through the institutionalization of research centers that generate accurate water data.

The laws, institutional arrangements, and mechanisms surrounding water evolved in distinct historical and international contexts. Reforms follow the country's paths of centralization under authoritarian rule, decentralization following the democratic shift, and the marketization mood of the global economy. To effect further reforms on institutional arrangements would require enormous investment in political capital to bring not just agencies with entrenched bureaucracies that may not want their mandates challenged but also water organizations and local communities on board. Legitimacy or widespread acceptance of policy changes is a good starting point.

References

ADB (Asian Development Bank). (2012). *The KALAHI-CIDSS Project in the Philippines: Sharing knowledge on community-driven development*. https://www.adb.org/sites/default/files/publication/29878/kalahi-cidss-project-philippines.pdf. Accessed 23 Nov 2016.

ADB (Asian Development Bank). (2013). *Philippines: Water supply and sanitation sector assessment, strategy and road map*. Mandaluyong City: ADB.

Almaden, C. R. C. (2014). Protecting the water supply: The Philippine experience. *Journal of Social, Political and Economic Studies, 39*(4), 467–493.

Araral, E., Jr., & D. Yu. (2013). *Comparative water law, policies and administration in Asia: Evidence from 17 countries*. Lee Kuan Yew School of Public Policy. Research Paper No. 13-18 IWP. http://ssrn.com/abstract=2338374. Accessed 4 Nov 2014.

Chia, P. L. G., Chua, K. C., Kim, F. C., Teo, S., & Toh, K. L.. (2007). *Water privatization in Manila, Philippines—Should water be privatized? A tale of two water concessionaires in Manila*. Economics and Management in Developing Countries (p. 21), INSEAD. http://www.circleofblue.org/wpcontent/uploads/2012/06/Insead_Water_Privatization_Manila_Philippines.pdf. Accessed 23 Nov 2016.

David, W. P. (2004). Water resources and irrigation policy issues in Asia. *Asian Journal of Agriculture and Development, 1*(1), 76–97.

Dayrit, H. (2000, March 30–31). *Role of interagency collaboration*. Paper presented at the 4th Multisectoral Forum on Watershed Management sponsored by UPLBCEFNR, Quezon City.

Duque-Piñon, C., Catacutan, D., Leimona, B., Abasolo, E., van Noordwijk, M., & Tiongco, L. (2010, June 28–July 1). *Conflict, cooperation, and collective action: Land use, water rights, and water scarcity in Manupali watershed, southern Philippines*. Paper presented at the CAPRi Workshop on Collective Action, Property Rights, and Conflict in Natural Resource Management, Siem Reap.

Ebarvia, M. C. M. (1997). *Pricing for groundwater use of industries in Metro Manila, Philippines*. EEPSEA Research Report. http://www.eepsea.org/pub/rr/10536132990ACF4D.pdf. Accessed 23 Nov 2016.

Elazegui, D. D. (2004). Water resource governance: Realities and challenges in the Philippines. In A. Rola, H. Francisco, & J. P. T. Ligaton (Eds.), *Winning the water war: Watersheds, water policies and water institutions* (pp. 84–104). Makati City: Philippine Institute of Development Studies and Philippine Council for Agriculture, Forestry and Natural Resources Research and Development.

Elazegui, D. D., Rola, A., & Allis, E. (2016). Enhancing institutional dynamics for multiple uses of water amidst climate-related risks: The case of Lake Buhi, Philippines. *Lakes & Reservoirs Research & Management, 21*(3), 224–234.

Greenpeace Southeast Asia. (2007). *The state of water resources in the Philippines*, Quezon City. http://www.greenpeace.org/seasia/ph/Global/seasia/report/2007/10/the-state-of-water-in-the-phil.pdf. Accessed 23 Nov 2016.

Hall, R., Lizada, J., Dayo, M. H., Abansi, C., David, M., & Rola, A. (2015). To the last drop: The political economy of Philippine water policy. *Water Policy, 17*(5), 946–962.

LWUA (Local Water Utilities Administration). (n.d.). *Sagana at Ligtas na Tubig Para sa Lahat (SALINTUBIG)*. http://www.lwua.gov.ph/salintubig/SALINTUBIG%20Program%20Briefer.pdf. Accessed 3 Sept 2016.

Malayang, B., III. (2004). A model of water governance in the Philippines. In A. Rola, H. Francisco, & J. P. T. Ligaton (Eds.), *Winning the water war: Watersheds, water policies and water institutions* (pp. 59–84). Makati City: Philippine Institute of Development Studies and Philippine Council for Agriculture, Forestry and Natural Resources Research and Development.

NEDA (National Economic and Development Authority). (2010). *The Philippine water supply sector roadmap* (2nd ed.). Pasig City: NEDA.

Paragas, V. S. (2012, January 26). *Water regulatory policies*. Paper presented at the Roundtable Discussion on Water Rights and Water Wrongs: Toward Good Water Governance for Development, Social Sciences Division, NAST PHL, Hyatt Hotel and Casino Manila, Manila.

Robinson, A. (2003). *Urban sewerage and sanitation: Lessons learned from case studies in the Philippines* (p. 6). World Bank Water and Sanitation Program. https://www.wsp.org/sites/wsp.org/files/publications/eap_urban_en.pdf. Accessed 23 Nov 2016.

Rola, A. C., & Elazegui, D. D. (2008). Role of institutions in managing agriculture-related climate risks: Angat reservoir case study, Bulacan, Philippines. *Journal of Environmental Science and Management, 11*(1), 26–39.

Rola, A. C., Pulhin, J. M., David, C. C., Wensley, C., Paragas, V. S., Tabios, G. Q., III, Lizada, J. C., Gazmen, P. G., & Dayo, M. H. F. (2012). Towards good water governance in the Philippines. *Transactions of the National Academy of Science and Technology (Philippines), 34*(2), 299–323. ISSN 0115-8848.

Rola, A. C., Pulhin, J. M., Tabios, G. Q., III, Lizada, J. C., & Dayo, M. H. F. (2015a). Challenges of water governance in the Philippines. *Philippine Journal of Science, 144*(2), 197–208.

Rola, A. C., Abansi, C. L., Arcala-Hall, R., Lizada, J. C., Siason, I. M. L., & Araral, E. K., Jr. (2015b). Drivers of water governance reforms in the Philippines. *International Journal of Water Resources Development, 32*(1), 135–152.

Rola, A. C., Abansi, C. L., Arcala-Hall, R., & Lizada, J. C. (2016). Characterizing local water governance structure in the Philippines: Results of the water managers' 2013 survey. *Water International, 41*(2), 231–250.

Silvertown, J. (2015). Have ecosystem services been oversold? *Trends in Ecology and Evolution, 30*(11), 641–648.

Singh, S. (2006). In *Paradigms of decentralization, institutional design and poverty: Drinking water in the Philippines. Philippine Institute for Development Studies (PIDS) discussion paper series 2006-19*. Makati City: PIDS.

Tabios, G. Q., III, & David, C. C. (2004). Competing uses of water: Cases of Angat reservoir, Laguna Lake, and groundwater systems. In A. Rola, H. Francisco, & J. P. T. Ligaton (Eds.), *Winning the water war: watersheds, water policies and water institutions* (pp. 105–132). Makati City: Philippine Institute of Development Studies and Philippine Council for Agriculture, Forestry and Natural Resources Research and Development.

The Water Dialogues. (n.d.). *The Philippine water situation*. www.waterdialogues.org/documents/PhilippinesCountryContext.pdf. Accessed 23 Nov 2016.

World Bank. (2005). Chapter 7: Water supply and sanitation from Philippines. In: *Meeting infrastructure challenges* (pp. 111, 113).

World Bank. (2013). *East Asia and the Pacific Region: Urban sanitation review*. http://documents.worldbank.org/curated/en/771821468036884616/text/842900WP0P12990Box0382136B00PUBLIC0.txt. Accessed 23 Nov 2016.

Wu, X., & Malaluan, N. (2008). A tale of two concessionaires: A natural experiment of water privatisation in Metro Manila. *Urban Studies, 45*(1), 207–229.

Dr. Rosalie Arcala Hall is a Full Professor at the College of Arts and Sciences, University of the Philippines Visayas (UPV). She completed her Master's degree in Political Science and PhD in International and Public Affairs (2002) at Northeastern University, Boston, Massachusetts. She completed research projects with The Asia Foundation, The Nippon Foundation, and Toyota Foundation concerning policy frameworks, their localization, resistance and competing narratives behind contestations in the areas of hard and soft security. She has contributed chapters to the books Rosalina Palanca-Tan ed. Nature and Culture: Environmental Issues in Asia (Ateneo de Manila Press, 2014), and Justine Vaz and Narumol Aphineves eds. Living Landscapes Connected Communities: Culture, Environment and Change Across Asia (Areca Books, 2014). She is co-author of two articles in Water Policy and International Journal of Water Resources Development. Currently, she is a member of the Philippine Commission on Higher Education Technical Committee on Political Science and the Philippine Political Science Association Board.

Dr. Corazon L. Abansi is Professor of the Institute of Management, University of the Philippines Baguio and holds a Ph.D. in Agricultural Economics major in Resource Economics. She is an ICM practitioner of UNDP-IMO facility "Environmental Management of the Seas of East Asia". She did research and development work on economics of water pollution, resource valuation and community mobilization while serving as officer and resource economist of the UNDP-IMO Program for the Prevention of Marine Pollution in East Asian Seas. Dr. Abansi is actively engaged in research in the Cordilleras focusing on the economics of hydropower generation and ecosystem services, and resource management of indigenous communities. As site coordinator and resource economist for the project, "Towards Good Water Governance for Development" she is working towards developing an adaptive-collaborative water governance mechanism for the Cordilleras and has co-authored two papers published in Water Policy and International Journal of Water Resources Development.

Dr. Joy C. Lizada is Professor at the College of Management, University of the Philippines Visayas (UPV). She holds a Ph.D. in Agricultural Economics (major in Resource Economics) from the University of the Philippines Los Baños. She completed her Diploma in Agricultural Economics from Massey University, New Zealand and Master of Management (major in Business Management) from UPV. She has done several research projects and publications on various topics such as institutional analysis, supply chain analysis, socio-economic studies, organizational development studies, environmental governance, economic valuation, and sustainability assessment. Dr. Lizada is actively involved in developing adaptive-collaborative governance mechanisms for water resource management in the Visayas region. She is co-author of two articles published in Water Policy and International Journal of Water Resources Development.

Chapter 4
Domestic Water Supply

Antonio R. De Vera and Rosalie Arcala Hall

Abstract The legal framework on domestic household water use and sanitation specifically mandates the Metropolitan Waterworks and Sewerage System to engage in capital infrastructure and water distribution in Metro Manila. On the other hand, the water districts and municipal-government-administered water works play a key role in the supply, distribution, and management at the local level. Historically, these water institutions have paid less attention to sanitation and have focused more on water provisioning. Insufficient public investments and mismanagement have led to serious gaps in performance. The patterns of water use in the domestic household sector follow the country's demographic distribution where urban and town centers are favored infrastructure-wise over rural and urbanizing areas. This chapter identifies the inefficiencies, the gaps in access by the poor, and the poor participation by consumers in water district schemes in urban areas. It also describes the challenges of rural villages distant from town centers that remain underserved or with crude water provisioning schemes. Urbanizing areas suffer from competing uses between households and small-scale industries and the attendant pollution arising from unregulated wastewater-dumping activities. Incipient attempts at reform either at the community or local government level toward improving domestic household water use and sanitation are mapped out.

Keywords Domestic water • Water service provider • Water district • Local government-run utility • Sanitation

A.R. De Vera (✉)
Subic Bay Water Regulatory Board, Subic Bay Freeport Zone, Philippines
e-mail: tonydvera@yahoo.com

R.A. Hall
Division of Social Sciences, College of Arts and Sciences, University of the Philippines Visayas, Miagao, Iloilo, Philippines

A.C. Rola et al. (eds.), *Water Policy in the Philippines*, Global Issues in Water Policy 8, https://doi.org/10.1007/978-3-319-70969-7_4

4.1 Sources of Domestic Water Supply

The Philippines obtains its water supply from rainfall, surface water resources (i.e. rivers, lakes, and reservoirs), and groundwater resources. It has 18 major river basins and 421 principal river basins as defined by the National Water Resources Board. Theoretically, the freshwater storage capacity and the high rate of precipitation assure the country an adequate supply of water for its agricultural, industrial, and domestic requirements. However, seasonal variations are considerable and geographic distribution is biased, often resulting in water shortages in highly populated areas, especially during the dry season.

Metro Manila is being served primarily by the Metropolitan Waterworks and Sewerage System (MWSS) through its two private concessionaires, the Maynilad Water Services, Inc. (MWSI) and the Manila Water Company (MWCI), and by some private companies serving subdivisions. The MWSS water supply comes mainly from surface water.

There are around 500 operational water districts (WDs) that mainly serve the towns and cities outside Metro Manila. In addition, about 1200 piped water systems are operated by municipal governments. Rural areas not covered by the WDs and municipal water systems are served by community-based organizations (CBOs) – e.g., rural water service associations (RWSAs), barangay water service associations (BWSAs) or village-level water systems, and water service cooperatives – or are not served at all (Manahan 2012).

Most utilities outside Metro Manila utilize groundwater (wells and springs). Only large utilities with more than 20,000 connections utilize surface water (rivers, lakes). Almost all community based utilities and LGU-operated systems, due to their small size, use groundwater sources.

4.2 Sector Framework

4.2.1 Sector Institutions and Their Roles

Institutions dealing with domestic water supply can be divided into water service providers (WSPs), regulators, sector planners at the national and local levels, program implementers at the local levels, funding institutions, and water-resource-related agencies.

There are a number of institutions with regulatory responsibilities in the domestic water supply sector. These include three primary regulatory agencies – the National Water Resources Board (NWRB), Local Water Utilities Administration (LWUA), local government units (LGUs) – and special regulatory units such as the Subic Bay Water Regulatory Board (SBWRB) created by the Subic Bay Metropolitan Authority (SBMA) and the MWSS Regulatory Office, which operates on contract-based regulation. All these existing regulatory structures have different regulatory

Table 4.1 Different regulatory practices for domestic water

Regulatory body	Regulatee/s	Tariff methodology	Monitoring system
NWRB	Private utilities and CBOs	Return on assets	Annual reports from utilities
LWUA	Water districts, RWSAs	Cash flows	Annual reports plus visits
LGU	LGU utilities	Cash and political considerations	Public complaints
MWSS-RO	2 private concessionaires of MWSS	Return on investment with appropriate discount rate	Monthly reports plus visits
Subic Bay Regulatory Board	One private utility	Return on equity	Monthly reports plus visits

Source: NEDA (National Economic and Development Authority) (2010)

practices, processes, and fees with cases of overlapping functions or jurisdictions. This environment suggests a fragmented regulatory framework and lack of coordination. Regulation of WSPs by these institutions is weak (ADB 2013). A summary of the different regulatory practices of these regulatory bodies is shown in Table 4.1.

4.2.2 Service Providers

Several types of WSPs exist nationwide, consisting of water districts; LGU utilities; CBOs such as rural water and sanitation associations (RWSAs), barangay water and sanitation associations (BWSAs), homeowners associations, and cooperatives); and private utilities. Only MWSS and the water districts are government corporations.

(a) *Water districts (WDs):* government-owned and -controlled water utility corporations created by LGUs in accordance with the law (PD 198). Water districts operate piped water systems.
(b) *LGU-operated systems*: directly owned and managed by an LGU. The LGU could be a province, city, town, or barangay. The LGU system may operate levels II or III systems or a combination of both.
(c) *Community-based organizations (CBOs):* groups of people who have banded together to own and operate water systems. CBOs are considered "private" and fall into three major categories:

- RWSAs/BWSAs: water systems run by a non-profit, non-stock association usually covering one or more barangays
- Water cooperatives (Coops): owned and managed by a cooperative whose members contribute to the cooperative equity base
- Homeowner associations (HOAs): owned and operated by associations of residents within gated communities

(d) *Private firms:* privately owned water systems operating within a given franchise area. Examples are the two Metro Manila concessionaires, Primewater, Balibago Waterworks, and systems still operated by subdivision developers.

Past sector studies have differed in terms of reported number of WSPs in the country. A World Bank report has an estimate of the entire range of WSPs (World Bank 2013a, b). However, a 2015 project called the "Listahang Tubig" of the NWRB gives the following number of utilities providing level III services.[1]

4.2.3 Service Levels

There are three domestic water service levels in the country (NEDA 2010):

Level I is a protected well or a developed spring with an outlet but without a distribution system, generally adaptable for rural areas where houses are thinly scattered.

Level II is a piped system with communal or public faucets usually serving 4–6 households within a 25-m distance.

Level III is a fully reticulated system with individual house connection.

Community-based organizations comprise the majority of WSPs in the Philippines, while LGU-run water utilities come to a close second. The number of CBOs decreased dramatically from 2003 to 2013, with the largest cut seen in BWSAs. In contrast, the number of LGU-run utilities increased by more than 20%, implying that, perhaps, the municipal governments took over the BWSAs (which are at the village level) that stopped operating (see Table 4.2). Table 4.3 provides estimates of Level **III** coverage in urban and rural areas. The urban bias in Level III

Table 4.2 Comparison of water supply providers (level III systems)

Type of WSP	2003	2013
Water districts	455	507
LGU utilities	1000	1277
MWSS	1	1
Community-based organizations (CBOs)	3800	2072
RWSAs	500	667
BWSAs	3100	1021
Cooperatives/HOAs	200	349
Unclassified	–	35
Private	900	863
Total	6156	4720

Sources: for 2003 data, World Bank (2013b); for 2013 data, NWRB (2015)

[1] This "list of water systems" is derived from a national survey (water register) of all WSPs covering all service levels in 2015. Data given as of 2013. NWRB gives the number of utilities providing level III services.

Table 4.3 Estimated level III water service coverage in the Philippines

	Population (million)(2010 Census)	Est. % of total population	Est. Level III coverage (% of total households)	Water supply provider(s)
Urban-NCR	11.9	13	88	Manila Water, Maynilad
Urban Outside NCR	33.0	36	50–65	Water districts LGUs Private operators
Rural	47.4	51	25	Coops, BWSAs, RWSAs
Total	92.3	100	42–48	

Sources: ADB (2013: 7)

BWSAs barangay water and sanitation associations, *est* estimated, *LGU* Local Government Unit, *NCR* National Capital Region, *RWSAs* rural waterworks and sanitation associations

Table 4.4 Consumption patterns (lpcd)

Utility model	2003	2013
Water districts	120	108
Private sector	144	129
CBOs	127	86
LGU systems	96	99

Source: NWRB (2015)

provisioning is self-evident; urban areas have three times more coverage than rural areas, which are mostly served by CBOs such as cooperatives, BWSAs, and RWSAs.

Planning bodies in the country advocate a design consumption of 20 liters per capita per day (lpcd) for Level I service and 40 lpcd for Level II. Since 1955, formal utilities have adopted a block consumption pattern as basis for designing water tariffs for Level III systems (MDGF-UNDP 2011).[2] Payment for water consumption is composed of a "minimum" charge and a "commodity" charge for each successive block. Since 1960s, the lifeline consumption or minimum volume had been set at 10 m^3 per month for a domestic connection by almost all formal utilities.

In 2014–2015, the NWRB, in coordination with the World Bank, started a national survey of all utilities in the country, including their 2013 operational performance metrics. The results for consumption patterns per person are shown in Table 4.4.

Consumption patterns dropped over time, attributed mainly to an increase in water tariffs. LGU systems hardly changed their tariffs and this is apparent in Table 4.4.

[2] Boracay, Subic Freeport, Clark, Baguio, Zamboanga, and Metro Manila.

4.3 Sector Performance Assessment

This section assesses the water and sanitation services in the country and presents its institutional strategy and its strengths and weaknesses in providing domestic water services to various sectors.

4.3.1 Institutional Strategy

Since the country's independence from America in 1946, the responsibility for waterworks, both urban and rural, was borne by the various LGUs. The provincial populace relied on springs and shallow wells as sources of water. Over time, the water supply situation did not improve and assistance had to be sought from the national government.

1. In 1955, the National Waterworks and Sewerage Authority (NAWASA) was established. This came about from the transformation of the Manila Waterworks Authority (created in 1878 to serve Manila) into a national facility with responsibilities for the entire country in cooperation with the LGUs. However, in 1971, the government discovered that the provincial urban systems (both LGU- and NAWASA-operated) were not able to keep up with demand and that the systems were in poor condition. The government then decided to return the management of the provincial water utilities to the LGUs and created three national offices for the water and sanitation sector. The NAWASA was transformed into the MWSS to service Metro Manila only.
2. The LWUA was formed in 1973 to assist LGUs create WDs in the provincial urban areas and to provide technical, financial, and institutional development assistance and regulations to them. The WD, a new management model for urban water supply, had a corporation-like organizational structure with a board distinct from the management team.
3. The Rural Water and Development Corporation (RWDC) was established (1980) to provide assistance to areas where neither WDs nor MWSS operates. The RWDC created RWSAs in order to construct, operate, and maintain their own water supply systems. In 1987, LWUA took over the work of the RWDC, which had been abolished.

Due to decreased central government funding for rural water supply, the new generation of rural projects tried to secure commitment and ensure sustainability by requiring capital contributions from communities and local governments as well as investing in "soft" components, such as institutional strengthening, capacity building, community-based planning, and health and hygiene promotion. In these projects, the national government typically provided grants equal to 90–100% of the total project cost. The remainder was provided by the community, through its water user association, in the form of voluntary labor, donated land, or cash contributions. Any costs associated with expanded levels of service were borne wholly by the community.

Responsibilities became more decentralized under the framework of the Local Government Code (LGC), which was promulgated in 1991 (Official Gazette 1991). This Code gave LGUs more power and more funding from the central government using national revenues. The LGC transferred powers and responsibilities from the central government to the LGUs, including the primary responsibility for developing water supply and sanitation services. The LGUs were required to share the installation costs for rural water supply systems, while the responsibility for system implementation was transferred from the central agencies to the LGUs.

This strategy also espoused the use of public-private partnership in various infrastructure undertakings of the government, including waterworks. The scope for private sector participation following the passage of the Water Crisis Act of 1995 and a Supreme Court resolution is reflected in the policy that WDs do not have exclusive franchises in their coverage areas where water service is still unavailable. These opened the door for private firms and CBOs, especially in areas of high demand.

The previously mentioned "Listahang Tubig"[3] gave a total of 863 private utilities operating in the country, composed of 742 private companies, 38 industrial locators, 63 real estate developers, and 20 peddlers. It does not mean that there are 863 companies involved inasmuch as one company may have more than one utility being operated. The private sector actively seeks opportunities to invest in domestic water provisioning mainly through the public-private partnership framework (Ndaw 2016). Private water companies are able to put forward proposals to provide water supply and sanitation services to local governments and local water utilities directly through PPP or joint venture schemes. This development signals the private sector's capability to professionalize the service and improve the level of service. More and more government agencies are turning to the private sector for performance improvement and/or efficiency gain.

4.3.2 Operational Performance

How well are the various utility models performing? Table 4.5 shows the performance of the utility models over time.

The WDs have relatively maintained their performance efficiency over time. Only the collection period has been lengthened. Although tariffs have increased by 26% over the last 10 years, this is less than the annual inflation rate.[4]

[3] Listahang Tubig is an ongoing study and results may differ from month to month due to the number of samples already included and tabulated. Except for WD and private firms, data from other WSPs have yet to be validated. Data listed herein are as of Feb 2017.

[4] San Fernando, La Union; Baliwag, San Jose del Monte, and Meycauyan, Bulacan; Cabanatuan, Nueva Ecija; Naga City, Camarines Sur, Dumaguete City, Bayawan, Negro Occidental: Cebu City; Malalag, Davao Sur; and five towns in Sarangani.

Table 4.5 Operational performance of various utility models

	Water districts		LGU systems		Private		CBOs	
Utility model year	2003	2013	2003	2013	2003	2013	2003	2013
Non-Revenue Water (%)	27	25	–	23.0	–	25	–	18.3
Collection period (mo)	1.2	2.06	2.1	1.73	1.7	2.22	2.3	2.31
Service hours/day	23	22.5	18.4	19.19	21.9	18.2	20	18.20
Average tariff (PhP/m^3)	17.82	26.0	7.22	13.0	15.37	22.7	7.99	13.1
Operating ratio	0.70	0.72	1.18	0.94	0.74	0.73	0.87	0.85

The LGU systems have shown improvements in their collection periods, service hours, and operating ratio. Tariffs have risen by only 12% over 10 years, the lowest among the four models; hence their operating ratio is still precariously high.

The private systems are good performers, given their NRW, collection period, and operating ratio performance. Although tariffs have risen by 23% over 10 years, just like the WDs, their operating ratio has gone down, indicating good control of expenditures.

The CBOs have also done well. Despite only a 14% increase in tariff, their operating ratio has gone down.

4.3.3 Sector Accomplishments

4.3.3.1 Piped Water Supply Coverage

The Philippine Water Supply and Sanitation Roadmap (PWSSR) states that, in 2000, 46.1% of the population had access to level II and III services. In 2003, the coverage of piped water had increased to 54% or 44.3 million (Table 4.6) (World Bank 2013b: 17).

A demand gap analysis for levels II and III was performed in 2011 for the Department of Public Works and Highways (DPWH) and the results indicated a 42.7% coverage of level III facilities. Although the percentage may have gone down, the population served has gone up due to a larger population base. To confirm this seeming reduction, the Joint Monitoring Program (JMP) figure of 40% piped coverage as of 2010 (Table 4.7) comes very close. The Joint Monitoring Programme (JMP) for Water Supply and Sanitation by WHO and UNICEF is the official United Nations mechanism that monitors progress towards the Millennium Development Goal (MDG) Target 7c, that of halving by 2015 the proportion of people without sustainable access to safe drinking water (and basic sanitation)

The JMP table also verifies that the Philippines may have already met said Target.

Table 4.6 Market share, by type of provider, 2003

<table>
<tr><td colspan="5">Access to formal service levels: 79%</td><td>No access: 21%</td></tr>
<tr><td colspan="3">Level III
44%
(36.1)</td><td>Level II
10%
(8.2)</td><td>Level I
25%</td><td>Private wells
Tankered
Vended water-</td></tr>
<tr><td>WDs
14%
(11.5)</td><td>Private
10%
(8.2)</td><td>LGUs/CBOs
20%
(16.4)</td><td colspan="2">LGUs/CBOs
35%</td><td>SSIPs and/or self-provision by households</td></tr>
<tr><td colspan="6">Complementary service provided by SSIPs or self-provision</td></tr>
</table>

Source: World Bank (2005)

Note: Figures in parentheses are added by the authors to reflect the population served (in million) as of 2003

SSIP Small scale independent providers. Most are real estate developers, homeowners' associations, local entrepreneurs, and mobile water truckers and haulers

Table 4.7 Estimates of JMP coverage (%)

Year	Total improved	Piped onto premises	Other unimproved	Surface water
1990	84	25	12	4
1995	85	29	11	4
2000	87	33	9	4
2005	89	36	8	3
2010	90	40	7	3
2015	92	43	5	3

Source: WHO-UNICEF (2015)

4.3.4 Sector Challenges

General policies in the water and sanitation sector concern issues such as decentralization of water supply provision, use of public-private partnership and meeting coverage targets. NEDA, in its Medium-Term Philippine Development Plan for 2011–2016, aimed to extend coverage of potable water, increase coverage of both level III and 24/7 services, with priority given to 400 barangays with poor water supply coverage (NEDA 2004).

4.3.4.1 WSP Service Performance

Results of a benchmark survey conducted by ADB/DILG covering 45 utilities revealed that performance of most utilities still fell under industry standards (NEDA 2010: 7). Among the WSPs covered, the LGUs had the most dismal performance.

The slow expansion and low quality of services and the inefficient operation of water utilities are attributed to the sector's weak regulatory and financing framework, lack of technical and managerial capacity, lack of access to financing for water and sanitation development, and dependence on subsidies for the majority of service providers. Thus, regulatory, financing, and utility reforms are imperative to improve WSP service performance.

Several constraints at the local level need to be urgently addressed: low LGU awareness and political will, inadequate information dissemination and development of human resources, low multistakeholder involvement, and inadequate financing schemes. These problems are exacerbated by the lack of local policies and programs on water supply and sanitation,resulting in low prioritization of water- and sanitation-related projects (specially sanitation and hygiene) and creating an upsurge in the incidence of diarrhea and other waterborne diseases in the community.

4.3.4.2 Regulation

The NWRB, by virtue of RA 2677 (Public Service Law), regulates private water providers. Despite its already extended regulatory mandate, in 2002, Executive Order 213 was issued, which transferred the regulation of WDs from LWUA to NWRB. This mandate was never carried out due to NWRB's lack of resources. In 2010, Executive Order 806 was issued transferring the administration of NWRB from the Office of the President to DENR.

Previously, LWUA was stripped of its regulatory functions over WDs; as a unit, it was transferred to the Office of the President, and then to DPWH (along with MWSS). This regulatory task (over WDs) was restored in 2010 and extended to include RWSAs.

Even with these regulatory mandates, LWUA's focus on financing means that RWSAs and other CBOs were ignored because of high transaction costs associated with these small water providers. Some regulation is extended over to other types of WSPs by other government agencies. Water cooperatives, for instance, register their existence as an organization with CDA, but the latter has to remit with respect to water fees collected and other management concerns. The DILG presumably has supervisory functions on LGU-run utilities and BWSAs, but these are not exercised at all in the absence of mechanisms. Technically speaking, NWRB regulates these WSPs by default, but the lack of resources and capacity, notably local presence, in effect means no regulation at all.

The sector's inability to respond to the water and sanitation needs of the population and derivative problems is rooted in the fragmented institutional environment, weak regulatory framework, and inadequate support for service providers and utilities, resulting in poor performance, limited access to financing and investments, low tariff and cost recovery, inadequate support for rural water supply, and lack of reliable and updated sector information needed for sector planning.

4.4 Sanitation Aspects

4.4.1 Public Health and Sanitation

Sanitation refers to a wide range of services and arrangements pertaining to the hygienic and proper management of human excreta (feces and urine) and community liquid wastes to safeguard the health of individuals and communities. It is concerned with preventing diseases by hindering pathogens or disease-causing organisms found in excreta and sewage from entering the environment and coming into contact with people and communities (DOH 2010). In a 2015 report, annual economic loss due to poor sanitation is about PhP 78 billion (L. Claudio, DENR, pers.commun.).

Fragmented institutional arrangements with no strong administrative mechanisms to guide policy implementation and to coordinate local program implementation show inadequate attention accorded to sanitation. Institutions mandated to construct, operate, and maintain sanitation and sewerage systems include MWSS for Metro Manila, the WDs, and LGUs for areas outside Metro Manila. The Clean Water Act of 2004 requires LGUs and WDs to create septage management programs in areas without sewerage systems. Four national agencies were given specific sanitation roles:

DPWH: administers the national government share through a National Sewerage and Septage Management Program (NSSMP) Office

DILG/DOH/DPWH: provides capacity building programs and assistance in developing local and regional plans and projects

DOH: sets guidelines and standards and issues environmental sanitation clearances

DENR: promulgates and enforces water and wastewater standards

In general, low priority has been given to sanitation at both national and local government levels. Most of the investments in sanitation have come from private investments in household toilets, housing estate wastewater treatment, and on-site wastewater treatment among commercial, industrial, and institutional establishments. Unfortunately, there has been little control or regulation of these private facilities and some are poorly designed and constructed.

The predominant sanitation technology in urban Philippines is the septic tank. According to a 2010 report, 84% of urban households discharge wastewater to a septic tank with another 10% with improved sanitation using other options (World Bank 2013a: 2). While several codes provide standards for design and construction of septic tanks, these are not often enforced. All too often, poorly maintained septic tanks discharge inadequately treated sewage and effluent directly into storm water drains, waterways, and streets, with serious consequences in terms of both water quality and public health. Usually desludging services are offered by private companies and the sludge is disposed of in local water bodies or solid waste dumpsites since septage treatment plants are relatively rare outside Manila.

Only about six of the 1500 provincial cities and towns in the Philippines contain functioning public sewerage systems.[2] As of 2015, the only LGUs with sewerage treatment facilities are Metro Manila, Zamboanga City, Baguio City, Boracay Island, and the Clark and Subic economic zones. Some LGUs have independent sewerage facilities serving only either housing developments or a small part of their business districts. Bacolod City has a sewerage system for two housing villages.

Few households are therefore connected to a sewerage network (less than 4%, by most estimates). There are reportedly a number of villages and condominiums with small treatment plants, but no national agency has this kind of information. The absence of communal sewer systems in urban areas has meant that storm water drains are frequently also used for wastewater disposal.

Only about 15 LGUs have built septage facilities, although most private malls, industries, and large hotels have their own septage treatment facilities.[3] Most of these LGU septage facilities have a design capacity from 30 to 60 m^3/day.[5]

4.4.2 Sanitation Coverage

In 2009, some 20 million Filipinos did not have access to improved sanitation, about 15 million shared toilets, and 9% still defecated in the open (DOH 2010). However, based on the 2015 WHO-UNICEF Joint Monitoring Program (JMP) report, total households with improved use of sanitation facilities reached 97% in the urban areas and 89% in the rural areas, a combined total of 92%.

The proportion of families with access to sanitary toilets has significantly increased from 71% in 1990 to 92% in 2015. Own toilets, shared toilets, and closed pits are considered sanitary in contrast to open pits, drop/overhang, pail system, and absence of access to a toilet facility (Table 4.8) (NEDA 2014).

Table 4.8 Total sanitation coverage (%) in 2015

	With access to sanitary facilities		No access to sanitary facilities	
Year	Improved	Shared	Other unimproved	Open defecation
1990	57	14	14	15
1995	60	15	11	14
2000	64	16	8	12
2005	67	16	6	11
2010	70	17	4	9
2015	74	18	1	7

Source: WHO-UNICEF (2015)

[5] One m^3/day is sufficient for one household, hence a 30-m^3/day facility can treat the sludge of 30–60 septic tanks daily, depending on the turnaround of the vacuum trucks.

Although there has been considerable progress in providing sanitation facilities in the Philippines from 1990 to 2015, it is also clear that a lot remains to be achieved, with about one-tenth of the population still not being served in 2015.

The government has some initiatives to improve sanitation. For instance, in 2008, a Supreme Court order to Philippine government agencies *"to clean up, rehabilitate, and preserve Manila Bay, and restore and maintain its waters ...to make them fit for swimming, skin-diving, and other forms of contact recreation"* triggered an extensive investment program by service providers (DENR 2008). This also opened the entry of private firms into areas not being served.

4.4.2.1 Metro Manila

Prior to the operation of two private concessionaires in 1997, the MWSS was serving only about 3–5% of the population in Metro Manila with sewer connections. In 2007, MWSS decided to privatize its operations via a concession mode to two private firms. The east zone would be serviced by MWCI and the west zone by MWSI.

A Sewerage and Sanitation Master plan prepared in 2005 by MWSS described a strategy that would provide sewerage coverage through combined systems (World Bank 2013a: 10).[6] The plan also envisioned a septage management program in combination with sewerage development. The target is to cover 100% of Metro Manila households by 2037.

In 2012, the sewerage connections and septic tank desludging service coverage was about 50% for MWCI and about 83% for MWSI. Manila Water currently operates a total of 39 septage treatment plants with a combined capacity of 142 million liters per day. A total of 160,860 households in the east zone now have access to full wastewater treatment. Manila Water currently operates two septage treatment plants capable of handling 1400 m^3 of septage daily. The other concessionaire, Maynilad, has a total capacity of 940 million liters per day for treating wastewater.

4.4.2.2 Outside Metro Manila

Before 1955, only two cities (Zamboanga and Vigan) had some limited sewerage facilities built during the 1920–30s by the Americans. From 1955 to 1991, no new sewerage or septage infra were built. The passage of the Local Government Code in 1991 gave more responsibilities to the LGUs with respect to sanitation. The NEDA Board passed several resolutions for the sector, which stated that LGUs shall primarily be the implementers of sanitation/sewerage programs, with the national government providing assistance to develop their capacities in certain areas. Several private proponents were thus able to secure contracts for concessions with some LGUs (Manila concessionaires, Subic, Clark, and Boracay) and were able to develop/improve sewerage facilities.

[6] The storm drainage systems would also be used to convey sewage.

With the passage of the Clean Water Act in 2004, some LGUs paid attention to their sanitation facilities and, consequently, some LGUs and/or their WDs were able to put up either a sewerage system (e.g., Baguio City through a grant from the Japan International Cooperation Agency) or septage facilities. A National Sewerage and Septage Management Program was prepared in 2011 by DPWH, which provided targets for urban cities with some grant funding (maximum of 40%) for sewerage projects. But, as of 2015, no LGUs had availed of the financing schemes offered under the program.

The urban poor generally remain excluded from sanitation services. Sewer networks, if any, do not reach the slum and squatter settlements found in most Philippine cities, and their inhabitants seldom have enough space or cash to construct private facilities such as septic tanks. Communal toilets have been provided in some densely populated low-income areas, but these typically serve small groups of households and are rarely well-managed. Instead, most urban poor rely on unsanitary toilets or people defecate in the open (Robinson 2003).

Non-poor urban households have responded by building their own sanitation facilities. Flush (or pour-flush) toilets are popular in the Philippines, and the majority of urban households have toilets connected to private septic tanks. Many private housing developments now construct small independent sewer networks, which serve those within the development area, and pipe their sewage to a communal septic tank. Whatever the system, be it individual or communal, septic tanks in the Philippines, are seldom de-sludged (Robinson 2003).

While LGUs are mandated to provide essential services for water and sanitation, 97% of the funds earmarked for water and sanitation go to water supply and only 3% for sanitation and wastewater treatment (CEC-Phil 2012; NEDA 2008).

4.4.3 Sanitation Constraints

The 2010 Philippine Sustainable Sanitation Roadmap identifies the lack of an effective national sanitation policy, including the lack of a clear policy on sanitation regulation, as one of the critical gaps in the sector. Policies on sanitation regulation should include national targets, a strategy for eliminating open defecation, and a strategy for facilitating localized sanitation improvement plans and budgets and national investment priorities and plans for sanitation.

4.4.3.1 Institutional Constraints

Due to the low priority given by WDs to sanitation, the main responsibility for sanitation lies largely with the LGUs. While LGUs are in the forefront of implementing, monitoring, and, to some extent, regulating sanitation programs and projects, they are generally not adequately informed about these regulations and standards.

The mandates for rural sanitation are not clearly defined between the LGU and DOH, which has adopted a policy objective to achieve open defecation-free status for all barangays, and universal access to sanitary toilets in cities and municipalities. DOH, thus far, has little or no funding allocated for implementation. While DOH plays a key role in the sector due to the health impacts of poor sanitation, its sanitation mandate is limited to policy formulation and monitoring of laws and policies.

There are also no plans, targets, or monitoring systems in place against which LGUs can be held accountable. Local private-sector providers of sanitation goods and services are neither organized nor regulated.

4.4.3.2 Financial/Sustainability Constraints

The operation of several septage facilities has shown the financial viability of such systems, but these are still not being taken up by many WDs or LGUs due to the political acceptability of tariffs. Even in Metro Manila, the component of the tariff allocated to sewerage and sanitation (20% of water tariff) is unlikely to cover costs, and sanitation and sewerage are effectively subsidized by the water supply component. The tariff collection and cost recovery for Dumaguete City's septage treatment facility could be one example that may be replicated. Under a Memorandum of Agreement between the LGU and the WD in Dumaguete, the LGU enacted the required legislation to obligate households to desludge septic tanks regularly. The WD manages the facility and collects user fees from service connections that are included in monthly water bills. Fees are collected from unconnected households directly. Dumaguete LGU/WD charges PhP 2/m^3 of water consumed, which is more than sufficient to meet operating and debt-servicing costs (PADCO 2006).

An increasing number of decentralized wastewater treatment systems are being constructed in cities to process wastewater from markets, slaughterhouses, hospitals, schools, universities, and housing developments (PADCO 2006).[7] Recurrent costs are part of the LGU or institution budget, but the sustainability of the approach is not clear.

The state of urban sanitation in the Philippines reflects a prolonged lack of activity or investment in the sector. In the last three decades, investment in urban sanitation in the Philippines totals only 1.5% of that spent on urban water supply (Robinson 2003). Sadly, urban sewerage and sanitation do not appear to be a priority of local governments.

4.4.3.3 Social Constraints

When discussing the reasons for the lack of progress in increasing coverage of sanitation and sewerage, all survey respondents point to lack of awareness of the need for sewerage and sanitation systems (Robinson 2003). One of the more important

[7] Use of anaerobic baffled chamber **to gravel filter**? and wetlands.

impediments to progress is the attitude that sanitation practice is a household issue and not the responsibility of government.

4.4.4 Sanitation Initiatives

4.4.4.1 National Sewerage and Septage Management Program

The NSSMP was prepared by DPWH in 2010 in consultation with an interagency steering committee and with technical assistance from the World Bank and Asian Development Bank. The primary focus of the NSSMP is sewerage and septage management infrastructure projects that will help cities/towns manage wastewater, and promote/create a supporting environment needed to make these projects successful. The goal of the NSSMP is to improve water quality and public health in urban areas of the Philippines by 2020.

However, due to the prevalence of septic tanks nationwide and the capital cost of building sewerage infrastructure, the government is increasingly focusing on septage treatment as a near-term sanitation solution. The objective of the NSSMP is for all LGUs to have septage management programs serving their urban barangays, with capital cost ranging from PhP 4 million to PhP71 million per project and per LGU.

For sewerage, NSSMP is initially targeting 17 highly urbanized cities (HUCs) outside Metro Manila, with capital cost estimates averaging P820 million per HUC. The LGUs and/or local WDs are expected to shoulder the cost of all septage management programs, while a 40:60% cost sharing scheme for sewerage systems will be implemented.

4.4.4.2 Philippine Sustainable Sanitation Roadmap

The PSSR consultation process started in 2007 and one of its milestone achievements is the formalization of the NEDA Infracom Sub-committee on Water Resources, which is tasked with oversight and coordination functions over the water supply and sanitation sector. This is an interagency body that monitors the implementation of the PWSSR and whose members are actively engaged as the Technical Working Group of the PSSR (DOH 2010).

The PSSR espouses five general strategies with each developed to address the multifaceted challenges besetting the sanitation sector. This is further translated into a cluster of related priority programs and activities directly supporting a specific policy directive. These programs will pursue the much needed policy reforms to enable sector institutions to perform their mandates effectively, ensuring that sanitation sector goals are achieved.

4.5 Water and Sanitation Sector Challenges

4.5.1 Lack of Centralized Regulatory Agency

There is no centralized economic regulatory agency for the water and sanitation sector. The existing national economic regulator, the NWRB, does not have the authority and resources to do economic regulation on water and sewerage affecting all the WSPs. Some agencies (LWUA, MWSS, and LGUs) have several functions that should not be housed in one agency—i.e. service provision and regulation, financing, supervision, and regulation. There are also no sanctions for government utilities that perform poorly.

There is no agency that sets and monitors coverage targets and operational standards of WSPs. The NWRB, LWUA, and LGU cover different WSPs and many more operate on "contract-based regulation," (MWSS and Subic Bay Water Regulatory Board (SBWRB)). Differences in regulatory practices, processes, and fees and cases of overlapping functions or jurisdictions have been observed.

LGUs have no regulatory capacity, except in granting business permits. LGU-run utilities are not required by any agency to submit regular reports. The DILG, which exercises authority over LGU-run utilities, is unable to monitor the performance of such water utilities mainly because of lack of resources.

No one is clearly accountable for implementing the reforms—there are a multitude of agencies involved in the sector. Each agency has its respective role in the sector and because the reform process cuts across the mandates of all agencies, they must all be involved. This means that all decisions are made by a committee, and the responsibility for implementation is often diluted. This lack of accountability for implementing reforms has affected most reform initiatives in the sector. For example, with respect to reform proposals on sector financing, no single agency has had the responsibility for ensuring that financing was available and that policies are in place to make this a reality.

4.5.2 Low Performance of WSPs

LGU-operated systems have the worst performance among all the utilities benchmarked. Water provision is simply politically motivated; thus, no emphasis is made on skills development, professional buildup, or financial sustainability. Most LGU systems are not ring-fenced so revenue is not linked to expenses. Dole-out mentality still exists. LGUs have also shown little interest in pursuing water supply projects due to leadership uncertainties brought about by the 3-year electoral term.

Water district tariffs are high because all their capital expenditure requirements are all funded from loans, whereas, most of the best performers among the types of utilities benchmarked have low coverage.

Water cooperatives need a lot of technical and financial support as CDA provides only administrative support. Community-based-organizations do not have access to commercial funds for expansion. They also need a lot of technical and financial support as no national agency seems to assist them.

Financing packages tend to cater only to creditworthy utilities with no concessional financing for the non-creditworthy or those still in the process of becoming creditworthy. Subsidy policies for LGU systems are not strictly implemented.

The government cannot possibly provide all the financing requirements. Support from Official Development Assistance, banks, and private sector providers is sorely needed.

4.5.3 Lack of Updated and Reliable Data

Planning at all levels is hampered by lack of reliable data and the absence of systematic and regular monitoring of sector activities at the LGU level. Many of the earlier provincial master plans and investment plans are based on decade-old designs without updated information on hydro-geologic and water resource conditions in the area. There are no updated local master plans; sector information are oftentimes conflicting or just simply missing.

Without data, it is difficult for the government and for users of services to assess critical aspects such as efficiency of services provided and quality of services given. The lack of such vital information makes it hard for the government to formulate policies, track progress over time, or hold agencies and service providers more accountable.

4.6 Policy Recommendations

There is a need to form a dedicated national agency for the water and sanitation sector. The first serious attempt was made in 2012 with the designation of the Secretary of DPWH as the "water czar" and placing MWSS and LWUA under the policy supervision of DPWH. This may be considered an interim step. The water czar has no dedicated unit for the sector within the department, aside from having numerous other duties as DPWH Secretary. Although there have been numerous attempts in the past to enact a law creating a national regulatory body, these have not been successful due to many factors. A draft of an interim body, called the National Water Resources Management Office (NWRMO)[8] had been made, needing only the signature of the President, but the draft has not been signed as of April 2016.

[8] The proposed National Water Resources Management Office (NWRMO) (Tabios and Villaluna 2012) is supposed to be mainly responsible for the management and protection of the country's

Given the 3-year terms of local executives, local governments alone cannot be relied upon to provide satisfactory and sustainable service for water and sanitation. The project cycle of water and sanitation infrastructure usually takes longer than 3 years. Tariff setting for LGU-run utilities also tends to be politically sensitive. As such, LGUs should be tasked with providing the services, but they should only become service providers themselves as a matter of last recourse.

There should be a cap on the percentage that LGUs can use from their IRAs to subsidize the operation of LGU water utilities (Internal revenue allotments (IRAs) are allocations from the national budget to the LGUs). Using IRA funds for this purpose does not motivate LGUs to improve service delivery and collect the proper tariffs. Most of the IRA funds should instead be used for capital expenditure.

There is a need to form a planning and monitoring body at the provincial level to oversee the water and sanitation sector within their boundaries. Having plans made at this local government level allows for nuancing, taking into account the specificities of hydrological conditions. It also brings greater accountability for WSP performance and subsequent decisions on public investments or financing.

References

ADB (Asian Development Bank). (2013). *Philippines: Water supply and sanitation sector assessment, strategy, and road map* (p. 7). Manila: ADB.

Center for Environmental Concerns-Philippines (CEC-Phils). (2012). *Confronting the ecological crisis: A situationer on Philippine environmental issues and struggles*. http://nafconusa.org/wp/wp-content/uploads/2014/11/State-of-the-philippine-environment-web041312.pdf. Accessed 28 Nov 2016.

DENR (Department of Environment and Natural Resources). (2008). *Supreme Court decision G.R. nos. 171947–48*. http://themanilabay.denr.gov.ph/supreme-court-decision/. Accessed 29 Nov 2016.

Department of Health. (2010). *Philippine sustainable sanitation roadmap (PSSR)*. Manila: Department of Health. https://www.scribd.com/document/52311529/Philippine-Sanitation-roadmap. Accessed 05 Apr 2016.

Manahan, M. A. (2012). *Focus at the alternative world water forum*. http://focusweb.org/content/bridging-gap-water-service-provision-philippines-new-roles-communities. Accessed 28 Nov 2016.

Millennium Development Goal Achievement Fund in the Philippines (MDGF). (2011). *Review of the national government-local government unit (NG-LGU) cost sharing for water and sanitation* (p. 6). Pasig City: MDG-F 1919 Program Management, NEDA. http://www.ombudsman.gov.ph/UNDP4/wp-content/uploads/2013/02/Review_of_the_NG-LGU_Cost-Sharing_Practices_for_Water_Supply_and_Sanitation.pdf. Accessed 15 Mar 2016.

Ndaw, M. F. (2016). *Private sector provision of water supply and sanitation services in rural areas and small towns: The role of the public sector* (p. 10). World Bank Group Water and Sanitation Program. http://documents.worldbank.org/curated/en/450101468179030315/pdf/104505-WPS-Box394877B-PUBLIC-Add-series-WSP.pdf. Accessed 10 May 2016.

water resources for domestic water supply, sanitation, irrigation, hydropower, fisheries, aquaculture, flood control, navigation and recreation, including the enhancement and maintenance of water quality, conservation of watersheds, control of water pollution, and environmental restoration, without compromising the natural ecosystems' functions and services.

NEDA (National Economic and Development Authority). (2004). *Medium-term Philippine development plan 2011–2016: Result matrices*. Manila: NEDA.
NEDA (National Economic and Development Authority). (2008). *Philippine water supply sector roadmap* (1st ed.). Manila: NEDA.
NEDA (National Economic and Development Authority). (2010). *Philippine water supply sector roadmap* (2nd ed.). Manila: NEDA.
NEDA (National Economic and Development Authority). (2014). *The Philippines: Fifth progress report- Millennium development goals*. Manila: NEDA.
NWRB (National Water Resources Board). (2015). *Listahang Tubig—A national water survey*. A28.
Official Gazette of the Philippines. (1991). *Republic Act No. 7160: An act providing for the Local Government Code of 1991*. Manila: Office of the President. http://www.officialgazette.gov.ph/downloads/1991/10oct/19911010-RA-7160-CCA.pdf. Accessed 25 Nov 2016.
Planning and Development Collaborative International Inc. (PADCO) (2006). *Local initiatives for affordable wastewater treatment in the Philippines/LINAW-1*. http://pdf.usaid.gov/pdf_docs/Pdaci905.pdf. Accessed 28 Nov 2016.
Robinson, A. (2003). *Urban sewerage and sanitation: Lessons learned from case studies in the Philippines* (p. 6). Final report. Water Supply and Sanitation Performance Enhancement Project. Quezon City.
Tabios G. Q, III, & Villaluna, R. (2012). *Development of the implementation and operational plan for the National Water Resources Management Office* (96 p). Submitted to the Interagency Committee on Water. Quezon City: NEDA.
WHO-UNICEF (World Health Organization-United Nations Children Fund). (2015). Joint Monitoring Program for Water and Sanitation: Estimates on the use of water sources and sanitation facilities, Philippines. https://www.wssinfo.org/documents/?tx_displaycontroller[type]=country_files. Accessed 01 Mar 2016.
World Bank. (2005). *Philippines: Meeting infrastructure challenges* (p. 111). http://siteresources.worldbank.org/INTEAPINFRASTRUCT/Resources/PHInfra.pdf. Accessed 28 Nov 2016.
World Bank. (2013a). *East Asia and the Pacific Region urban sanitation review: Philippine Country Study* (pp. 1, 2, 17, 36). Washington, DC: World Bank. http://documents.worldbank.org/curated/en/771821468036884616/pdf/842900WP0P12990Box0382136B00PUBLIC0.pdf. Accessed 09 Mar 2016.
World Bank. (2013b). *Developing the institutional framework for the water supply and sanitation sector and identifying investment plans and programs* (pp. 14, 15, 17, 31, 45–47). Report No: AUS151. Washington DC: World Bank. http://documents.worldbank.org/curated/en/984111468162537378/pdf/AUS15110WP0P1265570Box385184B00PUBLIC0.pdf. Accessed 09 Mar 2016.

Antonio "Tony" R. de Vera is a licensed civil engineer with completed academic units in Master's in Business Administration. He has over 40 years of water supply experience in the Philippine water sector, of which 26 years were spent with the Local Water Utilities Administration where he was Administrator and Vice Chair of the Board. He is currently Chairman of the Subic Bay Water Regulatory Board (2000–2020) and consultant in the sector for the Philippines and Asia, specifically on institutional frameworking, water regulation and capacity building. Among his accomplishments are the preparation of a three-volume manual (Design, Construction Supervision and Operations) for the Rural Water Sector, his participation in several reform studies funded by various Official Development Assistance (ODAs) and his assistance to several government water sector agencies in improving the performance of rural water associations and cooperatives.

Dr. Rosalie Arcala Hall is a Full Professor at the University of the Philippines Visayas (UPV). She completed her Master's degree in Political Science and PhD in International and Public Affairs (2002) at Northeastern University, Boston, Massachusetts. She completed research projects with The Asia Foundation, The Nippon Foundation and Toyota Foundation concerning policy frameworks, their localization, resistance and competing narratives behind contestations in the areas of hard and soft security. She has contributed chapters to the books Rosalina Palanca-Tan ed. Nature and Culture: Environmental Issues in Asia (Ateneo de Manila Press, 2014), and Justine Vaz and Narumol Aphineves eds. Living Landscapes Connected Communities: Culture, Environment and Change Across Asia (Areca Books, 2014). She is co-author to two articles in Water Policy and International Journal of Water Resources Development. Currently, she is a member of the Philippine Commission on Higher Education Technical Committee on Political Science and the Philippine Political Science Association Board.

Chapter 5
Industrial Water Use and the Associated Pollution and Disposal Problems in the Philippines

Veronica P. Migo, Marlo D. Mendoza, Catalino G. Alfafara, and Juan M. Pulhin

Abstract This chapter characterizes the state of Philippine water policy as it relates to the industrial sector and compares it with those of ASEAN countries. The water consumption profiles (source and quantity) of major industries are described, with a focus on those with large water footprints (e.g., oils and chemicals, food and beverages) to assess relative potential to cause water pollution. From water generation and pollution load profiles of the industrial sector, the wastewater characteristics of major industries are evaluated. A review of government regulations related to the prevention and control of water pollution, their state of implementation, and the degree of compliance/non-compliance by the industrial sector are presented. Using three environmental water pollution events as cases (nickel and chromium mining in Sta. Cruz, Zambales; gold mining in Padcal, Benguet Province, and the biophysical deterioration of the Marilao-Meycauayan-Obando river system due to factory effluents), the difficulties in the implementation of environmental laws, which lead to water pollution and great public health impacts, are explored.

Keywords Industrial water use • Water pollution • Philippine Clean Water Act of 2004 • Water quality • Effluents

---

V.P. Migo (✉) • C.G. Alfafara
Department of Chemical Engineering, College of Engineering and Agro-Industrial Technology, University of the Philippines Los Baños, College, Laguna, Los Baños, Philippines
e-mail: vpmigo@up.edu.ph

M.D. Mendoza
Department of Environment and Natural Resources, Field Operations and Staff Bureau, OIC Undersecretary, Quezon City, Philippines

J.M. Pulhin
Department of Social Forestry and Forest Governance, College of Forestry and Natural Re-sources (CFNR), University of the Philippines Los Baños, Los Baños, Laguna, Philippines

A.C. Rola et al. (eds.), *Water Policy in the Philippines*, Global Issues in Water Policy 8, https://doi.org/10.1007/978-3-319-70969-7_5

5.1 Introduction

The Philippines is naturally endowed with abundant freshwater resources to fulfill its need for water supply, industrial usage, as well as for irrigation. From about 2360 mm of annual precipitation, a third fills the country's inland waters, flows to the ocean as natural runoff, and enters underground aquifers through percolation. The country has 59 lakes covering some 200,000 ha. There are about 130,000 ha of artificial reservoirs and 126,000 ha of freshwater swamps. Water flows reach the sea through 421 principal rivers, with drainage basins over 40 km^2 each (Matsumura et al. 2003).

Based on an assessment by the Water Environment Partnership Asia (WEPA), there are about 85,000 manufacturing industries in the Philippines (WEPA n.d.). Basically, all manufacturing activities consume significant amounts of water during some point in the production process. Metro Manila is reported to be the main industrial region and accounts for about 52% of total manufacturing industries. As with many countries in the world, the rapidly developing industrial sector in the Philippines is a major consumer of water and a major contributor to water pollution. In highly urbanized and highly industrialized areas, industrial wastewater contributes about 30% of pollution discharges, while 60% comes from domestic sources. While industrial wastewater contributes less to pollution discharges than domestic wastewater, its adverse environmental impact is often greater due to the nature and concentration of pollutants.

This chapter looks at the industrial sector scenario with respect to Philippine water policy. As the industrial sector consumes and generates wastewater, the aspects of water allocation/consumption as well as wastewater generation/regulation policies will be discussed. Case studies related to industrial wastewater pollution are presented to highlight the challenges and to show a success story. Recommendations are also presented.

5.2 Sectoral Framework

5.2.1 Institutions

5.2.1.1 Regulation and Allocation of Water Resources

The National Water Resources Board (NWRB) operates under the mandate of the Philippine Water Code and regulates the allocation of water resources for the different sectors (including the industrial sector). Regulation is usually in the form of water permits, as would be described in the succeeding subsections.

5.2.1.2 Regulation for Industrial Wastewater Pollution Control

The Department of Environment and Natural Resources (DENR) is the primary agency responsible for the protection of the Philippine environment and natural resource sector (land, water, air, biodiversity, etc.). Under the DENR, pollution issues are usually handled by the Environmental Management Bureau (EMB) and the Pollution Adjudication Board (PAB). The EMB supports the DENR on matters related to environmental management, conservation, and pollution control. It also conducts environmental policy studies and takes the lead in setting environmental standards (DBP 1999). The PAB conducts public hearings on violations of pollution control laws and issues "cease and desist" orders to companies discharging effluents that do not comply with environmental standards or those that pose an immediate threat to public safety. The EMB serves as secretariat to the PAB.

5.2.1.3 Management of Water Quality Issues within Local Jurisdictions Under the Philippine Clean Water Act

Under the Philippine Clean Water Act of 2004, local governing boards of DENR/NWRB-declared water quality management areas (WQMA) are tasked with the formulation of strategies and coordination of policies needed for the implementation of the Act and the creation of a multi-sectoral group to oversee effective water quality surveillance and monitoring. A WQMA can be any water body (watershed, river, or water regions) with similar hydrological, meteorological, or geographic conditions that generate pollution reaction and transport in the water body or "share a common interest or face similar development programs, prospects, or problems." An industrial sector with potential pollution issues on a declared WQMA can be under the local jurisdiction of local water quality management boards.

5.2.2 *Water Allocation Permits Within the Context of the Industrial Sector*

Presidential Decree No. 1067, also known as the Philippine Water Code, deals with both aspects of water consumption and wastewater management in the Philippines. The law recognizes that even with the seemingly bountiful natural resources of the country, water supply is limited and should be used responsibly as well as sustainably. The Water Code covers the appropriation of water among the different sectors, while citing regulations against water pollution. The sectors of interest to this chapter are power generation (i.e., utilization for electrical or mechanical power) and industrial (i.e., utilization of water in factories, industrial plants, mines, etc.). In times of water shortage, these two sectors only come in third (for power generation), and sixth (for industrial use) in priority.

The implementing rules and regulations (IRR) of the Philippine Water Code focus on details such as permit requirements, registration process, administration, enforcement, resolution of conflicts, fines, and penalties. The general water permit requirements for industrial use in the Philippines are (1) proof of land ownership; (2) certificate of registration from the Department of Trade and Industry (DTI), Cooperative Development Authority (CDA), or Securities and Exchange Commission (SEC); (3) vicinity map; (4) brief description of the project; (5) well drilling data; and (5) environmental compliance certificate (for projects considered as environmentally critical). The Water Code and its IRR do not include the definition of or criteria used to identify an environmentally critical project or area; instead, the definition can be found in Presidential Proclamation No. 2146.

The process of approval for an industrial water permit in the Philippines starts with a review of the application submitted to NWRB for a period of 30 days. During the review period, notices are posted to the local government unit overseeing the jurisdiction of the projects, as well as to other agencies such as the Department of Public Works and Highways (DPWH). Seven days are dedicated to the investigation of environmental effects, land use, and project impact on existing public and private interests. There is also a dedicated period for "protests," scheduled until the 30th day after the last day of posting of the notice by NWRB. During this period, other parties may file objections against the upcoming project. If there are no impediments and upon approval of NWRB, the board conducts a periodic inspection of the project, in coordination with deputized agents. The product of the final inspection is a report, submitted not later than 10 days after the inspection date. For hydraulic power applications, the DPWH shall forward the application records and notice to the National Power Corporation central office. Nevertheless, apart from the registration process and a bit brushing on the regulations for industrial waste, both the Water Code and its IRR tend to be generic in their approach to water protocols for industrial use.

5.2.3 Government Regulations and Implementation Related to Prevention of Water Pollution

Complementary to the general approach of the Philippine Water Code toward pollution, the government provides extensive regulations for the protection of its water resources against pollution in the form of republic acts and administrative orders. The focus of this section is to review and assess these laws in light of industries and power generation. The first law to be tackled is Republic Act (RA) 9275, also known as the Philippine Clean Water Act of 2004. Section 8 of RA 9275 requires agencies and concessionaires to connect sewage lines of industrial complexes to sewage systems. Currently, there is paucity of data on the sewage systems for industrial complexes. However, for the National Capital Region (NCR), there are available data from EMB-NCR on sewage discharges tagged to industrial and commercial establishments. The data imply that these sewages are connected to the domestic sewage system.

RA 9275 also mandates that fees and charges will be applied for wastewater discharges or effluents and that industrial complexes shall be subjected to environmental impact assessments. Linkages are required between the business/industry sectors and agencies such as the Laguna Lake Development Authority (LLDA), DPWH, etc. Meanwhile, section 23 of RA 9275 deals with the requirement of record keeping and providing access to documents. Pollution control officers of respective pollution sources (such as industries) are required to submit reports to the concerned public departments. These records are to be made available to the public. In spite of the law being clear on the availability of records to the public, as mentioned, data on inorganic and toxic substances are restricted. Republic Act 9275 also gives incentives to industrial wastewater treatment facilities in the form of fiscal and non-fiscal measures. Industrial wastewater treatment and collection facilities, equipment, and parts presumably enjoy tax and duty exemptions.

Republic Act No. 3931 created the National Water and Air Pollution Control Commission. This Act covers the powers of the commission in issuing permits, setting standards, conducting public hearings on projects, etc. Presidential Decree 984 goes hand in hand with RA 3931 and is known as the Pollution Control Law, which requires permits to be secured before discharging industrial wastes to air, land, and water resources.

On the other hand, Republic Act No. 4850 created the LLDA, which is tasked to guide, support, and regulate industries in terms of water usage and disposal. RA 4850 is quite general in its approach to water quality management. Hand in hand with this law is Resolution No. 33, Series of 1996, the rules and regulations implementing the environmental user fee system in the Laguna de Bay region. The resolution details the effluent variable fee amounting to PhP 5 (USD 0.11) per kg biological oxygen demand (BOD) if within the 50 mg/L BOD limit, and PhP 30 (USD 0.64) per kg BOD for levels above 50 mg/L. The law also imposes a penalty of PhP 5000 (USD 106.4) for refusing entry of authorized representatives inside industrial premises. As mentioned, this resolution focuses on one pollutant only, BOD, and does not lay out the fees for other inorganic, toxic pollutants.

Corresponding sets of laws provide important measurable standards for various water pollutants, on top of BOD. The laws detailing water quality and effluent standards are DENR Administrative Order (DAO) No. 34 (Revised Water Usage and Classification) and No.35 (Revised Effluent Regulations of 1990) (DENR 1990a, b). The water body classifications provided by DAO No. 34 are straightforward; the interests of this chapter are water sources class C or industrial water class I (for manufacturing process after treatment) and classes D and SD or industrial water class II (for cooling, etc.). Both classes C and D are for fresh surface waters, while class SD is for coastal and marine waters. So far, the bodies of water discussed in this chapter (Laguna Lake, Marilao-Meycauayan-Obando river systems (MMORS), and NCR rivers) are all under class C.

DAO No. 34 also specifies the water quality criteria for conventional and other pollutants, contributing to aesthetics and oxygen demand both for fresh waters and coastal/marine waters. In other words, these standards are meant for ambient water not industrial effluents. As such, they are much lower in values. Parameters such as color, temperature, pH, minimum dissolved oxygen, 5-day BOD, total suspended solids, etc. are listed. The complete criteria are shown in Table 5.1 for industrial

Table 5.1 Water quality criteria for conventional and other pollutants contributing to aesthetics and oxygen demand for fresh waters and coastal/marine waters, for industrial water classes[a]

		Fresh water		Coastal/ marine
Parameter	Unit	Class C	Class D[b]	Class SD
Color	PCU	[c]	[c]	[c]
Temperature[d] (maximum)	°C rise	3	3	3
pH (range)		6.5–8.5	6.0–9.0	6.0–9.0
Dissolved oxygen[e] (minimum)	% sat'n mg/L	605.0	403.0	502.0
5-day 20 °C BOD	mg/L	7(10)	10(15)	–
Total suspended solids	mg/L	[f]	[g]	[g]
Total dissolved solids	mg/L	–	1000[h]	–
Surfactants (MBAS)	mg/L	0.5	–	–
Oil/grease (petroleum ether xtract)	mg/L	2	5	5
Nitrate as nitrogen	mg/L	10[i]	–	–
Phosphate as phosphorus	mg/L	0.4[j]	–	–
Phenolic substances as phenols	mg/L	0.02[k]	–	–
Total coliforms	Most probable number (MPN)/100 mL	5000[l]	–	–
Or fecal coliforms	MPN/100 mL	–	–	–
Chloride as chlorine	mg/L	350	–	–
Copper	mg/L	0.05[m]	–	–

Adapted from DAO No. 34, 1990–2015

[a]Except as otherwise indicated, numerical limits in Tables 5.1 and 5.3 are yearly average values. Values enclosed in parentheses are maximum values

[b]For irrigation purposes, Sodium Adsorption Ratio should have a minimum value of 8 and a maximum value not to exceed 18. Boron should not exceed 0.75 mg/L

[c]No abnormal discoloration from unnatural causes

[d]Allowable temperature increase over the average ambient temperature for each month. This rise shall be based on the average of the maximum daily temperature readings recorded at the site but upstream of the mixing zone over a period of 1 month

[e]Sampling taken between 9:00 AM and 4:00 PM

[f]Not more than 30 mg/L increase

[g]Not more than 60 mg/L increase

[h]Do not apply if natural background is higher in concentration. The latter will prevail and will be used as baseline

[i]Applicable only to lakes or reservoirs and similarly impounded water

[j]When applied to lakes or reservoirs, thephosphate as Pconcentration should neither exceed an average of 0.05 mg/L nor a maximum of 0.1 mg/L

[k]Not present in concentrations to affect fish flavor/taste

[l]These values refer to the geometric mean of the most probable number of coliform organism during a 3-month period and that the limit indicated shall not be exceeded in 20% of the samples taken during the same period

[m]Limit is in terms of dissolved copper nil—extremely low concentration and not detectable by existing equipment—Means the standard of these substances are not considered necessary for the present time, considering the stage of the country's development and DENR capabilities, equipment, and resources

Table 5.2 Water quality criteria for toxic and other deleterious substances for fresh waters and coastal/marine waters (for the protection of public health), for industrial water classes

		Fresh waters		Coastal/marine
Parameter	Unit	Class C	Class D	Class SD
Arsenic[a]	mg/L	0.05	0.01	–
Cadmium[a]	mg/L	0.01	0.05	–
Chromium[a] (hexavalent)	mg/L	0.05	–	–
Cyanide	mg/L	0.05	–	–
Lead[a]	mg/L	0.05	–	–
Total mercury[a]	mg/L	0.002	0.002	–
Organophosphate	mg/L	nil	nil	–
Aldrin	mg/L	–	–	–
DDT	mg/L	–	–	–
Dieldrin	mg/L	–	–	–
Heptachlor	mg/L	–	–	–
Lindane	mg/L	–	–	–
Toxaphane	mg/L	–	–	–
Methoxychlor	mg/L	–	–	–
Chlordane	mg/L	–	–	–
Endrin	mg/L	–	–	–
PCB	mg/L	–	–	–

Adapted from DAO (No. 34, from 1990 to 2015)

Limiting values of organophosphates and organochlorines may in the meantime serve as guidelines in the interim period pending the procurement and availability of necessary laboratory equipment. For barium, cobalt, fluoride, iron, lithium, manganese, nickel, selenium, silver, and vanadium, see the 1978 NPCC rules and regulations (NPCC 1978)

[a]Do not apply if natural background is higher in concentration. The latter will prevail and will be used as baseline

waters. Note that the 5-day 20 °C BOD (our pollutant for study) limit for industrial water sources is between 7 mg/L and 15 mg/L.

In addition, DAO No. 34 provides water quality criteria for toxic and other deleterious substances such as arsenic, cadmium, and other heavy metals—again for both fresh waters and coastal/marine waters. Table 5.1 shows the specifications for toxic pollutants for industrial waters (classes C, D and SD). Again, it is important to emphasize that the standards set by DAO No. 34 are for ambient waters, not industrial effluents. Note that there are no toxic substance criteria set for class SD (Table 5.2). Unfortunately, no ambient water quality data are collected from the agencies.

While DAO No. 34 deals with quality standards for ambient water sources, DAO No. 35 details the standards for effluents or waste discharges. Tables 5.3 and 5.4 present the standards for industrial effluents that will be discharged from various receiving water classes. The label OEI stands for "old existing industries," while NPI for "new and proposed industries." Naturally, NPI presents much stricter limits as compared with OEI with the expectation that newer industries are able to streamline their waste management processes ahead of time. The data from LLDA, EMB-MMO

Table 5.3 Effluent standards: Toxic and other deleterious substances (maximum limits for the protection of public health)[a]

Parameter	Unit	Protected waters Category I[b] (Classes AA & SA)		Protected waters Category II (Classes A, B, & SB)		Inland waters Class C		Marine waters Class SC		Marine waters Class SD	
		OEI	NPI	OEI	NPI	OEI	NPI	OEI	NPI	OEI	NPI
Arsenic	mg/L	–	–	0.2	0.1	0.5	0.2	1.0	0.5	1.0	0.5
Cadmium	mg/L	–	–	0.05	0.02	0.1	0.05	0.2	0.1	0.5	0.2
Chromium (hexavalent)	mg/L	–	–	0.1	0.05	0.2	0.1	0.5	0.2	1.0	0.5
Cyanide	mg/L	–	–	0.2	0.1	0.3	0.2	0.5	0.2	–	–
Lead	mg/L	–	–	0.2	0.1	0.5	0.3	1.0	0.5	–	–
Mercury (total)	mg/L	–	–	0.005	0.005	0.005	0.005	0.005	0.005	0.05	0.01
PCB	mg/L	–		0.003	0.003	0.003	0.003	0.003	0.003	–	–
Formaldehyde	mg/L			2.0	1.0	2.0	1.0	2.0	1.0	–	–

Adapted from DAO No. 35 (from 1990 to 2015)

[a]Except as otherwise indicated, all limiting values in Table 5.3 are maximum and therefore shall not be exceeded

[b]Discharge of sewage and/or trade effluents is prohibited or not allowed

Notes:

1. In cases where the background level of total dissolved solids (TDS) in freshwater rivers, lakes, reservoirs, and similar bodies of water is higher than the water quality criteria, the discharge should not increase the level of TDS in the receiving body of water by more than 10% of the background level
2. The COD limits in Tables 5.3 and 5.4 generally apply to domestic wastewater treatment plant effluents. For industrial discharges, the effluent standards for COD should be on a case-to-case basis considering the COD-BOD ratio after treatment. In the interim period that this ratio is not yet established by each discharger, the BOD requirements shall be enforced
3. There are no effluent standards for chloride, except for industries using brine and discharging into inland waters, in which case, the chloride content should not exceed 500 mg/L
4. The effluent standards apply to industrial manufacturing plants and municipal treatment plants discharging more than 30 m^3/day

Table 5.4 Effluent standards: Conventional and other pollutants[a]

		Protected waters					
		Category I[b]		Category II		Inland waters	
		(Classes AA & SA)		(Classes A, B & SB)		Class C	
Parameter	Unit	OEI	NPI	OEI	NPI	OEI	NPI
Color	PCU	–	–	150	100	200[c]	150[c]
Temperature (max rise in °Celsius in RBW)	°C rise	–	–	3	3	3	3
pH (range)		–	–	6.0–9.0	6.0–9.0	6.0–9.0	6.5–9.0
COD	mg/L	–	–	100	60	150	100
Settleable solids (1 h)	mg/L		–	0.3	0.3	0.5	0.5
5-day 20 °C BOD	mg/L	–	–	50	30	80	50
Total suspended solids	mg/L	–	–	70	50	90	70
Total dissolved solids	mg/L			1200	1000	–	–
Surfactants (MBAS)	mg/L			5.0	2.0	7.0	5.0
Oil/grease (Petroleum ether extract)	mg/L	–	–	5.0	5.0	10.0	5.0
Phenolic substances as phenols	mg/L	–	–	0.1	0.05	0.5	0.1
Total coliforms	MPN/100 mL			5000	3000	15,000	10,000

Adapted from DAO No. 35 (from 1990 to 2015)

[a]Except as otherwise indicated, all limiting values in Tables 5.3 and 5.4 are 90th percentile values. This is applicable only when the discharger undertakes daily monitoring of its effluent quality, otherwise, the numerical values in the tables represent maximum values not to be exceeded once a year

[b]Discharging of sewage and/or trade effluents is prohibited or not allowed

[c]Discharge shall not cause abnormal discoloration in the receiving waters outside of the mixing zone

Notes:

1. In cases where the background level of total dissolved solids (TDS) in freshwater rivers, lakes, reservoirs, and similar bodies of water is higher than the water quality criteria, the discharge should not increase the level of TDS in the receiving body of water by more than 10% of the background level
2. The COD limits in Tables 5.3 and 5.4 generally apply to domestic wastewater treatment plant effluents. For industrial discharges, the effluent standards for COD should be on a case-to-case basis considering the COD-BOD ratio after treatment. In the interim period that this ratio is not yet established by each discharger, the BOD requirements shall be enforced
3. There are no effluent standards for chloride, except for industries using brine and discharging into inland waters, in which case, the chloride content should not exceed 500 mg/L
4. The effluent standards apply to industrial manufacturing plants and municipal treatment plants discharging more than 30 m^3/day

Region III, and EMB-NCR discussed earlier in Tables 5.2, 5.3, and 5.4 follow the class C standard for inland waters since the water resources in these regions are also being used for fisheries, boating, and other recreational activities, apart from industrial use. The old standard is 80 mg/L for BOD but was lowered to 50 mg/L for NPIs.

The standards set by DAO Nos. 34 and 35 are comprehensive and at par with other ASEAN countries such as Singapore and Malaysia. The industries surrounding Laguna Lake and the MMO river system in Region III more or less comply with the 50 mg/L BOD limits based on data from their respective agencies. As mentioned, NCR is more challenged in meeting the regulations. What is lacking in the assessment is that of the other pollutants—as there are no readily available data from the agencies. There are stockrooms and warehouses of data, but they are yet to be encoded. This, on top of the restrictions set by industries, adds to the data problem.

As of May 2016, the DENR, through Administrative Order No. 2016-08, issued the new Water Quality Guidelines and General Effluent Standards. In the new guideline, the water body classification was modified for both fresh and marine waters, so that it does not include the industrial water supply classes. The previous industrial water classes C, D, and SD were modified to fishery water, recreational, agricultural, irrigational, and livestock watering (class C), navigable fresh waters (class D), and navigable marine waters (class SD). There is a note, however, indicating that, for unclassified water bodies, "classification shall be based on the beneficial use as determined by the EMB. The new standards indicate new values for the limits of the different water quality guidelines for primary, secondary, organic, and inorganic parameters. The updated guideline is shown in Table 5.5. One key feature of the new water quality

Table 5.5 Water quality guidelines for primary parameters for classes C, D, and SD

		Water body classification		
Parameter	Unit	C	D	SD
BOD	mg/L	7	15	n/a
Chlorine	mg/L	350	400	n/a
Color	TCU	75	150	150
Dissolved oxygen[a] (minimum)	mg/L	5	2	2
Fecal coliform	MPN/100 mL	200	400	400
Nitrate as NO_3-N	mg/L	7	15	15
pH (range)		6.5–9.0	6.5–9.0	6.5–9.0
Phosphate	mg/L	0.5	5	5
Temperature[b]	°C	25–31	25–32	25–32
Total suspended solids	mg/L	80	110	110

Adapted from DENR Administrative Order 2016–08 (as of 2016)

Notes:

MPN/100 mL most probable number per 100 milliliters, *n/a* not applicable, *TCU* true color unit

[a]Samples shall be taken from 9:00 AM to 4:00 PM

[b]The natural background temperature as determined by EMB shall prevail if the temperature is lower or higher than that under water quality guidelines; provided that the maximum increase is only up to 10% and that it will not cause any risk to human health and the environment

guideline is the addition of a table on "significant effluent quality parameters per sector," where the main pollutants have been identified per industry sector.

5.3 Sector Performance

5.3.1 Water Allocation Based on Permits

Data on Philippine water allocation for 2014 were obtained from NWRB to check water consumption for the industrial and power generation. Figure 5.1 shows the water allocation for consumptive use for both groundwater and surface water (NWRB 2014). Consumptive water use is defined as the water taken from a water resource without returning to a water resource system; an example of which is water that is made part of a commercial product. On the other hand, non-consumptive water use such as the water used in hydroelectric power plants, is water that is returned to the resource system after use. Industrial water usage is 10.76% of the total (Fig. 5.1). In addition, there is no allocation for the power sector in the consumptive area. Meanwhile, a large chunk of the non-consumptive water use is taken by the power sector (57.72%) (Fig. 5.2). This is highly expected since hydroelectric power plants utilize large volumes of water while returning the water to rivers and lakes. Combining this value with industrial usage results in 62.27% usage for both power and industrial sectors, in the non-consumptive area.

The values in Figs. 5.1 and 5.2 highly depend on the data submitted by water permit applicants, as they were based on water application grants. The data do not cover informal and illegal water users in the different sectors, which continue to be a gray area in the regulations. In addition, upon reviewing the raw data used to gen-

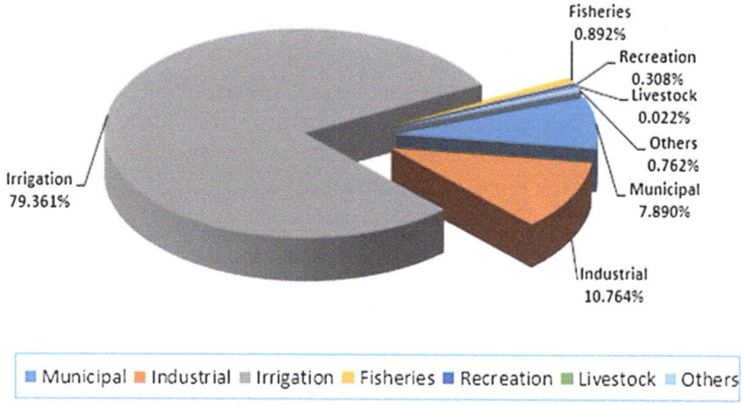

Fig. 5.1 Water allocated for groundwater and surface water consumptive use (as of December 2014) (Adapted from NWRB 2014)

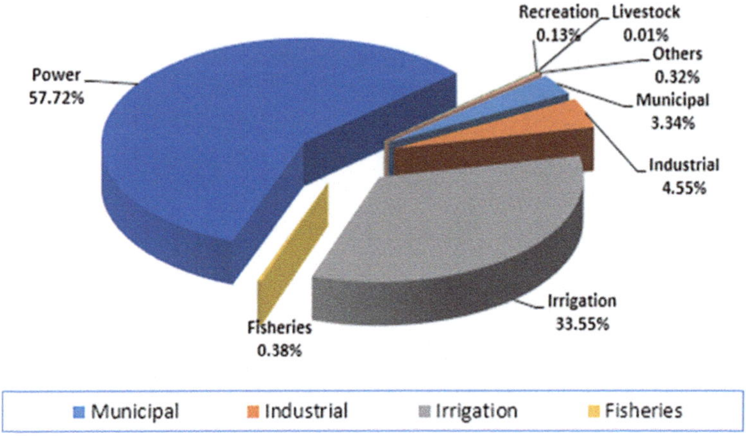

Fig. 5.2 Water allocated for groundwater and surface water non-consumptive use (as of December 2014) (Adapted from NWRB 2014)

erate the figures, it appears that there are still regions and companies in the country that are not included. Nevertheless, the available data can provide an estimate of water consumption profiles in the Philippines.

It is a known fact that water-permit applicants tend to "under declare" their water usage plans in order to ease on the registration fees. "Under declaration" is difficult to quantify. Thus, the accuracy and appropriateness of the records heavily depend on the investigations and surveillance by NWRB, as well as by the other involved agencies. Furthermore, there is no subclassification within industry (i.e., petroleum, mining, textile, etc.). The different sectors in the data remain to be categorized based on the generic categorization from the Philippine Water Code, as there is no provision to classify the industries further. However, Manila Water Company, a private concessionaire, provides some details on industrial water consumption for the east zone of Metro Manila. Table 5.6 presents the average consumption rates in 2015 of the different industries in the area (Manila Water 2015). The highest consumers of water belong to the oil and chemical industries at an average of 36.7 m^3/day, followed by construction material industries at 29.8 m^3/day, and electronic parts and supplies at 25.6 m^3. Wood furniture and automotive lag behind in their water consumption at an average of less than 4 m^3/day.

Singapore also has a more straightforward pricing when it comes to water, imposing an industrial tariff of up to SGD 0.66 (equivalent to USD 0.47) per cubic meter of water consumed (if over 40 m^3 per month) (Singapore's National Water Agency 2017). On the other hand, there is a huge gap in water cost between Metro Manila and the countryside in the Philippines as NWRB delegates the distribution of quality water to private concessionaires. In the countryside, the industrial rate can go as low as PhP 2 (USD 0.04) per cubic meter (LWUA 2000), while blowing up to PhP 86.62 (USD 1.84) in Metro Manila (Maynilad 2015), for consumption of more than 1000 m^3 per month. Nevertheless, NWRB standardizes these companies

Table 5.6 Average consumption (m^3/day) for industries in the east zone of Metro Manila as of 2015

Industry subsector	Average consumption
Oil and chemicals	36.7
Construction material	29.8
Electronic parts and supplies	25.6
Plastic, rubber, glass, styrofoam	21.9
Food and beverages	21.1
Pharmaceutical	20.2
Soap and cosmetic products	8.8
Printing press	7.6
Garment	5.2
Paper products and packaging	4.6
Wood furniture	3.2
Automotive	3

Adapted from Manila Water (2015)

through economic regulation and sets their fees and charges not exceeding 12% of their return on investment.

Aside from differences in water pricing schemes, Singapore also posts heavy fines (USD 743) on offenses such as making illegal water connections, contaminating water, and even as simple as wasting water. On the other hand, in the Philippines, charges are set on a daily basis, with a fine of PhP 500 (USD 10.7) for illegal diversion and a charge of up to PhP 1000 (USD 21.4) per day for grave offenses such us destruction of hydraulic works, etc.

5.3.2 Wastewater Generation and Pollution Load Profiles of the Industrial Sector in the Philippines

Industrial waste water generation and pollution load profiles are based on data taken from three agencies, where information could be obtained: (1) LLDA), (2) the National Capital Regional Office of the Environmental Management Bureau (EMB—NCR), and (3) Region III Office of the Environmental Management Bureau (EMB-Region III), where the MMORS is located. Data for only one pollutant parameter are available, which is the 5-day BOD (BOD5), where the regulation set for this parameter is 50 mg/L for class C water bodies (waters used for fishery, recreation and industrial water supply), where most industrial wastewaters are discharged (DENR, Administrative Order No. 35).

The different industries that release wastewater to Laguna Lake were subcategorized and summarized (Table 5.7). Since the BOD loading data provided are in kilograms per year, the data were processed to convert the units to mg/L to make comparisons with the regulation value of 50 mg/L. This was done by converting

Table 5.7 Wastewater generation data of industries around Laguna Lake as of 2014

Industry	Establishments (no.)	Total BOD loading (kg/year)	Total volume (m^3/day)	BOD of effluent (mg/L)
Restaurants and fast food	492	1,070,953	4920	596.32
Recycling, waste management and dumpsites	75	88,669	917	265.03
Water utilities	73	3,869,579	211,370	50.16
Textile and garments	116	43,148	2474	47.79
Memorial parks	23	2462	148	45.56
Food and beverage manufacturing	490	514,363	31,041	45.40
Automotive	816	59,060	4598	35.19
Power	38	11,279	1057	29.24
Hotels, resorts, condominiums	1192	1,183,084	111,570	29.05
Hospitals	152	67,433	7586	24.35
Wood furniture	17	1715	197	23.85
Industrial parks	31	247,922	32,637	20.81
Agriculture	256	54,434	7435	20.06
Schools, universities, laboratories	106	34,912	4945	19.34
Plastic, rubber, glass, styrofoam	120	6434	930	18.96
Paper products and packaging	109	24,834	3704	18.37
Others (laundry services, jewelry, etc.)	85	23,785	3719	17.52
Oil, chemicals, fertilizer and soap	291	13,183	2296	15.73
Electronic parts and supplies	189	132,147	24,686	14.67
Construction material	444	52,503	11,013	13.06
Mining	51	56	47	3.25

Adapted from LLDA (2014)

total BOD loading rate (kg/year) to (kg/day), and dividing this value with the volumetric wastewater generation rate (m^3/day). The value obtained in (kg/m^3) was then multiplied by 1000 to convert the units to mg/L.

Restaurants and fast food establishments were seen to have non-compliant BOD levels at 596 mg/L, followed by recycling, waste management and dumpsite facilities at 265 mg/L. Textile/garments, memorial parks, and food/beverage manufacturing establishments were just at the 50 mg/L limit.

The BOD for mining is the lowest among industries, at only 3.25 mg/L. As one would expect, the mining industry will deal more with inorganic waste manage-

Table 5.8 Wastewater generation data for industries and commercial establishments in Bulacan Province, 2014

Industry	Average BOD of effluents from 69 companies (mg/L)	BOD reduction (kg/year)
Paper products and packaging	70.00	4.55
Hotels, resorts, condominiums	50.00	89.00
Others	50.00	226.50
Oil, chemicals, fertilizer and soap	45.50	6.61
Construction material	44.00	36.00
Agriculture	43.75	429.16
Recycling, waste management and dumpsites	43.75	189.75
Textile and garments	37.48	530.13
Food and beverages	37.35	437.08
Automotive	35.43	29.12
Warehousing	32.00	1018.00
Electronic parts and supplies	28.00	142.00
Hospitals	26.00	47.34
Commercial malls, markets	25.85	1480.70

Adapted from EMB-MMO Region III (2014)

ment, particularly heavy metals and toxic wastes. Unfortunately, there are no data found for these pollutants to enable analysis. Public access to complete pollution parameters by establishments seem to be restricted, thus a comprehensive analysis of pollution profiles of industries could not be made.

An environmental user's fee is imposed on establishments that discharge their wastewater into Laguna Lake. It is the sum of a fixed fee (based on volumetric rate of discharge) and a variable fee (based on "unit pollution load," which is only the BOD of the wastewater). The computation of this fee does not consider other pollutants, which may produce negative environmental impacts on Laguna Lake.

The EMB also has its own method of classifying establishments that generate industrial wastewater. Table 5.8 shows the BOD data for 69 companies and commercial establishments in Bulacan Province, releasing wastewater to the MMORS. The different companies were subcategorized into industries. Since BOD data are already available in mg/L, the BOD loadings of the different companies belonging to the same category were averaged. In contrast with the data from Laguna Lake, the industry with the highest BOD rate in the MMORS is the paper product and packaging industry at 70 mg/L, failing the standard set by DENR. The rest of the industries are able to comply with the standard.

Surprisingly, the waste management industry, which comes in second place in Laguna Lake as the highest BOD contributor, only comes in seventh place in the case of MMORS – it has a BOD loading of 43.75 mg/L. The explanation is that the waste management industry in the area is composed mostly of lead-recycling industries, thereby increasing the toxicity of the effluents and preventing organisms to thrive. This may cause low BOD detections. The data also show BOD reduction per year. Commercial malls, markets, and warehousing industries reveal BOD reductions of more than 1000 kg. Textile and tannery industries, which are heavily present in this region, meet regulations and at the same time has a high BOD reduction of 530 kg/year. The noteworthy compliance of many industries may be credited to the MMORS rehabilitation actions, which began in 2008, with DENR declaring the river system as a WQMA. With funding from the Japan International Cooperation Agency (JICA) and support from NGOs such as the Blacksmith Institute, a New York-based non-for-profit organization, the project jumpstarted the compliance monitoring of the effluents released by the said industries (more information in Sect. 5.4). But then again, for some industries with effluents containing toxic and hazardous substances, BOD may not be a reliable parameter for gross pollution.

The 69 companies and establishments in Table 5.7 may only represent a small fraction of the whole industrial establishments. For example, there are no records for the restaurant and fast food industry, which is an industry that is expected to account for one of the highest BOD. Thus, a more comprehensive and systematic information-gathering system may be necessary.

Another region for study is NCR, Metro Manila, which comprises the cities of Manila, Caloocan, Las Piñas, Makati, Malabon, Mandaluyong, Marikina, Muntinlupa, Navotas, Parañaque, Pasay, Pasig, Quezon, San Juan, Taguig, Valenzuela, and Pateros. Here, domestic and industrial effluents drain into the following bodies of water: Navotas-Malabon-Tenejeros-Tullahan River system, San Juan River, Paranaque River, Manila Bay, Pasig River, Meycauayan River, and Las Pinas River. Some of these waterbodies are shown in Fig. 5.3.

A limited list from EMB-NCR is available; however, data tagged as industrial or commercial BOD for most of the companies are not available. Sewerage data tagged to the industries have been excluded. The data were processed to reflect the unit of mg/L for BOD by dividing kg BOD per day with the volume of discharge in m^3/ day.

Surprisingly, the industry with the highest BOD is the electronic parts and supplies at 4780 mg/L, followed by the automotive industry at 1511 mg/L. The BOD content in these effluents may be attributed to possible use of chemicals with high organic content during processing. The BOD values were also higher than those observed in other regions by LLDA and EMB-Region III (MMO). The restaurant and fast food industry's BOD loading for this region came in third at 1117 mg/L. This value is twice that reported by LLDA for Laguna Lake (596 mg/L). Another surprising information from Table 5.9 is the BOD value for power generation industries, which was 165 mg/L (as against Laguna Lake's 29 mg/L BOD). Since power generation has a non-consumptive use for water, the presence of BOD from these effluents merits further study.

In general, the industry BOD data from EMB-NCR shows that this area is not quite as compliant as the other regions. In fact, of the 13 industries reported in NCR, only five pass the set limit of 50 mg BOD/L. Furthermore, the highest BOD value attained in NCR is in thousands, while it is only in hundreds for Laguna Lake and less than a hundred for the MMOS.

Fig. 5.3 The major river systems flowing through the main cities and municipalities in Metro Manila with respect to locations of Manila Bay and Laguna de Bay (Source: Adapted from Map data from Google Maps)

As mentioned previously, only the BOD loading data are available for NCR-EMB and MMO Region III. Despite the presence of mining, tannery, and other industries, which heavily release inorganic and toxic substances to water resources, information remains restricted from study. Another difficulty in the data provided is the immediate disparity of units against the standards, thereby forcing a secondary processing of the data. One would think that, in order to assess compliance properly, the data must be as straightforward as possible. Because of this, harmonization in data acquisition and data validation among the environment-related government agencies is recommended.

Table 5.9 EMB wastewater generation data for industries in NCR, 2013

Industry	Volume of discharge (m^3/day)	BOD of effluents (kg/day)	BOD of effluents from 81 companies (mg/L)
Electronic parts and supplies	4.73	22.61	4780.13
Automotive	20.05	30.31	1511.38
Restaurant and fast food	162.74	181.90	1117.71
Others	19.53	3.38	172.86
Power	7.02	1.16	165.00
Paper products and packaging	31.32	3.54	113.15
Recycling, waste management and dumpsites	38.18	3.45	90.35
Food and beverages	3010.00	199.74	66.36
Plastic, rubber, glass, styrofoam	4.60	0.23	50.00
Construction material	103.21	4.96	48.02
Textile and garments	900.85	38.20	42.40
Commercial malls, markets	2567.86	74.47	29.00
Hotels, resorts, condominiums	175.65	3.15	17.93

Adapted from EMB-NCR (2013)

5.4 Case Studies Highlighting the Challenges

The following cases illustrate the challenges related to formulating a water use policy for the industrial sector in the Philippines. The first case on backyard smelting and other polluting industries around the MMORS demonstrates the need to establish reliable baseline data on which subsequent monitoring of water pollution can be based. The second case, involving mining companies in Sta Cruz, Zambales, and Padcal, Benguet Province, illustrates the challenge of imposing legal and political remedies for violations of industrial standards to minimize water pollution and associated environmental destruction. Finally, the last case, which is a success story about agro-recycling of industrial wastewater, shows the importance of tapping the support of a research institution in finding technical solutions to the wastewater problem. It also highlights the need to give incentives for water reuse in the industrial sector and for scaling up the initiative using economic instruments.

5.4.1 The Case of Backyard Gold Smelting and Other Industries Around MMORS

Located 19 km north of Metro Manila, Meycauayan, Bulacan, has an area of 32 km^2 with an estimated population of more than 190,000. The residents are distributed in 26 villages (barangays). The Meycauayan River extends 17.2 km in length and connects to Manila de Bay. This river is part of the Marilao-Meycauayan-Obando river system, mainly used for fishponds, domestic, and industrial purposes. In 2007, the New York-based organization, Blacksmith Institute, named the MMORS as one of the top 30 "dirtiest rivers of the world" (Blacksmith Institute 2007). DENR stated in 2009 that, in order to save Manila de Bay, which is the largest port in Metro Manila, the MMORS in Bulacan has to be cleaned first.

DAO No. 34 classifies the MMORS as Class C, deemed for fishery, recreational, and industrial uses. While the locals depend on this river for their livelihood, they continue to dump their wastes into the river, an act done together with the surrounding industries in the area. The orientation of the river also proves to be a disadvantage because located upstream is industrial and agricultural land, while the catchment or downstream area are surrounded by settlements, small commercial establishments, and fishponds. As toxic effluents from the industries are deposited downstream, the effluents continue to endanger the aquaculture industry—especially the *bangus* (milkfish) industry that grosses up to USD 80–100 million in sales yearly. Note that Bulacan Province is one of the largest producers of bangus in the whole country.

There are 567 registered large business establishments in Bulacan. Out of this number, the top industries in Bulacan (aside from the milkfish industry) include lead recycling, tanneries (for leather products), and gold smelting industries. Jewelry-making, which is linked to gold smelting, flourishes in this region; in fact, there are 107 registered jewelers and 36 gold smelters in the municipality. This number does not include the numerous unregistered or "informal" gold smelters. These industries are the main sources of toxic, heavy metal-rich effluents, which spoil the MMORS.

According to the Blacksmith Institute study, even the small-scale, backyard gold smelting done by the residents when grouped collectively, can be considered as one sizeable industry itself. Gold smelting continues to thrive as buyers and makers of jewelry make the industry profitable. Inappropriately, these independent manufacturers did not have the required waste treatment facilities, as well as the right level of awareness of the impact of their wastes. The waste products from the gold smelting process include lead, arsenic, cadmium, nickel, copper, as well as large volumes of sulfuric and nitric acid—all of these pollutants released to the MMORS.

To solve the MMO pollution problem, Blacksmith Institute stepped in in 2005 and began its multistakeholder meetings with representatives from Region III, offices of the EMB, the Bureau of Fisheries and Aquatic Resources, and the Department of Health; the local governments of Marilao, Meycauayan, and Obando;

the provincial government of Bulacan; the industry sector, as well as NGOs and socio-civic organizations. It was not until 2008 when DENR officially declared the MMORS as a WQMA when true accomplishments of the awareness campaign began. The MMORS was selected as a pilot site for Luzon in taking the Clean Water Act into action, getting international support from JICA and the Asian Development Bank (ADB).

The Blacksmith Institute, funded by ADB, organized a team of scientists, human ecologists, and development communication experts from the University of the Philippines Los Baños (UPLB) in an almost 8-year long project in Bulacan. The following are the main players and their respective tasks (ADB 2013):

- The College of Human Ecology (CHE) from UPLB was tasked to carry out a health risk assessment and knowledge-attitude-perception study among industry players and community members.
- The College of Forestry and Natural Resources from UPLB conducted a policy review analysis of the municipalities' local environmental codes, together with the CHE. The College was also involved in exploring phyto-remediation in sequestering inorganic pollutants and planting of mangrove species along the river banks.
- The College of Development Communication from UPLB was tapped to design an information, education, and communication (IEC) strategy and develop the IEC materials.
- The UPLB-National Institute of Molecular Biology and Biotechnology (BIOTECH) was involved in the conduct of several baseline studies, particularly the mass balance and mapping studies. They were also involved in identifying approaches to improve engineering designs of the gold smelting to reduce pollution at source.
- The Department of Science and Technology-Industrial Technology and Development Institute was involved in the design and testing of cost-effective technologies to abate heavy metal pollution from the gold refineries.

During the project's interview with the residents, they found that only a few could identify the commonly used chemicals for each industry and that most are unsure of their view regarding the gold smelting business, despite knowing its harmful effects on their health and environment. From the study, only 7% felt positive toward the gold smelting industry while a whopping 40% felt the opposite. The 7% of the population in favor of the gold smelting industry recognized its contribution to the economic and cultural growth of the region. Those not in favor of the industry complained of fumes released during the gold smelting process, saying that the odor makes it difficult to breathe. The gold smelting process begins with extractive purification, using either silver (for yellow gold) or lead (for white gold). The next step is thermal treatment where two layers are formed, an impurity-rich molten layer that is decanted and a gold-rich molten layer that is further quenched by nitric acid and water. The final product of this process is gold granule.

Aside from silver, lead, and nitric acid contaminants, water samples in public areas around the region also showed contaminations of copper and chromium, while plant samples showed the presence of manganese, lead, and copper. These contaminants are further washed throughout the public areas due to frequent flooding. Another industry, lead recycling, affected not only the water source in the areas but also the soil quality of the region.

The Blacksmith Institute gathered the baseline data along the Meycauayan river system to determine the presence and level of heavy metals using four types of sampling: surface water, groundwater, sediment, and through selected fishes and shellfishes. The results of their investigation in different stations along the MMORS were compared with the limits set by DENR DAO No. 34 for Class C water (Table 5.5). Note that the original measurements in the report were in parts per million (ppm); by assuming the density of water to be 1000 kg/L, the units were converted into mg/L as presented below.

Arsenic. Two sampling stations exceeded the limit of 0.05 mg/L during the dry season at 0.1 and 0.07 mg/L, respectively. However, measurements did not exceed the standards during wet season.

Cadmium. Two stations exceeded the limit set of 0.01 mg/L during the dry-season monitoring. The highest concentration recorded was 0.06 mg/L. During the wet-season monitoring, most of the stations exceeded the limit.

Copper. Most of the stations exceeded the limit of 0.02 mg/L, the highest was at 0.15 mg/L near an effluent pipe.

Chromium. Two stations were found to exceed the set limit of 0.05 mg/L during the dry season and the highest recorded value was 0.19 mg/L near the highway.

Lead. In general, lead measurements exceeded or were near the set limit of 0.05 mg/L. The highest value recorded was 0.48 mg/L, followed by 0.32 mg/L.

Manganese. Manganese was consistently present in all stations and exceeded the set limit of 0.2 mg/L. The highest value recorded was 40 mg/L, followed by 2.01 mg/L.

Mercury. Mercury was found in five surface water samples, exceeding the set limit of 0.002 mg/L during the dry-season monitoring. The highest value recorded was 0.01.

Nickel. One of the stations exceeded limit of 0.2 mg/L, registering a significantly high concentration of 0.82 mg/L.

Zinc. Zinc was detected in almost all stations during the wet- and dry-season monitoring, but their concentrations were found way below the limit of 2 mg/L. However, there were two stations that exceeded the limit, registering 2.9 and 2.06 mg/L.

Sediment monitoring also showed a consistent presence of manganese, copper, lead, mercury, zinc, and chromium. As for ground water, all stations did not register concentrations of arsenic, copper, chromium, lead, and mercury. However, cadmium, nickel, zinc, and manganese were present, exceeding the limits set by the Philippine drinking water standards. Shellfish and fish were also checked for contamination as samples of paros, prawn, tilapia, tahong, and bangus were analyzed. Heavy metals were found in the fish and shellfish samples and these exceeded the limits set by BFAR.

The project staff presented their results to the stakeholders and conducted a series of seminars for the locals. Blacksmith Institute continues to monitor the fishes and shellfishes cultured in the MMORS through another project funded by HSBC.

5.4.2 The Cases of Nickel and Chromium Mining in Sta. Cruz, Zambales and Gold Mining in Padcal, Benguet Province

Sta. Cruz, a town in Zambales Province located in Central Luzon, Philippines, has an area of 438 km^2 and a population of more than 54,000. Popular for its chromite and nickel reserves, Sta. Cruz considers mining as one of its most profitable industries since 2007.In fact, DENR declared nine mineral-rich areas for excavation in this region, and the country boasts of Zambales' chromite reserve as one of the largest in the world. According to the Mines and Geosciences Bureau, there are 15 existing mineral production sharing agreements (MPSAs) approved between 1991 and 2010, seven of which are in Sta. Cruz (Sazon 2016). Sazon reported the effect of mining in Sta. Cruz's ecosystem as the town has at least six MPSAs, giving permission for large-scale mining activities in the municipality. The MPSAs cover more than 10,878 ha of land, corresponding to 24.81% of the municipality's total land area. The buyer of the minerals (nickel and chromite) is China; the companies transport their products through local seaports. The effect of almost a decade of mining is now visible as ponds and farmland surrounding the area experience changes in color and higher levels of heavy metals.

The town made headlines in 2016, as residents of Candelaria and Sta. Cruz barricaded the roads preventing trucks from transporting mined ores to the local port. The 30 police officers and a hundred workers who escorted these trucks were not enough against a group of 200 protesters, who, in turn, barricaded the road for 2 weeks. It is said that this protest was ultimately triggered by the red mudflows that flooded Sta. Cruz in October 2015, an aftermath of Typhoon Lando (international name Koppu). The red mud killed two persons and many farm animals and destroyed residents' houses (Badilla 2015). The red mud is believed to be composed of nickel laterite, which also silted the fishing sites in the area and thereby renderedthe sites unproductive.

Right after this incident, residents of Sta. Cruz filed a petition under the "Writ of Kalikasan" to the Supreme Court against the mining companies, accusing five mining firms of: "destruction of the ecosystem in Santa Cruz, Zambales, and its neighboring municipality of Candelaria, Zambales; water, air, and soil pollution; heavy laterite siltation of river systems, coasts, farmlands, fish ponds, and residential areas; forest denudation, resulting in soil erosion, exacerbated flood problems during typhoons and heavy rains; and destruction of irrigation system in Santa Cruz, Zambales, which severely reduced the palay production of the rice granary of Zambales, and heavily affected the livelihood of residents" (Datu 2016). As of June 2016, the Supreme Court decision on the case is still pending and the mines remain under temporary environment protection order (TEPO). Because of the suspension, more than 2000 mine workers lost their jobs, impacting the economy of the region. Due to the loss of jobs, the region is now struggling against petty theft, which rose from 15 to 20% (Campos 2016). The month after, DENR ordered the suspension of all large-scale mining companies in Zambales.

Complaints filed against the Zambales miners are not new. In 2013, the Supreme Court issued TEPO against 94 small-scale mines (Bondoc 2013). Prior to this, the Zambales governor signed a petition to allow small-scale mining permits outside the officially declared mining zone. It has been known that these mines are fronts of giant nickel miners from China, "all tainted by bribery reports, they set up the five "small mines" through Filipino dummies" (Bondoc 2013). Several of these informal small-scale mines are run by residents and middlemen, and most do not follow mining and environmental standards and good practices. Because of the water pollution from mining, fishermen in the area are forced to sail far out into the sea. However, upon reaching 124 miles west of Zambales at the shoal called Panatag, the Chinese military vessels chase them out (Bondoc 2013). Aside from water pollution, residents also suffer from acute respiratory infections due to the worst air pollution, raising the morbidity rates in the area (Bondoc 2013). Unfortunately, there are no data on the water quality profiles of the surrounding water systems in the area, as well as the quality of the effluents from the mines.

On the other hand, Benguet is a mountainous province and home to over 440,000 people, spread out in 2770 km^2 of land. Because the region is made of highlands and is gifted with naturally cool weather (ideal for growing vegetables), it is named the "Salad Bowl of the Philippines." Benguet is mostly settled by the indigenous Igorot people whose major means of livelihood are agriculture, mining, and tourism. Its main mineral reserves are gold, copper, chromite, and silver; but it is most popular for gold, even being tagged as the country's leading gold producer. Silver smithing is also popular in Benguet, producing silverworks that are much cheaper compared with Manila.

One of the biggest mining companies in Benguet, Philex Mining Corporation, started operation in 1955. Philex's most valuable mining site is located in Sto. Tomas, also called the Padcal mine. "A Gift of God, A Work of Man" served as the company motto for 50 years. As the Padcal mine covers the ancestral lands of Igorot tribes, 95 has in Benguet are subject to royalties paid to indigenous claim holders, Philex asserts. The company is currently a holder of three MPSAs for gold, copper and chromite, which cater to Japanese companies. There are other mining companies present in Benguet, but Philex remains the biggest and most controversial of all. As of December 2014, its main stockholders are Asia Link B.V (20.71%); Social Security System (20.53%); PCD Nominee Corp. (20.46); and Two Rivers Pacific Holding Corp. (14.96%) (PMC 2015).

In August 2012, the Padcal mine made headlines as a massive mining spill caused the release of 20.6 million tons of toxic tailings into the local water systems. The incident happened shortly after the arrival of Typhoon Gener (international name Saola). While the company claims that the leakage is due to the abnormally large volume of rainfall brought about by the typhoon, further independent investigations showed negligence on the part of Philex as a significant contributor to the disaster (Boquiren 2013). The report showed that the root cause of the spill is poor planning on the part of the company, as the tailing storage is way past its lifetime. Boquiren also concluded that there is nothing abnormal with the volume of rainfall brought by the typhoon. The design of the tailing storage should last only up to 2010 to 2012, but Philex decided to extend the use of the facility. Even though the mine spill from Padcal is less toxic than in the other mine spills, its sheer volume was the biggest concern as the spill made a heavy impact on the surrounding ecosystem.

The Mines and Geosciences Bureau decided to suspend Philex's mining operations the day after the massive spill. However, the suspension was lifted 2 years after, in 2014, declaring that Philex handled the disaster adequately and paid their fines accordingly (over PhP 1 billion or USD 21.2 million). Furthermore, the company announced in 2015 that they will be extending their mining activities in Padcal until 2020 as recent studies show an additional 110.9 million tons of mineral sources available in the area (ABS-CBN 2015).

This was not the first time that a disaster struck Padcal. In the 1980s and 1990s, two tailing storage ponds have also collapsed. Furthermore, small spills have recurred before the massive spill in 2012 (Boquiren 2013). The repeated tragedies have taken its toll on the nearby water systems, the Balog creek and the Agno River. Previously tagged as class A water bodies, the said creek and river suffered greatly due to heavy metal contamination, including zinc, arsenic, and copper. Unfortunately, there are no data available on ambient water quality in the surrounding rivers, as well as on effluent quality. Hence, imposing legal sanctions on environmental violation has not been pursued.

5.4.3 A Success Story on Agro-recycling of Industrial Wastewater

In 1999, fish kills in Lian River, Batangas, were reported and this was attributed to an alleged disposal of distillery wastewaters. A distillery in Batangas was given a "cease and desist order" because of its barging operations. Then DENR Secretary Cериles called for a meeting with stakeholders and other concerned parties (distillery officials, sugarcane farmers, NGOs, members of the academe, scientists, government officials, and community representatives) to come up with a win-win solution on recycling of distillery wastewater as fertilizer for sugarcane. UPLB's BIOTECH spearheaded the research on liquid fertilization project.

The land application of distillery wastewater has been practiced in Brazil, China, Indonesia, Australia, and Malaysia. The distillery effluent is a molasses-based wastewater which is a byproduct of alcohol fermentation. The distillery wastewater are of two types: fresh slops that come directly after alcohol distillation and biodigester effluents that are generated after anaerobic digestion (production of methane). The fresh slops are highly colored, with 9–12% total solids and high BOD and COD of about 50,000 ppm and 100,000 ppm, respectively. In addition, the effluent smells like rotten eggs because it comes from anaerobic digestion. At that time, results from local studies in the Philippines on the use of fresh slops as fertilizer were from greenhouse experiments only. These greenhouse studies showed that the slops have no adverse effects on the early measurable growth characters of sugarcane. During that period, none yet has been done on large-scale applications in sugarcane plantations.

Because the fresh slops are still useful to distilleries for the production of methane (used to fire their boilers), the main problem is the disposal of the biodigester effluents. The Fertilizer and Pesticide Authority, which is the government agency handling the registration of fertilizers, tasked BIOTECH to conduct efficacy trials on the use of biodigester effluents as liquid organic fertilizer for sugarcane. BIOTECH studies showed that, at the rate of 200 m^3/ha, the biodigester effluents can either supplant or supplement mineral fertilizers as source-applied nutrients without negative effects on the soil properties. Simultaneous with the efficacy trials, large-scale applications were conducted in the different sugarcane plantations in the Batangas area using biodigester effluents from two distilleries. Farmers have testified to the positive effects from either sole application of the biodigester effluent or its use in combination with conventional fertilizers.

Encouraged by the early success of the fertilization project in Batangas, another distillery in Pampanga also conducted large-scale utilization of distillery effluents as fertilizer for sugarcane in lahar-laden areas. The biodigester effluents increased organic matter content, improved soil structure, and favored the growth of soil microflora in lahar-laden soil. Several studies followed on the use of biodigester effluents in Pulupandan and La Carlota, Negros Occidental.

Agro-recycling of biodigester effluents through land application produces the following benefits:

1. Because the fertilizer is liquid in nature, it is able to provide not just the nutrient requirements but also the partial moisture requirements of the soil. Thus, there is less dependence on surface and groundwater, conserving vital water resources.
2. It promotes sustainable agriculture due to less dependence on synthetic fertilizers, which have negative effects on soil characteristics when used for long periods of time.
3. Aside from NPK, the biodigester effluents contain a wide array of nutrients such as calcium, magnesium, sodium, chloride, copper, iron, manganese, and zinc, which are essential for plant growth and development.

5.5 Recommendations

5.5.1 *Establish Systematic and Harmonized Data/Information Infrastructure*

It is quite difficult to make a full assessment of the status of water policy implementation among industries in the Philippines because of the lack of data. All that we know are the environmental disasters and continuous pollution taking place all over the country – and we need water quality reports for stronger implementation of the law.

The available data are incomplete and disorganized because there is no single government agency analyzing, studying, consolidating, and reporting them. Reliance on private groups is needed to get more data. At the same time, only one pollutant makes it on record (BOD), whereas data on other parameters and toxic substances remain unavailable or restricted to the public. The unit being used by the government agencies in their measurement of the BOD parameter does not match the unit indicated in the standards. All these create an impression that the implementation of DENR DAO No. 34 and 35 is severely wanting. Improvements are needed in the implementation framework of the policy.

Interviews with the respective government agencies reveal a deficiency in terms of adequate personnel to do the legwork in collecting and analyzing data. Because of this, there are serious gaps in the monitoring of the various environmental management sites. While reports may exist, most of these are stored in a warehouse and are not yet encoded for analysis. There is also no harmonization in producing quality data among the different government agencies.

It is recommended that a national data system dealing with environmental standards be formed. Under this system, there should be a dedicated analyst who will generate the necessary reports for action. With the help of other agencies, the existing knowledge database should be utilized and these agencies need to collaborate in handling/processing their data.

5.5.2 *Strengthen the Monitoring of Water Pollution Policies*

While the Philippines has an existing water allocation policy and while the government recognizes the need to regulate and conserve the nation's water resources, major gaps still exist, which need to be filled in the area of power generation and industrial sectors. Due to lack of personnel, there is inadequate monitoring of industrial projects. There is no clear accountability in monitoring and implementation as several agencies are involved in such tasks. Moreover, there is difficulty in assessment because the data heavily rely only on water permit applicants and grantees and there is no aggressive monitoring of illegal water use.

Evidently, a more systematic and effective monitoring system needs to be instituted to better implement the water pollution policies in the industrial sector. This would primarily require a clear delineation of the monitoring functions and accountabilities of the different agencies involved such as the NWRB, LLDA, EMB, LGUs, and others. Once these are clarified, it is crucial to satisfy the required number and competencies of the personnel dedicated to do effective monitoring by hiring new ones and instituting a continuing capacity-building program. Corresponding budgetary support should be allocated to conduct regular monitoring activities. To strengthen current monitoring efforts, it is also important to implement the provision of the Local Government Code in forming a multipartite monitoring team to be led by the Department of Interior and Local Government from the national to the municipal level, which shall cover, among others, the monitoring of compliance with established standards, guidelines, systems, and procedures. Likewise, it is important to tap the support of universities and research institutions in establishing benchmark data for monitoring and to involve them in the monitoring process.

5.5.3 Give Incentives to Water Reuse by Developing Tools for Economic Assessment of Pollution Load and Clean-Up Costs

The Philippine government, through the Department of Agriculture (DA) Administrative Order 26, Series of 2007, supports the reuse of industrial effluents for land application, either as fertilizer or irrigation water. The DA, through its implementing agencies and bureaus, provided guidelines for the safe reuse of wastewater for irrigation and other agricultural purposes. In addition, according to Section 6c of RA 9367 (Biofuel Law), effluents such as but not limited to distillery slops from the production of biofuels used as liquid fertilizer and other agricultural purposes are considered "reuse" and are therefore exempt from wastewater charges.

In spite of existing policies that encourage the reuse of industrial effluents, particularly in the agricultural sector, there are no sufficient incentives for wider application of these policies. There is therefore the urgent need to promote the reuse of water in the industrial sector by providing necessary incentives such as tax rebates, exemptions from wastewater charges, etc. for those who will practice it. To achieve this, it is necessary to develop tools for economic assessment of pollution load and clean-up costs to serve as basis in determining these incentives.

5.5.4 Compel Institutional Action to Address Pollution Through the Writ of Kalikasan and Writ of Continuing Mandamus

Existing political-legal mechanisms such as the "Writ of Kalikasan" should be employed to compel institutional action on the part of industries to address water pollution and the associated problems brought about by violations of environmental regulations. This writ is a legal remedy under Philippine law for persons whose constitutional right to "a balanced and healthful ecology" is violated by an unlawful act or omission of a public official, employee, or private individual or entity (Republic of the Philippines Supreme Court 2010). As illustrated in the case of five mining firms in Sta. Cruz, Zambales, the petitioners can include residents of the affected areas who can file a case to the Supreme Court for environmental violations of the firms contributing to environmental damages such as water pollution and heavy laterite siltation of river systems, coasts, farmlands, fishponds, and residential areas and their associated livelihood impacts.

Similarly, a petition for continuing mandamus should be filed against LGUs or any government agency or officer that unlawfully neglects the performance of official duty concerning the enforcement or violation of an environmental law, rule, or regulation (Republic of the Philippines Supreme Court 2010), particularly the inability of the agency to address water pollution problems. Once judgment is pronounced by the Court in favor of the petitioner, support should also be provided to the latter to monitor compliance of the former to perform its obligations until the required action is fully satisfied.

The employment of the writs of kalikasan and continuing mandamus requires an informed and dynamic civil society and citizenry who are vigilant and can mobilize and pursue civil action in case of environmental violations. Continuing IEC programs are therefore necessary to teach local communities about their environmental rights so they can be more active in their pursuit of environmental justice and in addressing water pollution problems. Moreover, strategic partnerships with members of civil society such as people's organizations, nongovernment organizations, media, and environmental groups should be forged to enable citizens to become vanguards of environmental protection.

References

ABS-CBN. (2015). *Philex mining: Padcal mine life may be extended*. http://news.abs-cbn.com/business/03/20/15/philex-mining-padcal-mine-life-may-be-extended. Accessed 5 Dec 2016.

ADB (Asian Development Bank) and Blacksmith Institute. (2013). *Reduction of mercury and heavy metal contamination resulting from artisanal gold refining in Meycauayan, Bulacan River System (report card)*. https://www.adb.org/results/water-pda-reducing-mercury-and-heavy-metals-contamination-meycauayan-river. Accessed 10 Jan 2017.

Badilla, N. S. (2015). Four firms blamed for mud flows. *The Manila Times*. http://www.manilatimes.net/4-firms-blamed-for-mud-flows/225602/. Accessed 10 Jan 2017.

Blacksmith Institute. (2007). *"Dirty Thirty"—List of the world's worst polluted places for 2007.* Unpublished report. www.worstpolluted.org/reports/file/2007%20Report%20updated%20 2009.pdf. Accessed 5 Dec 2016.
Bondoc, J. (2013). SC stops Zambales mines; Chinese 'invaders' socked. *The Philippine Star.* http://www.philstar.com/opinion/2013/07/24/1005781/sc-stops-zambales-mines-chinese-invaders-socked. Accessed 5 Dec 2016.
Boquiren, A. (2013). *Philex mine spill: Not due to Typhoon Saola (Gener), it is a test on responsible mining.* https://file.ejatlas.org/docs/ATM_Art_Boquiren_On_the_Philex_Mine_Spill_of_August_2012__written_7_May_2013_.pdf. Accessed 5 Dec 2016.
Campos, O. V. (2016). Closure of Zambales mines affects local economy. *Manila Standard.* http://www.thestandard.com.ph/business/202253/closure-of-zambales-mines-affects-local-economy-.html. Accessed 5 Dec 2016.
Datu, R. (2016). *Writ of Kalikasan filed vs mining firms, Zambales, DENR officials.* http://www.rappler.com/nation/134153-writ-kalikasan-mining-firms-zambales-denr. Accessed 5 Dec 2016.
DBP (Development Bank of the Philippines). (1999). *Handbook of environmental standards.* Manila: DBP.
DENR (Department of Environment and Natural Resources). (1990a). *Administrative order no. 34: Revised water usage and classification/water quality criteria amending section nos. 68 and 69, Chapter III of the 1978 NPCC rules and regulations.* Manila: DENR.
DENR (Department of Environment and Natural Resources). (1990b). *Administrative order no. 35: Revised effluent regulations of 1990.* Manila: DENR.
DENR (Department of Environment and Natural Resources). (2016). *Administrative order no. 2016-08. Water quality guidelines and general effluent standards of 2016.* Manila: DENR.
EMB-NCR (Environmental Management Bureau-National Capital Region). (2013). *EMB NCR waste water generation data.* Unpublished.
EMB-Region III (Environmental Management Bureau-Region III). (2014). *Water pollution for industries and commercial establishments.* Unpublished.
LLDA (Laguna Lake Development Authority). (2014). *LLDA waste water generation data of industries as of 2014.* Unpublished.
Local Water Utilities Administration (LWUA). (2000). *Manual on water rates and related practices.* Retrieved from http://www.lwua.gov.ph/downloads_10/LWUA_water_rates_manual.pdf.
Manila Water Company. (2015). *Average consumption per day in cubic meters for industries in east zone of Metro Manila as of 2015.* Unpublished.
Matsumura, M., Nakao, N., Alfafara, C. G., & Migo, V. P. (2003). The current situation and problems of water treatment technologies in the Philippines [in Japanese]. *Journal of Water and Waste, 45*(10), 69–76.
Maynilad. (2015). *Maynilad tariff schedule.* http://www.mayniladwater.com.ph/uploaded/Dec_16_TARIFF_table_for_2016_web.jpg. Accessed 10 Jan 2017.
NPCC (National Pollution Control Commission). (1978). *Revised water usage and classification.* http://open_jicareport.jica.go.jp/pdf/11948882_17.pdf. Accessed 5 December 2016.
NWRB (National Water Resources Board). (2014). *Water allocated for groundwater and surface water consumptive use.* Unpublished.
Philex Mining Corporation. (2015). *Top 20 stockholders.* http://www.philexmining.com.ph/investor-relations/share-information/top-20-stockholders. Accessed 5 Dec 2016.
Republic of the Philippines. (2010). *Administrative matter no. 09-6-8-SC. Rules of procedure for environmental cases.* Manila: Supreme Court, Republic of the Philippines.
Sazon, R. R. (2016). *Environmental impacts of mining operations on the riverine ecosystem of Sta. Cruz, Zambales, Philippines.* University of the Philippines Los Baños. Unpublished.
Singapore's National Water Agency. (2017). Water price. https://www.pub.gov.sg/watersupply/waterprice. Accessed 10 May 2017.
Water Environment Partnership in Asia (WEPA). (n.d.). *State of water: Philippines.* Retrieved November 15, 2017, from http://www.wepadb.net/policies/state/philippines/overview.htm

Dr. Veronica P. Migo is a Professor at the Department of Chemical Engineering, College of Engineering and Agro-Industrial Technology of University of the Philippines Los Baños (UPLB). She obtained her BS Agricultural Chemistry, MS Soil Science and Ph.D. Agricultural Chemistry from UPLB. Her field of specialization is Analytical and Environmental Chemistry. Her research interests include remediation of water and sediments in aquaculture farms, river quality monitoring, treatment of industrial wastes and agro-recycling of distillery effluents. She has in-depth experience in putting up analytical chemistry laboratories and is a technical assessor for ISO 17025 accreditation for laboratory competence. She was a Fulbright Scholar at the Department of Chemistry, University at Buffalo, New York where she worked on the detection of emerging contaminants in the environment.

Prof. Marlo D. Mendoza is an Associate Professor at the College of Forestry and Natural Resources (CFNR), University of the Philippines Los Baños (UPLB). He earned his B.S. Forestry from UPLB and Master in Development Management from the Asian Institute of Management (AIM). His fields of expertise are development management, forest and natural resources governance, watershed management, ecotourism, and strategic planning. He was a former Director of the Forest Management Bureau and Assistant Secretary and Undersecretary of the Department of Environment and Natural Resources. He was also a Country Coordinator of the Blacksmith Institute Philippines, a New York-based not-for-profit organization undertaking action research from site assessment, project design and remediation work of toxic places in developing countries. He was the Program Director of the Bantay Kalikasan (Environment Watch) of ABS-CBN Foundation, Inc., a leading media-based environmental organization in the Philippines. He is currently involved in action-research and actual project implementation of various watershed and environmental restoration work in the Philippines.

Dr. Catalino G. Alfafara is an Associate Professor of the Department of Chemical Engineering, College of Engineering and Agro-Industrial Technology (CEAT), University of the Philippines Los Baños College, Laguna (UPLB). He earned his B.S. Chemical Engineering (*magna cum laude*) from the University of San Carlos, Cebu City and Masters in Environmental Engineering at the Asian Institute of Technology, Bangkok, Thailand. His Ph.D. is in Fermentation Technology from Osaka University Japan. He further pursued a Post-Doctoral Fellowship in Eco-Environmental Engineering in University of Tsukuba, Japan. His major fields of expertise are in Environmental Engineering and Fermentation Technology. His research studies include physico-chemical treatments of industrial wastes from artisanal used-gold-jewelry smelting plants, sugar refinery spent ion-exchange-process (SIEP) effluent, distillery slops and other highly-colored effluents employing ozonation and electrolysis. He has an extensive research experience in biogas production from molasses-based distillery slops. He has a number of peer-reviewed scientific publications both local and international.

Dr. Juan M. Pulhin is full Professor and former Dean of the College of Forestry and Natural Resources, University of the Philippines Los Baños (UPLB). He earned his Bachelor of Science and Master of Science in Forestry degrees in UPLB and Ph.D. degree in Geographical Sciences from The Australian National University. He was a Visiting Professor at The University of Tokyo for four times and has more than 30 years of experience in natural resources education, research and development. He has authored more than 100 technical publications on various aspects of natural resources management and climate change. He was a Coordinating Lead Author of the Intergovernmental Panel on Climate Change (IPCC) Fifth Assessment Report and a Lead Author of the Fourth Assessment Report. He has been involved in an interdisciplinary research project on water governance for development of the University of the Philippines and in numerous watershed development planning projects since 1998.

Chapter 6
Agricultural Water Management Issues in the Philippines

Arlene B. Inocencio, Dulce D. Elazegui, Roger A. Luyun Jr., and Agnes C. Rola

Abstract This chapter discusses the role of water management and governance in agriculture as it affects irrigation and agricultural productivity. The governance framework discusses the various actors involved in water in agriculture and the policy environment, which, in turn, affect the irrigation development performance. The country's performance in agricultural water management is assessed in terms of area irrigated by public investments, cropping intensity, collection efficiency, and productivity impact. To improve performance and agricultural productivity, the major challenges facing national and communal irrigation systems are identified. These challenges require policy changes to address poor performance, unsustainability of systems, and weak management capacities.

Keywords Irrigation • Cropping intensity • Firmed up service area • Irrigation service fee • Agricultural productivity

A.B. Inocencio (✉)
School of Economics, De La Salle University, Manila, Philippines
e-mail: arlene.inocencio@dlsu.edu.ph

D.D. Elazegui
Center for Strategic Planning and Policy Studies (CSPPS), College of Public Affairs and Development, University of the Philippines Los Baños, College, Laguna, Los Baños, Philippines

R.A. Luyun Jr.
Institute of Agricultural Engineering, Land and Water Resources Division, College of Engineering and Agro-Industrial Technology, University of the Philippines Los Baños, College, Laguna, Los Baños, Philippines

A.C. Rola
Institute for Governance and Rural Development (IGRD), College of Public Affairs and Development, University of the Philippines Los Baños, College, Laguna, Los Baños, Philippines

A.C. Rola et al. (eds.), *Water Policy in the Philippines*, Global Issues in Water Policy 8, https://doi.org/10.1007/978-3-319-70969-7_6

6.1 Introduction

Agriculture uses about 80.5% of water that is allocated for consumption (Luyun 2016). Irrigation investment drives higher agricultural productivity. However, recent data from the Philippines show that even with huge irrigation investments, cropping intensity (effective crop area) was not increasing in a similar pace (Inocencio 2016). Several factors were hypothesized to cause this, including lack of water. Water scarcity is seen to be a very real problem at the irrigation-system level (Nguyen 2015; Clemente 2015) and the literature has pointed to water governance[1] as the culprit. Natural occurrences such as shifts in historical temperatures and rainfall patterns that alter the timing and quantity of annual water flows pose ever increasing challenges to irrigation. The current short cycle of both drought and flood events cause significant agricultural damage. The "new normal" conditions are recognized, but policy and institutional responses have been slow to adjust to achieve agricultural resilience (Rola 2016).

Irrigation development and management in the Philippines has historically been the single biggest expense item in agriculture, accounting for about a third of the total since the 1960s. In the 1970s and early 1980s, as well as in recent years when world rice prices rose at unprecedented levels, this ratio was even higher at close to half of total public expenditures for agriculture. In recent years, irrigation has taken up from one-third to close to half of the Department of Agriculture (DA) budget (Inocencio et al. 2016).

Despite these efforts, irrigated area in the country is still small, thus jeopardizing food security. Siltation problems in canals of the national irrigation system (NIS) cause reduced flow capacities that deprive the downstream portion from adequate water supply (Clemente 2015). The reason, according to Clemente (2015), is lack of maintenance especially of unlined canals and poor watershed management, which results in upland erosion and siltation of the rivers and canals downstream. Siltation in the dams and canals is considered a major problem by communal irrigation systems (CIS) (Luyun 2015). Sediments occur due to forest denudation in the catchment, slash-and-burn agriculture (locally termed as *kaingin*), upland agriculture, quarrying, and lack of silt control devices in the dam.

Among the factors that cause inadequate system-level irrigation water supply is conveyance inefficiency (Clemente 2015; Luyun 2015). Conveyance inefficiency is caused by physical degradation of the system that can be due to storms/typhoons and faulty designs. It can also be caused by siltation as well as lack of maintenance brought about by limited funds, shortage of personnel, and lack of capacity to do maintenance work.

This chapter focuses on the role of water management and governance in agriculture as it affects irrigation and agricultural productivity.

[1] Over the past decade, the terms 'water governance' and 'water management' have entered into the standard vocabulary of professionals and academics engaged with the water sector. The term water governance highlights a shifting state-society relationship in which the state has altered its responsibilities and/or activities related to water management and water service provision.

6.2 Governance Framework

6.2.1 Actors in the Agricultural Water Management Sector

There are many actors involved in water use in agriculture. For irrigation design, development/construction, management and operation of bigger systems, the National Irrigation Administration (NIA) and the irrigators' associations (IAs) are the key players, with NIA being responsible for 80% of total irrigated area. The Bureau of Soils and Water Management (BSWM) and the regional field offices (RFOs) of the DA are the other players that are responsible for small systems. These agencies established small water irrigation system associations (SWISA) to manage and operate the fully turned-over systems. Both NIA and BSWM collaborate with local government units (LGUs), which play a role in communal and barangay irrigation (Fig. 6.1).

The mandate of NIA is to plan, design, construct, and/or improve all types of irrigation projects and to operate, maintain, and administer all NIS. It could also delegate the partial or full management of NIS to duly organized cooperatives or associations. NIA also has the authority to supervise the operation, maintenance, and repair, or otherwise, administer temporarily, all communal (CIS) and pump irrigation systems wholly or partially with government funds. As a government-owned and–controlled corporation (GOCC), NIA has autonomy and authority to contract domestic and foreign loans with government absolute and unconditional guarantees. NIA is exempt from payment of all forms of taxes, duties, fees, imposts, and import restrictions.

NIA has 14 regional irrigation offices (RIOs) and irrigation management offices (IMOs) consisting of 40 clustered provinces; nine district offices, and two dam and reservoir divisions, the Upper Pampanga River Integrated Irrigation System (UPRIIS) and the Magat River Integrated Irrigation System (MRIIS). The IAs are the direct beneficiaries of completed irrigation projects/systems. An IA is an association of farmers within a contiguous area served by an irrigation system (IS).

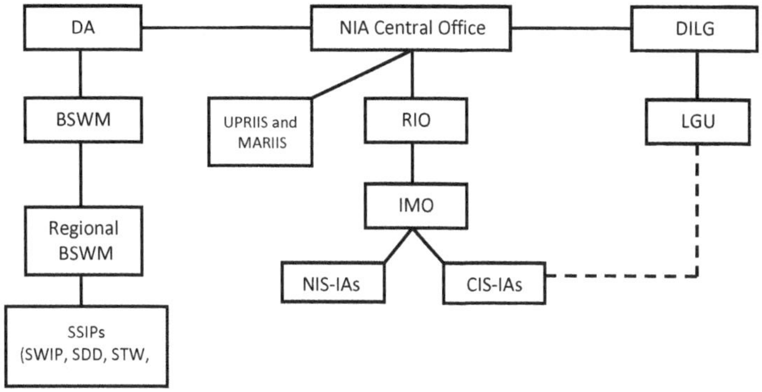

Fig. 6.1 Philippine irrigation water governance administration (*Source*: Rola 2015)

It is a non-profit organization that has to register with the Securities and Exchange Commission (SEC) to apply for a water permit from the National Water Resources Board (NWRB) and to enter into contract with NIA for irrigation development. An IA operating as a cooperative has to register with the Cooperative Development Authority (CDA).

The BSWM is responsible for the design and implementation of small-scale irrigation projects (SSIPs) with LGUs and RFOs of DA and for formulating measures and guidelines for effective soil, land, and water resource utilization. SSIPs include techniques for rainwater harvesting such as small water impounding projects (SWIPs) and small farm reservoirs (SFRs); diverting flowing water by gravity such as diversion dams (DD); using stone/earth/brushwood or concrete structures (run-of-river systems); creating a reservoir to harness spring through spring development (SD) projects; and lifting water by mechanical power (e.g., shallow tubewell or STW and pump irrigation using open sources or PISOS); and the use of renewable energy (ram pumps, solar pumps, and wind pumps) (Tejada et al. 2015).The Philippines' Agriculture and Fisheries Modernization Act (AFMA) of 1997 provides for the devolution of CIS to LGUs and directs the DA to devolve the planning, design, and management of CISs, including the transfer of NIA's assets and resources in relation to the CIS, to the LGUs.

From 1974 to June 2015, the SSIPs contributed about 11% to total irrigation development. SSIPs are being operated and managed either individually or by groups of farmers within adjacent farms. Unlike the NIA systems, the SSISs, specifically SWIPs and DDs, are being operated and maintained by SWISA. SWISA is registered with either SEC or the Department of Labor and Employment (DOLE). The BSWM and DA-RFOs provide support to capacitate and strengthen the SWISAs (Tejada et al. 2015).

6.2.2 The Policy Environment

6.2.2.1 Multiple Goals and Corporate Form

NIA has multiple and seemingly conflicting objectives. Specifically, it is tasked with the social objective of assisting farmers through provision of irrigation service. And yet, as a corporate body, it is expected to generate income and recover costs and maximize returns. World Bank (1995) has indicated that one of the reasons government corporations perform poorly is that "no one has a clear stake in generating positive returns because there is no identifiable owner since the government can be the ministries, or parliament, or the general public." The preferential treatment of GOCCs may be contributing to the essentially weak or absent accountability (SEPO 2006). On the other hand, as a corporate body, NIA enjoys greater flexibility, particularly in its investment activities compared with line agencies. As a corporation, it has been given authority to borrow with national government guarantee. The borrowing spree had the approval of government and can be clearly seen from the 1970s to the mid-2000s with foreign funding dominating the fund sources (SEPO 2006).

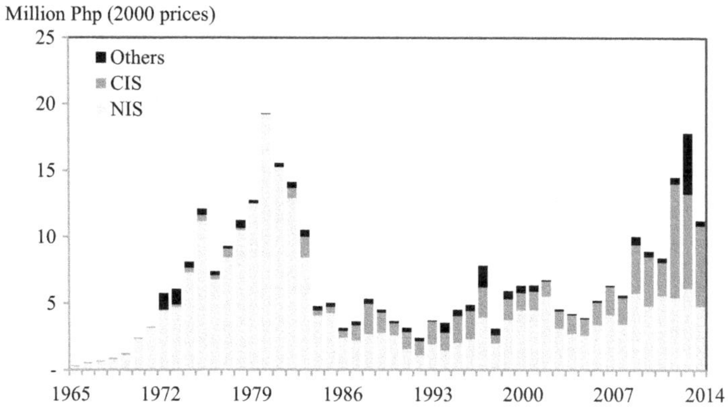

Fig. 6.2 Irrigation investment trends for national, communal, and pump irrigation systems at 2000 prices, 1965–2014 (*Source*: Inocencio 2016)
Notes: *CIS* Communal irrigation system, *NIS* National irrigation system

Irrigation investment remains the most important policy instrument that the government uses to increase productivity in agriculture and to achieve its food staple sufficiency objective. Figure 6.2 presents the trends in total public expenditure for irrigation in 2000 prices. Over the past four decades, irrigation investments peaked in the late 1970s to the early 1980s and rose again in more recent years. The increase in world rice prices in the 1970s, together with the introduction of modern rice varieties suited to irrigated conditions, raised the marginal rates of return for irrigation investments. As world commodity prices declined, yields of modern rice varieties leveled off, and as the cost of irrigation expansion increased, public expenditures declined.

In terms of trends in type of project, shifts in investments from largely NIS to more CIS in early 2000 and in more recent years (David and Inocencio 2012) are discernible. These changes are consistent with the delayed enforcement of the provisions in the AFMA (supposedly to begin in 1997) to give more attention to smaller systems and promote participation of LGUs in developing CIS.

Investment shifted from new development to rehabilitation projects, which were reported to give higher returns to investment. The rise in investments in 2008 was a response to the increase in world rice prices in 2007. This trend continued with the Aquino administration's food self-sufficiency program. More systematic analyses indicate that levels of public investments respond to short-run changes in world rice prices, as these affect marginal rates of return to irrigation investments and adoption of rice self-sufficiency instead of consideration of long-term benefits and costs (Kikuchi et al. 2001).

Also, the funding source shifted from predominantly foreign in the 1970s and 1980s to more local in recent years (Fig. 6.3). This pattern is reflective of the country's development strategy at the earlier time, when foreign funding was the key element. That period was characterized by debt-driven growth. The poor fiscal position of the country in early 2000 led it to foreign funding for irrigation projects.

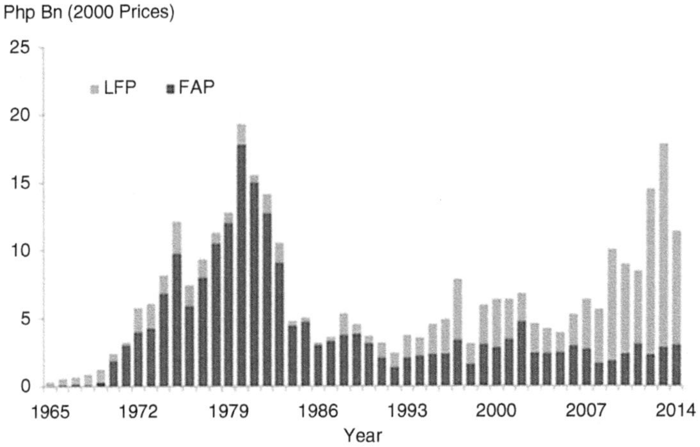

Fig. 6.3 Trends in irrigation investments, by source of funding, 1965–2014 (*Source*: Inocencio 2016)
Notes: *FAP* Foreign-assisted projects, *LFP* Locally funded projects

The recognition of the need to reduce foreign exchange risks directed borrowings back to local sources. This fiscal policy had implications for the quality of projects and their implementation. Foreign funding appeared to have certain implementation rules and standards from the conception of a project up to completion. Local funding, however, appears to have had fewer requirements and less stringent rules and accountability.

6.2.2.2 Capital Cost and O&M Recovery

NIA's mandate to develop and construct irrigation systems requires huge capital and yet, it has not recovered the cost of investment, except for CIS. In lieu of recovering capital costs for NIS, the government funds the capital requirements of NIA through the annual national allocation provided in the General Appropriations Act. Where internally generated funds mostly composed of ISF are insufficient, the national government also subsidizes NIA's operating costs (Fig. 6.4). Because of the capital needs of NIA, it enjoys guaranteed loans. All these policies create a moral hazard problem since there is no incentive to be efficient and collect all dues as there is practically no threat of bankruptcy. Also, NIA is in a bind because it cannot exclude farmers who do not pay the ISF from getting water even if the unpaid irrigation fees or administration charges are considered as "preferred liens, first, upon the land benefited, and then on the crops raised thereon. NIA has preference over all other liens except for taxes on the land which cannot be removed until all fees or administration charges are paid or the property is levied upon and sold by the National Irrigation Administration." The seemingly inconsistent policies contribute to poor performance of NIA.

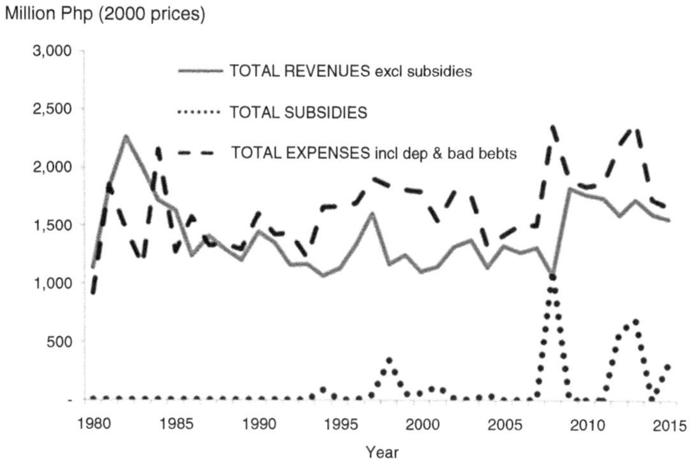

Fig. 6.4 NIA corporate income and expenditures at 2000 prices, 1980–2015 (*Source* of basic data: NIA 2016a)

In the case of SSIPs, the BSWM does not recover any capital cost. For O&M, the SWISAs collect from members.

The present policy on subsidy is biased in favor of NIS. The government fully subsidizes the construction and about half of the O&M costs of these systems. In the case of CIS, farmers are required to finance the operation and maintenance as well as contribute 10% of the cost of construction and repay the balance without interest within 50 years. Another option is to pay 30% of the chargeable cost (in cash or in the money value of labor, materials, and supplies) during the construction or rehabilitation period and then the CIS is turned over to the IA.

The heavy subsidy on NIS and CIS has been the primary disincentive to the judicious choice of irrigation technologies. Such inequity in subsidy increases farmers' demands for NIS relative to the more cost-effective irrigation technologies and discourages private sector participation in irrigation development.

6.3 Agricultural Water Management Performance Assessment

6.3.1 Irrigation Development

NIA's development program largely focuses on NIS and CIS. The NIS are irrigation systems with minimum service areas of 1000 ha, the biggest reaching over 130,000 ha. Management of NIS systems is by NIA and IAs. There are 232 NIS with a total service area of 815,308 ha (or 732,774 ha firmed-up service area) benefitting a total of 595,325 farmers as of December 2015. The three largest NIS with massive reservoirs operated in combination with run-of-the-river gravity irrigation systems account for about a third of total NIS service area. The remaining two-thirds are mostly run-of-the-river gravity irrigation, except for six medium and large-sized NIS that pump water from large rivers.

The CIS are those whose service areas fall below the 1000-ha threshold. The total service area of CIS continues to be sizeable at about 603,106 ha or a firmed-up service area of 578,455 ha, benefitting a total of 391,918 farmers as of December 2015. While many CIS started as private initiatives, most of them have been receiving government support for the cost of rehabilitation and new construction. Most CIS are constructed by NIA, but the IAs are responsible for the management and maintenance of the systems. At least 95% of CIS are gravity systems obtaining water from rivers or streams. However, few have been given funding support for medium sized pumps to also abstract and distribute water from rivers. As of end of 2015, the national systems comprise 43.6% of total irrigated area; the communal systems, 35.6%; private irrigation systems, 10.8%; and other government-assisted systems, 10%.

While the government started to promote the adoption of pump irrigation in the 1950s, recent data indicate that at least 90% of pump irrigation used by rice farmers have been purchased privately, and even more so for pumps used for non-rice cultivation.

Table 6.1 gives the potential irrigable area, by region, as of December 2015. The estimates indicate that there is still much potential left. However, it should be noted that these estimates only considered contiguous areas of 100 ha or more with slopes not exceeding 3%. Included are areas that have been converted into non-agricultural uses. Important determinants of irrigation potential such as water availability and existing land use were not considered. If water availability and economic feasibility will be factored in, these estimates would likely be overestimates. From Luyun (2015), many CIS are already in areas with slopes between 3% and 8%. This information implies that there are potential irrigable areas in steeper slopes.

David (2009) reported the rapid deterioration of the gravity irrigation system service area in the Philippines. The deterioration rate of about 70,000 ha/year in the total NIS and CIS service areas during the pre-AFMA years (1992–1996) earlier reported by David (2003) had increased to about 134,000 ha/year during the post-AFMA years of 1998–2004 (David 2008, 2009). This trend accounted for the very slow annual rate of increase of only about 10,000 ha/year in the actual NIS and CIS service areas. This is in spite of massive efforts of rehabilitating an average of 124,597 ha/year and constructing new irrigation facilities at 19,285 ha/year during 1995–2005. Ella (2015) reported the same findings. He noted that the net increase in total irrigated areas from 1985 to2014 is only 294,939 ha or just 10,170 ha/year. This rate increased for the period 2009–2014, showing a net increase in total irrigated areas of 168,130 ha or 33,626 ha/year. However, for the same period, new area generated is only 224,316 ha and total area restored is 298,840 ha. The sum of new areas generated and areas restored minus actual irrigated areas for the same period

Table 6.1 Status of irrigation development and potential based on 3% slope, as of December 2015 (in '000 ha)

Region	Estimated total irrigable area (1)	Firmed-up service area					Irrigation development (2)/(1) (%)	Potential area for development (1)–(2)
		NISs	CISs	PISs	Others	Total (2)		
CAR	97	14	50	25	4	93	96	5
1	263	48	54	21	51	174	66	89
2	457	152	57	46	21	275	60	182
3	481	196	70	9	20	295	61	186
4A	86	21	19	6	2	49	57	37
4B	139	19	35	14	12	80	58	59
5	239	23	74	25	16	138	58	102
6	190	47	37	15	15	114	60	76
7	46	12	26	4	2	44	95	3
8	84	24	38	6	3	71	84	13
9	75	17	24	2	4	47	62	28
10	114	26	26	6	4	62	54	52
11	147	36	26	1	3	66	45	81
12	286	64	36	3	10	113	39	173
ARMM	156	26	20	0.09	0.295	46	29	110
CARAGA	159	30	26	3	7	65	41	94
Total	3020	755	616	188	173	1731	57	1288

ARMM Autonomous Region of Muslim Mindanao, *CARAGA* Caraga Administrative Region or Region 13
Source: NIA (2016b)

showed total areas which deteriorated to be 355,026 ha or 71,005 ha/year. Despite huge irrigation investments, the data showed that the rate of deterioration is faster than the establishment of new systems.

6.3.2 Cropping Intensity

Irrigation increases cropping intensity, which is affected by seasonality of water supply. Availability of water the whole year round allows for a second or third cropping. Water scarcity leads to a decline in irrigation water supply, thus affecting cropping intensity. This is caused by several factors that may be beyond the control of NIA – e.g., sectoral water allocation (priority for domestic use over irrigation in cases of water crisis); watershed degradation, and climate change.

The national average cropping intensity is defined as the ratio of area actually irrigated for all seasons (including the 3rd and ratooning) to firmed-up service area (FUSA). CIS shows some increase from 122% in 2012 to 128% in 2015. The NIS, on the other hand, shows the same modest growth from 147% to 156% (Table 6.2). Given the investments poured into NIS, CIS growth appears better.

Table 6.2 Cropping intensity (%) in the Philippines, by irrigation system

	2010	2011	2012	2013	2014	2015
NIS	147	112	159	147	146	156
CIS	ND	ND	122	129	134	128

Sources: NIA (Various years)

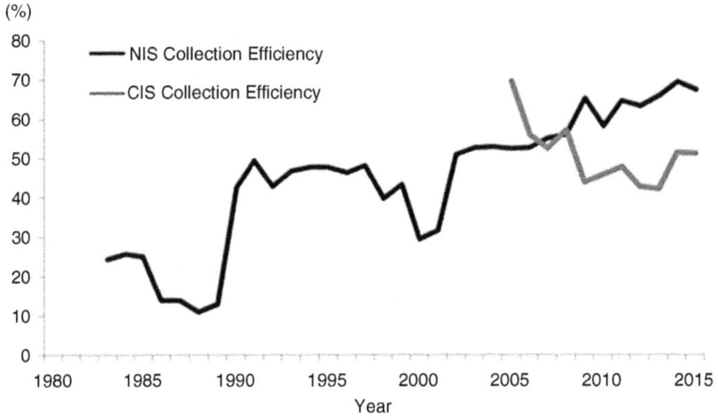

Fig. 6.5 Trends in collection efficiency in NIS and CIS, Philippines, 1965–2015 (*Sources*: NIA Various years)

Low cropping intensity is due to over assessment at the design stage of the individual system service area relative to what is actually irrigated and available water supply (World Bank 1992). Overestimation of service area results from failure to account for built-up areas, flooded areas during rainy season, and elevated areas that cannot be reached by gravity irrigation. In more recent years, flooding during the wet season has reduced substantially the irrigated area for that season and, correspondingly, cropping intensity (Inocencio et al. 2016). The lack of water supply due to declining water intake from degraded watersheds has largely reduced the irrigated area in the dry season (Clemente 2015). Rivers and creeks, the major water sources of CIS, produce extremely low flow or at times no flow at all during the dry season.

6.3.3 Collection Efficiency in Irrigation Systems

The collection efficiency (Fig. 6.5) is defined for NIS as the ratio of current ISF collections to current collectibles, while for CIS, it is the ratio of current amortization payments to current collectibles; these show opposite trends. The collection efficiency for NISs was below 50% until the early 2000s, but it started rising in 2003 to more than 60% in more recent years. ISF collections are used for O&M of the systems and to partly cover the operations costs of the IMOs and the RIOs. For the CIS, the amortization collection rate has been declining since 2004, which should be alarming, given the declining share of total revenues from CIS. This pattern could imply much reduced efforts at collection and/or a less effective collection strategy.

The trends in actual cost of NIS O&M compared with recommended/desirable levels and with ISF collectibles and collections at 2000 prices revealed two key concerns: (a) the recommended O&M cost is much higher than the current collectible ISFs, implying that, even with 100% collection efficiency, the ISFs would not be able to cover the recommended O&M; and (b) the actual O&M seems to stay close to the actual ISFs collected – i.e., actual collection seems to determine how much O&M will be done rather than the desired level of O&M. The historically poor O&M spending may partly explain why the increasing investments in irrigation development, FUSA, and actual irrigated area and irrigation intensities hardly improved over time.

The ISF collection from NIS for the 1983–2015 period averaged only 43% of the total collectible. It is interesting to point out that this was approximately the same as the percentage of the service area actually irrigated during the dry seasons of the same period. It is quite clear that NIA should improve the efficiency and performance of its NIS if it has to increase its ISF collection.

6.3.4 Irrigation System Governance

Irrigators' associations are supposed to govern the irrigation system, for both NIS and CIS. The primary aim of the two systems is to provide irrigation water to association members at the time it is most needed.

The performance of IAs is based on NIA's functionality rating based on parameters related to O&M performance, organization, financial performance, and organizational discipline. Results of the functionality survey are used in the search for outstanding IAs at the provincial, regional, and national levels. This provides good motivation to IAs and their members. It also helps NIA in identifying appropriate strategies to enhance IA capabilities. The rating is done through discussions/consultation with IAs. The overall rating of IA functionality is an aggregation of four major factors: O&M with 40% weight; financial performance, 30%; organization, 15%; and organizational discipline, 15%. O&M indicators include O&M planning and implementation, and performance such as annual cropping intensity, irrigated area vs programmed area, status of irrigation facilities and structures, yield, and collection efficiency. Financial performance includes income generation and fund utilization and viability index. Organization includes information on membership, meetings, and recording/filing system. Organizational discipline includes attendance in meetings and group work, holding of regular elections, conflict resolution, and imposition of discipline.

6.3.4.1 NIS Irrigators' Associations

The NIS scheme has been established and is being maintained by the NIA. In this type of irrigation system, farmers have to pay ISF to cover O&M expenditures. The ISF is payment for the delivery of irrigation water services rendered by NIA to be paid by the beneficiaries of NIS. This fee is used primarily to finance the continuous operation of the IS. Rates are based on the IS development scheme (run-off-the-river, reservoir, and pump), the crops planted, and season (Nguyen 2015).

Rice farms with yields of 40 cavans/ha or less are exempted from paying the ISF. The fees collected by NIA should cover the costs for O&M. NIS farmers are not paying for capital outlay. Consistent with the current trend of devolved management of irrigation systems in other countries and to make viable the rationalization or staff reduction of the agency, NIA has been preparing the IAs to take on greater responsibilities through an irrigation management transfer (IMT) scheme for NIS. The IMT transfers the responsibility and authority for the management of irrigation systems from NIA to IAs. This process may include transfer of decision-making authority (governance) or transfer of ownership of infrastructure (considered in privatization), or transfer of water rights from government to water users' associations (FAO 2001). The IMT may also be turning over partial management responsibilities – e.g., water delivery, canal maintenance, and paying for irrigation services such as that followed in the Philippines. The goal is for the NIS under the IAs to be self-managed just like the CIS.

There are four models of IMT or transfer of NIS O&M responsibilities from NIA to NIS IAs (Table 6.3). Operation and management of turnouts and farm-level facilities is the inherent responsibility of the IA. As of 2014, IMT accomplishments involved models 1 and 2 (95% of IAs) and were very minimal for models 3 and 4 (Table 6.3). This rate raises concern if devolving management of NIS has reached a sufficient depth since Model 1 is limited only to maintenance of canals and Model 2 to management of lateral canals, remote from the system's complete turnover.

For NIS, a study (Clemente 2015) in Luzon revealed that most IAs have moderate performance levels; only 12% show high performance, and these are found at the upstream parts, which receive adequate water supply. Downstream IAs show low performance. The study cites that, even without much technical data on flows that are included in the analysis, water delivery is one major factor causing low performance.

6.3.4.2 CIS Irrigators' Associations

CIS has been established either by farmers or by the NIA. CIS farmers pay for the capital outlay and irrigation fees for maintenance. Management is turned over by NIA to the IA for O&M (Lauraya and Sala 1995; Jopillo and de los Reyes 1998). CIS farmers use the less formal and customary rules in governing irrigation water. The legal mandate that AFMA provided was for the LGU to take charge of the CIS investments and oversee the CIS IA operations.

Table 6.3 Status of irrigation management transfer (IMT) of NIS as of October 31, 2014

IMT model	Description	IAs involved (no.)	Percentage
Model 1	Maintenance of canals delegated to IA; IA is compensated based on canal area maintained and existing labor rate	1192	49.69
Model 2	Turnover of management of lateral canals to IA; IA gets a share of ISF collected; typical ISF sharing: NIA: 70%, IA:30%	1103	45.98
Model 3	Turnover of management of main and lateral canals to IA federation (headworks/dam not included); IAs get a share of ISF collected; typical ISF sharing: NIA: 70%, IA:30%	77	3.21
Model 4	Complete turnover of irrigation system to IA; IA pays NIA a rental fee at a rate of 75–100 kg of dry palay per ha per year	27	1.13
Total		2399	82.98
Total NIS IAs organized		2891	

Source: NIA (2016c)

The AFMA mandated that the DA, particularly through the NIA and the Agricultural Training Institute (ATI), in collaboration with the Department of Finance (DOF) and the Department of Interior and Local Government (DILG), conduct a capability-building program to enable LGUs to independently and successfully sustain the CIS. This program for the LGUs was supposed to incorporate components for technical and financial assistance, logistical support, and training. This task, however, was not accomplished due to lack of coordinated efforts to improve the technical capacity of the LGUs and the lack of political will to implement the AFMA provisions. There has really been limited capacity building for LGUs and IAs. For CIS, 80% of the IAs organized had achieved complete turnover of irrigation systems (Table 6.4).

Among functionality indicators, a study (Elazegui 2015) in Luzon indicated that O&M and finance are common weak points among CIS IAs. O&M involves different activities such as minor repair, routine maintenance, emergency repair, and annual repair. IAs do not include all of these in their collection targets, as they usually refer to canal cleaning for their O&M activity. Moreover, NIA's financial assistance to CIS projects is for the main diversion and main conveyance facilities. Farm-level facilities (e.g., turnouts) are not included in the project cost and have to be developed by the farmers. Thus, even if collection efficiency is high relative to amortization payments, the IAs find it difficult to collect other dues from their members (Elazegui 2015).

Table 6.4 Status of CIS turnover, as of October 31, 2014

Mode	Description	IAs involved (no.)	Percentage
CIS Turnover			
Complete turnover of irrigation system to IA for O&M	Generally, IA contributes 10% equity during the construction period and pays NIA amortization fee of the direct cost of construction of the project for a period not exceeding 50 year; IA has the option to contribute equity equivalent to 30% of the project cost, and the rest of the project cost is considered fully paid	4267	80.18
Total CIS IAs organized		5322	

Source: NIA (2016c)

6.3.5 *Irrigation and Rice Productivity*

Irrigation increases yield and cropping intensity. With water available all year-round, some farmers can have a second or third crop. Figure 6.6 shows the comparison between rainfed and irrigated rice in terms of area, yield, and production for the 1970–2014 period. Figure 6.6a shows decreasing hectarage of rainfed rice and increasing irrigated rice areas for the same period. On average, irrigated rice has higher yields than rainfed rice, by about a ton per hectare (Fig. 6.6b). The increasing area and high yield lead to greater rice production for irrigated rice compared with rainfed rice (Fig. 6.6c). The sharp decline in production and rice area in 1998 has been attributed to the worst El Niño episode in 1997, which affected the harvest in 1998. The development of irrigation in the 1970s has resulted in substantial increases in crop yield. It has coincided with the introduction of high-yielding varieties, particularly for rice, in the Green Revolution era.

6.4 Policy Challenges

6.4.1 *Low Performance of Irrigation Systems*

Concerned with the continuous deterioration of the NIS and the declining water yield during dry seasons as well as the millions of pesos spent annually for desilting, the NIA proposed the inclusion of the catchment management program to address the said problems under the World Bank-assisted Water Resources Development Project (World Bank 1998). The study found that another main reason for the low performance of the systems is sedimentation in the storage area of reservoirs and dams and along canals. Sources of sediments are sidehills, drainage/creeks, side slopes of irrigation canals, and the catchment. Canals passing through foot of hills

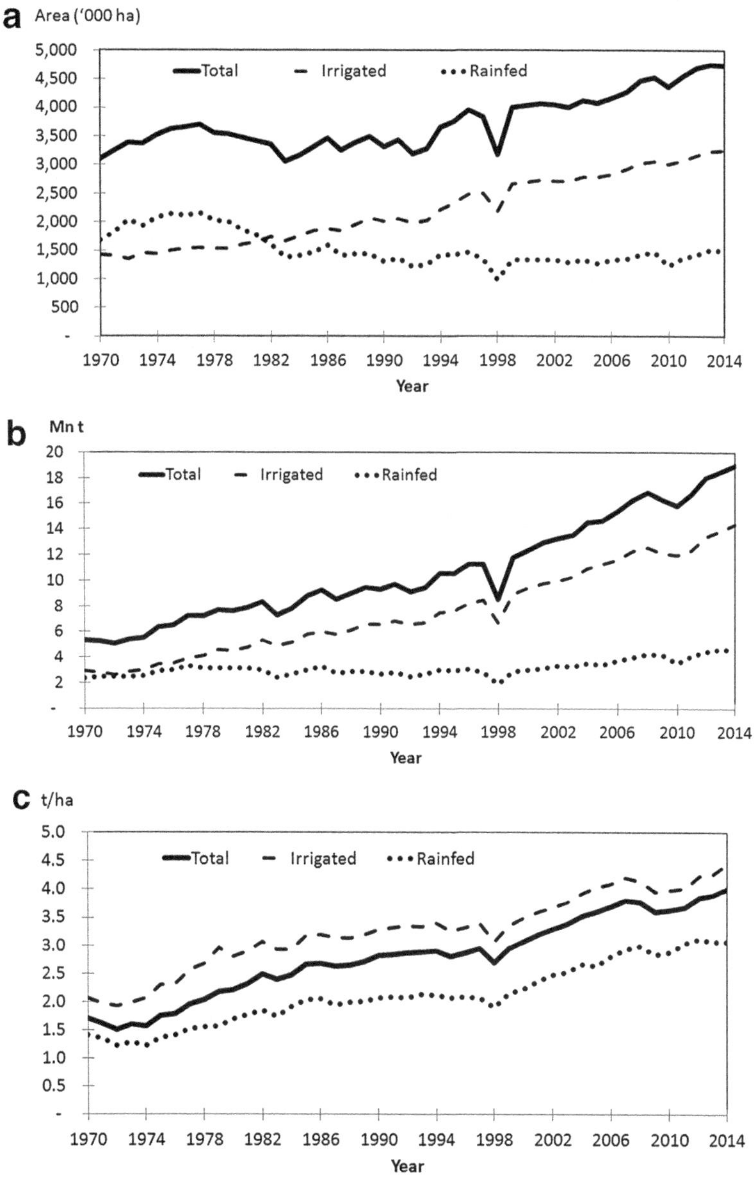

Fig. 6.6 Trends in irrigated vs rainfed area, yield, and production, 1970–2014 (**a**) Palay area (**b**) Palay production (**c**) Palay yield. (*Source*: PSA 2015)

planted with crops such as corn have reduced flow capacities or are rendered useless after being completely filled up by eroded or collapsed soils. Some canals passing through built-up areas serve as drainage of roads and communities. When shallow creeks are tapped for additional water, sediments are directly added into the system. However, the catchment is where the bulk of sediments came from, especially when the forest cover has been denuded or *kaingin* was practiced and the upland area is cultivated. In combination with defective sluice gates and absence of silt control devices, this leads to severe reduction of storage capacity of most dams.

The decreasing dependable flow of surface water sources and the inability to control sedimentation are exacerbated by the prevalence of damaged or dilapidated dams, headworks, and control structures. Some of these problems are caused by typhoons but others are simply due to faulty design and poor O&M. In a study of several NIS, CIS and SWIPs in Ilocos Norte, David et al. (2012a, b) showed that the poor performance of the irrigation systems studied can be traced to their design shortcomings and undesirable design features. The very low dry-season irrigation intensity are due to design shortcomings at the headworks, including underestimation of flood flows and sediment loads, inadequate provisions for sediment control, and underestimation of reservoir inflow and outflow hydrographs. These problems are more evident in the case of the CIS, where most dams are already old, with exposed rock cores, damaged spillways, and silted storage area. Possibly due to limited funds, there is apparent neglect in the estimation of dependable flow and sediment discharge, relying on old design criteria or adopting design parameters from other systems.

6.4.2 Non-optimal Cropping Intensity Due to Water Supply Seasonality

The lack of water during dry season was one of the main reasons cited by Luyun (2015) for the low performance of CIS. Of the 66 CIS surveyed, only 22 CIS or 33% have river sources that were deemed capable of providing irrigation even during dry seasons. Four of these rivers are large enough to provide water even to large NIS and to the CIS through pumps. In 14 CIS, creeks are the major source of water. These creeks have adequate water flow for small areas during the wet season but produce extremely low flow or at times no flow at all during the dry season. Springs are the principal water sources of most CIS in mountainous regions, which, due to their usually low discharges, are supplemented by flows from creeks and runoff. Three other CIS rely on excess runoff either as the main source or in support of other principal water sources. These CIS are equipped with costly storage reservoir dams to impound water, but they can only support small areas for irrigation.

The country also has an extensive groundwater reservoir with shallow and deep well areas of about 50,000 km^2 and 123,000 km^2, respectively. However, these are relatively underutilized for irrigation with only a 25% share of the 3737 million m^3 water permits granted by the NWRB in 2013. In most of the irrigation systems, farmers have shallow tubewell (STW) pumps serving as supplemental water source in times of inadequate or intermittent irrigation from the canal systems or as a primary water source for vegetables or other high-value crops. Most STWs are purchased under the farmers' own initiatives.

Reuse of wastewater for irrigation could help address water supply and improve cropping intensity. There are existing guidelines for this. Watershed degradation has posed problems on sustainability of irrigation systems. IAs apparently have none or a limited role in watershed management despite their complaints about activities in the watershed affecting their irrigation system. IAs can serve as partners in watershed management programs in collaboration with the Department of Environment and Natural Resources (DENR), the lead agency in watershed concerns. There is also the Convergence Program between DENR, DA, and LGUs. They can lobby for representation in appropriate bodies. For example, Irrigation committees may be formed within sub/watershed management councils where IAs may express their concerns.

6.4.3 Need for Sustainable Source of Operations and Maintenance Funds

The national government, in wanting to provide service to a larger populace, is less stringent on loan repayments for service-oriented cooperatives and does not impose stiff fines on delinquent and erring end-users (such as NIA farmer beneficiaries). This exacts a moral hazard problem: these service providers and end-users have no incentive to shape up and improve efficiency since the national government is on standby to bail them out and would not allow them to fail. This, in turn, leads to poor collection efficiency.

The poor financial condition of NIA arises from operational factors and inconsistent policy objectives of government. NIA is mandated to provide irrigation service, which has some social objective, which results in losses. In this case, the government is obliged to subsidize its operations. NIA's operations require huge capital investments that the government cannot afford, given its budgetary constraints, so it contributed to foreign borrowing.

Moreover, NIA's financial assistance to CIS projects is for main diversion and main conveyance facilities. Farm-level facilities, e.g., turnouts, are not included in the project cost and have to be developed by the farmers, according to the CIS Manual (NIA 1985). O&M is not included in the CIS project cost. Thus, IA collection mainly goes to amortization payments and it is difficult to collect fees for other activities. Between 2009 and 2014, average actual collection efficiency did not improve at around 40% level until 2014 when it rose to 60%. Low collection efficiency is due to attitude and perception of farmers. They claim that providing irrigation is the responsibility of the government; for NIS, farmers are paying ISF, not the cost of irrigation project. Some politicians also committed to pay the project cost but failed to do so. Other IAs claimed that there is not enough water, so they will not pay (Elazegui 2015).

Projected yields are also overestimated but water use efficiency has declined over the years. Another cause is lack of investments in recurrent costs associated with O&M activities once construction is completed (Ostrom 1990). Donors normally restrict their involvement to design and construction and view O&M as a responsibility of the recipient of the system. Routine maintenance is delayed until deterioration of the system is large enough to require rehabilitation.

6.4.4 Capacity Needs for Irrigation Water Management

NIA is mandated to provide needed capacity-building/training activities to the IAs before transferring O&M. Under its IMT program, it shall also gradually transfer the management and O&M of NIS, wholly or partially, to duly organized IAs. Likewise, NIA develops communal or small irrigation projects (CIP) or rehabilitates existing CIS with IA participation, then completely transfers to the IA the O&M of the completed CIP or rehabilitated CIS. NIA assists the IAs in establishing linkages with other agencies (both government and private) that provide support services and other assistance programs, including livelihood/small business ventures to increase farm productivity and family income. The above mandates give responsibility to the NIA to ensure that the IAs have the capacity to govern their systems. However, it does not have an explicit mandate to regulate the activities of the IAs.

The IMT program offers various schemes (models) to IA for partial or full responsibility in O&M, depending on the IA's capacity. The original target of NIA is to roll out its rationalization or organizational streamlining plan side-by-side with IMT implementation. The slow IMT implementation in 2010 was attributed to the seeming reluctance of some IAs to accept O&M responsibility unless irrigation facilities are rehabilitated or restored to operable conditions. Others argued that there are some unclear provisions in the IMT policy guidelines, notably: (1) vagueness of the concept "fair sharing of burdens and benefits," (2) computation of break-even point as basis for ISF collection sharing, (3) IAs' lack of start-up capital or seed fund to assume management, and (4) extent of IA share in the collection of back accounts (Ofrecio 2016).

6.5 Policy Recommendations

To address the abovementioned policy challenges, the following policy recommendations are being proposed to improve agricultural water management in the Philippines toward better productivity and sustainability.

6.5.1 *Exploring Private-Public Sector Partnerships in Future Irrigation Development*

Given the perennial problem of government in funding O&M and rehabilitation and problems with the management and sustainability of irrigation systems, a study on public-private partnerships (PPPs) in irrigation was commissioned. Benabderazik and Inocencio (2013) explored the viable and effective business options for PPPs in developing and modernizing the irrigation sector in the country. Adopting the PPP scheme in irrigation will require specific answers to the following questions: (a) how PPP contractual arrangements can help to address the usual problems in public irrigation; (b) how PPPs can enhance financial viability; (c) how PPPs can avoid political interference in setting irrigation fees; and (d) how PPPs can increase the useful life of irrigation systems and break the vicious cycle of low collection of ISFs, poor maintenance, network degradation, and expensive rehabilitation. Any PPP proposal will have to make the case that the financial costs of privately funded projects are on par with those of publicly funded projects.

The efficiency gains from private sector participation stem from private experience and capacity to manage construction more efficiently and to provide more effective O&M. Its ability to structure contracts where finance and performance are entwined to promote efficient outcomes is generally superior to standard public procurement. Publicly funded projects are constrained by procurement, supervision, and decisionmaking processes. Budgeting and appropriation of allocations are complex political tasks that can reduce the ability of the public sector to react quickly and manage efficiently. In terms of O&M, the flexibility of the private sector in setting wages to attract skilled staff is generally an efficiency factor not permitted in the public sector.

Benabderazik and Inocencio (2013) proposed some options for the government to break out of that vicious cycle and provide potentially more efficient and sustainable irrigation service. The options cover financing of new and rehabilitation projects and making O&M sustainable. Specifically, in pursuing PPP projects for irrigation, the government should (a) begin with some experiments or pilots; (b) use PPPs; and (c) use the dam nexus, an expanded role of IAs, the rice policy, and social policy as entry points.

The use of experiments in irrigation sector schemes, whether technical or managerial, is proposed in order to provide more substance to the reform process. Several variations or mixes of activities can be pursued (Benabderazik and Inocencio 2013): (a) NIA in a service contract; (b) a PPP with innovative techniques for conveyance, distribution, and collection; (c) a pilot scheme to measure actual O&M costs; (d) a pilot PPP-IA project with volumetric pricing; (e) a pilot PPP in high-value crops; (f) a PPP with agricultural extension services; and (g) expansion of the role of IAs. The experiments or pilots will help develop realistic and workable guidelines, assess the likely success of proposed options, identify logistical and operational problems and uncover potential ones, determine what resources (funds, materials, staff) will be required, and convince other stakeholders of the potential of PPPs for delivering better irrigation service.

Another option is to exclude the cost of dams from the irrigation investment cost. In this case, the cost of dams would be treated as pure public infrastructure. For instance, in multipurpose dams that will benefit not just farmers but also other water users, the dams and reservoirs can be fully subsidized, but the cost can be shared with the LGU and/or water districts. In the case of hydropower, NIA can collect the full value of the benefits from its users while remaining in the business of administering NIS.

If PPPs are to be pursued right away, the government can begin with a management contract. This track can help improve the technical abilities of IAs in the devolution process through a mandatory alliance between a private partner and the IA. The management contract will be between NIA and the private partner. In this case, the transfer of irrigation management from NIA to the IA will be more rapid than the current format, where NIA directly trains the IA. In order to realize such a transfer, NIA could finance the operation of the private partner for the first few years. The selection of the private partner can be done through open bidding. Once the period of full subsidy is over, a reduction in the contribution from NIA would allow a progressive transfer to the IA that is being assisted by the private partner.

6.5.2 Increasing Cropping Intensity with Strategies to Reuse Water

Pursuant to Republic Act 9275, the Philippine Clean Water Act of 2004, the DA through Administrative Order No. 26 (2007) shall allow reuse of wastewater for irrigation, fertilization and aquaculture, and other agricultural uses, on the condition that generators of wastewater shall secure a discharge permit from the DENR and submit a certification of safe wastewater reuse to the DA (2007). Wastewater generators shall include but not be limited to the different primary users of livestock, agriculture, and food industrial processes (e.g., food handling, processing and manufacturing plants, sugar mills, refineries and distilleries, including biofuel production, slaughterhouses, and poultry dressing plants), aquaculture, domestic and animal sewage, and other industrial and commercial establishments.

Requirements for the permit include the quantity of wastewater to be reused; wastewater characteristics, plan on the reuse of wastewater, and baseline conditions of the surface water, groundwater and soils in the area.

Wastewater shall be subjected to a treatment process, as may be required, to achieve the quality limits before use for irrigation, fertilization, and aquaculture as secondary uses. General requirements are based on standard analytical procedures of wastewater quality for irrigation, fertilizers, and aquaculture stated in established standards in the country and abroad (i.e., DAO 34 series of 1990: Revised Water

Usage and Classification; DAO 35 series of 1990: Revised Effluent Regulations; Standard methods of examination of water and wastewater from American Public Health Association [APHA], American Water Works Association [AWWA], Water Environment Federation [WEF, and the U.S. Environmental Protection Agency [US-EPA]).

Wastewater quantity shall be determined based on requirements specific to the land application and site-peculiar conditions. The method of wastewater application shall not, in anyway, result in negative impacts to the environment. However, the DA recognizes that there are many facilities that practice reuse of wastewater that are undocumented, unregulated, and probably unsafe. These pose risks to human and animal health, crop production, and the environment.

6.5.3 Sourcing O&M Funds

For NIS, farmers do not pay ISF because water delivery is not sufficient or maybe the timing of delivery is off-schedule. The problem of high siltation and quality of canals could be a reason for this. The solutions to the O&M issue may start with the design and the quality of materials of canals and other structures. This implies that even during extreme events, the structure will be strong. This means that the IAs will make sure that during turnover, the appropriate design and materials of the IS are within or even beyond standard. This will minimize O&M in the subsequent years.

For dams that are defective, rehabilitation may be requested to minimize O&M activities. If farmers cannot afford the technical advice of NIA personnel, it can tap other institutions such as state colleges or universities in the area.

Among the CIS, O&M is the primary role of the IA. However, IAs still need technical advice when the structure is damaged. Other community-based participatory irrigation water governance should be promoted in the community; for instance, residents can minimize throwing solid wastes into canals. The community should also be vigilant on thieves of staff gates and other equipment. Some successful CIS IAs have sufficiently done their O&M because they have high collection efficiency. They also pay the water master, who, more often, is also a farmer. Silt removal can also be a community-based activity.

NIA needs to exert more effort toward increasing its ISF collection. This may be done by equitable, adequate, and timely delivery of irrigation water and improved delivery of irrigated agriculture support services for higher yield and cropping intensity. Delineating the NIS irrigation service areas down to realistic levels will go a long way toward increasing unit command area water supply and improving irrigated agriculture support services. Strengthening IAs by giving them better incentives to take charge of the repair and O&M of secondary and tertiary canals will also help reduce O&M costs and improve ISF collection.

The present policies governing ISF rates should be carefully reviewed. The present rates, which were approved in 1974 and implemented starting in 1975, now fall short of the cost of repair, O&M, and rehabilitation. These rates were partly based on the capacity of farmers to pay. During times of poor harvest (less than 2.0 time/ha). farmers are exempted from the payment of ISFs. The same farmer-beneficiaries are not, however, charged higher fees during period of good harvests.

When confronted with the low level of ISF collection, officials tend to support non-enforcement on social grounds: the farmers are deemed too poor to pay the ISF. In this case, it is better to adopt an explicit exemption for poor farmers than to accept a poor collection rate, with its detrimental effect on the quality of irrigation service. This policy could be applied at the IA level, with a general assembly approbation of the farmers to be exempted and a partial increase in the ISF to recover the loss in revenue. The policy could be applied by NIA based on a poverty assessment to be done by a special committee composed of social departments, the DA, LGUs, and NIA. An alternative would be to use the *Pantawid Pamilyang Pilipino* Program of the government. With this transfer in place, all users would pay the ISF, while the poorest would receive benefits from the conditional cash transfer (CCT) program. If they do not pay their ISF, CCT benefits will be cut off.

6.5.4 Capacity Building for Sustainable Water Management

NIA's IMT program for NIS is a work in progress. The current IMT models (1 to 4) are an incomplete devolution because NIA effectively still plays significant roles from management to financing rehabilitation and repairs. Giving an expanded role to IAs beyond the current IMT framework can lead to systems that are more responsive to farmers' needs or are financially sustainable. Specifically, NIA can make the IAs accountable for the rehabilitation of existing systems. A progressive transfer of rehabilitation responsibilities can be facilitated through an agreed-upon schedule of reduction in the share of costs.

With the expansion of IA roles, NIA can focus on the higher level role of supervising the devolution, managing the headwork (reservoirs, dams, and main canals), implementing volumetric charges to IAs at the head gate, and providing technical support to IAs. The IAs will take care of transferred assets, collect water fees to cover their O&M costs, and manage water efficiently and equitably. The empowered IAs will become real service providers for their members and can even hire professional personnel. For this option to succeed, NIA has to provide financial support for asset rehabilitation, if not done before the transfer; and technical support for O&M. As a transition arrangement, the IAs may need to hire professional support staff.

References

Benabderazik, H., & Inocencio, A.B. (2013). *Public-private partnership (PPP) options for irrigation investment in the Philippines*. Final Report to World Bank (June), Manila. Unpublished.

Clemente, R. (2015). *Technical and institutional evaluation of selected national irrigation systems*. Makati: Philippine Institute for Development Studies.

DA (Department of Agriculture). (2007). *Administrative Order No. 26. 2007. Guidelines on the procedures and technical requirements for the issuance of a certification allowing the safe re-use of wastewater for purposes of irrigation and other agricultural uses*. Quezon City: DA.

David, W. P. (2003). *Averting the water crisis in agriculture*. Policy and Program Framework for Irrigation Development in the Philippines. Quezon City: University of the Philippine Press and Asia Pacific Policy Center.

David, W.P. (2008). Chapter 6: Irrigation. In R. T. Dy, L. A. Gonzales, M. F. Bonifacio, W. P. David, J. P. E. De Vera III, F. A. Lantican, G. M. Llanto, L. O. Martinez, E. E. Tan (Eds.), *Modernizing Philippine agriculture and fisheries: The AFMA implementation experience* (pp. 85–128). Manila: University of Asia and the Pacific.

David, W. P. (2009). Impact of AFMA on irrigation and irrigated agriculture. *The Philippine Agricultural Scientist, 91*(3), 315–328.

David, C., & Inocencio, A. B. (2012). *Irrigation policy and performance indicators in the Philippines*. Report submitted to the Philippine Institute for Development Studies (PIDS) as part of the Monitoring and Evaluation of Agricultural Policy Indicators Project. Makati City: PIDS.

David, W. P., Delos Reyes, M. F., Villano, M. G., & Fajardo, A. L. (2012a). Design shortcoming of the headwork and water distribution and control facilities of the canal irrigation systems of Ilocos Norte, Philippines. *The Philippine Agricultural Scientist, 95*(1), 64–78.

David, W. P., Delos Reyes, M. F., Villano, M. G., & Fajardo, A. L. (2012b). Faulty design parameters and criteria on farm water requirements result in poor performance of canal irrigation systems in Ilocos Norte. *The Philippine Agricultural Scientist, 95*(2), 199–208.

Elazegui, D. D. (2015). *Establishment of technical and institutional baseline information and preliminary evaluation of socio-economic impacts of communal irrigation system*. Makati: Philippine Institute for Development Studies.

Ella, V. (2015, November 24). *Irrigation development in the Philippines: Status, challenges and opportunities*. Paper presented at the Asia Rice Foundation's Annual Rice Forum: Water in Agriculture. Los Baños: DOST-PCAARRD.

FAO (Food and Agriculture Organization). (2001). *Overview paper: Irrigation management transfer, sharing lessons from global experience*. International E-mail Conference on Irrigation Management Transfer. www.fao.org.pdf. 21 May 2015.

Inocencio, A. B. (2016). Water in agriculture: Key challenges and opportunities for the Philippines. In S. Banta (Ed.), *Water in agriculture: Status, challenges and opportunities*. Papers presented at The Asia Rice Foundation Annual Rice Forum, 24 Nov 2015. Los Baños: DOST-PCAARRD.

Inocencio, A. B., Ureta, C., Baulita, A. L., Baulita, A. R.,Clemente, R., Luyun, R., Jr., & Elazegui, D. D. (2016). *Technical and institutional evaluation of selected national and communal irrigation systems and characterization of irrigation sector governance structure* (Philippine Institute for Development Studies (PIDS). Discussion paper no. 2016-12). Makati City: PIDS.

Jopillo, S. M. G., & de los Reyes, R. P. (1998). *Partnership in irrigation: Farmers and government in agency-managed systems*. Quezon City: Institute of Philippine Culture, Ateneo de Manila University.

Kikuchi, M., Maruyama, A., & Hayami, Y. (2001). *Investment inducements to public infrastructure: Irrigation in the Philippines and Sri Lanka since independence*. Manila, Sri Lanka: International Rice Research Institute and Colombo, International Water Management Institute.

Lauraya, F. M., & Sala, A. L. R. (1995). *Performance determinants of irrigators associations in national irrigation systems in Bicol, the Philippines: Analysis*. Colombo: IWMI.

Luyun, R.A., Jr. (2015). *Technical assessment of communal irrigation systems in Luzon*. Policy notes. Makati: Philippine Institute for Development Studies.

Luyun, R.A., Jr. (2016). Water resources in the Philippines. In S. Banta (Ed.), *Water in agriculture: Status, challenges and opportunities*. Papers presented at The Asia Rice Foundation Annual Rice Forum, 24 Nov 2015. Los Baños: DOST-PCAARRD.

Nguyen, M. (2015). *Characterizing the governance structure of the irrigation sector in the Philippines*. Makati: Philippine Institute for Development Studies.

NIA (National Irrigation Administration). (2016a). *National income accounts*. Quezon City: Accounting Division, NIA.

NIA (National Irrigation Administration). (2016b). *Irrigation development and potential*. Quezon City: Corporate Planning Office, NIA.

NIA (National Irrigation Administration). (2016c). *Status of irrigation management transfer of NIS*. Quezon City: Institutional Development Division, NIA.

NIA (National Irrigation Administration). (Various years). *National irrigation system performance (NISPER) and communal irrigation system performance (CISPER) data*. Quezon City: System Management Division, NIA.

NIA (National Irrigation Administration). (1985). *Communal irrigation systems manual*. Quezon City: NIA.

Ofrecio, B. P. (2016). Irrigation management transfer and the future of national irrigation systems. In S. Banta (Ed.), *Water in agriculture: Status, challenges and opportunities* (pp. 147–166). Laguna: The Asia Rice Foundation.

Ostrom, E. (1990). *Crafting irrigation institutions: Social capital and development*. Workshop in Theory and policy analysis. Bloomington: Indiana University.

PSA (Philippine Statistics Authority). (2015). *Production, area and yields data for Palay*. Quezon City: PSA.

Rola, A. C. (2015). *Characterizing the governance structure of the irrigation sector in the Philippines: National, regional and irrigation management office level analysis*. Makati: Philippine Institute for Development Studies.

Rola, A. C. (2016). *Improving irrigation water governance for a resilient agriculture*. Paper presented at the 2nd Annual Public Policy Conference, Risks, Shocks, Building Resilience, 22 Sep 2016, Mandaluyong City.

SEPO (Senate Economic Planning Office). (2006). *A profile of selected Philippine—Government-owned and-controlled corporations*. Manila: Senate of the Philippines.

Tejada, S. Q., Sandoval, T. S., & Contreras, S. M. (2015, November 24). *The Department of Agriculture-Bureau of Soils And Water Management Small-Scale Irrigation Project Development Program (DA-BSWM-SSIP Program)*. Paper presented at the Asia Rice Foundation's Annual Rice Forum: Water in Agriculture. Los Baños: SEARCA, University of the Philippines Los Baños.

World Bank. (1992). *Philippines: Irrigated agriculture sector review* (Report No. 9848-PH). Washington, DC: World Bank.

World Bank. (1995). *Bureaucrats in business: The economics and politics of government ownership*. World Bank policy research report. New York: Oxford University Press.

World Bank. (1998). *Water Resources Development Project document*. Washington, DC: World Bank.

Dr. Arlene B. Inocencio is full professor at the School of Economics, De La Salle University (DLSU) in Manila. She was connected with the International Water Management Institute (IWMI) in Pretoria, South Africa and later in Penang, Malaysia before joining De La Salle. Before IWMI, she was Research Fellow at the Philippine Institute for Development Studies (PIDS) in Makati, Philippines. Her research work includes assessments of the rice sector performance and budgetary allocation, the application of landscape planning and monitoring in agriculture, the Philippine irrigation program, the economic and poverty impacts of the government national greening program, the potential of public-private partnership in irrigation, reviewing the national irrigation service fee policy, establishing payments for ecosystem services under the sustainable financing of protected areas project, monitoring and evaluation of agricultural programs, and benchmarking of performance of river basin organizations. She teaches micro and macro-economics, environmental economics and resource valuation, public finance, and agricultural economics.

Ms. Dulce D. Elazegui is a University Researcher of the Center for Strategic Planning and Policy Studies (CSPPS), College of Public Affairs and Development, University of the Philippines Los Baños. She finished the MA Program in Technology Policy and Innovation Management at the State University of Limburg (now University of Maastricht), Netherlands and Master's program in Agricultural Development Economics at the Australian National University. She also attended the Summer Certificate Course on Sustainable Environmental Management at the University of California-Berkeley, U.S.A. She has been involved in collaborative research and written publications related to water, watershed management, climate change adaptation, irrigation, with areas of interest in policy and institutions. She has worked on evaluation of irrigation programs, institutional evaluation of irrigation systems, policy study on small scale irrigation systems; and capacity building of communal irrigator's associations.

Dr. Roger A. Luyun Jr., is an Associate Professor at the College of Engineering and Agro-industrial Technology (CEAT), University of the Philippines Los Baños (UPLB), where he is the current Chair of the Land and Water Resources Division, Institute of Agricultural Engineering. He obtained his bachelor and master's degrees in Agricultural Engineering at the UPLB and his doctoral degree in Agricultural Engineering (Water Resources Engineering) from the United Graduate School of Agricultural Sciences, Kagoshima University, Japan. His fields of specialization include water resources engineering, surface and groundwater hydrology, irrigation and drainage engineering, hydrologic modeling, hydraulics and hydraulic structures, and agrometeorology. He has been or is currently involved in collaborative researches related to irrigation, water resources assessment, water management, climate change adaptation, seawater intrusion and artificial recharge of aquifers. He is a member of the technical committee which drafted the Philippine Agricultural Engineering Standards for Irrigation. He has worked on water balance and loss assessment of the two largest irrigation systems in the Philippines (UPRIIS and MRIIS) as well as technical evaluations of small scale irrigation systems, including capacity building of communal irrigator's associations.

Dr. Agnes C. Rola is full Professor at the University of the Philippines Los Baños (UPLB), former Dean of the College of Public Affairs and Development, UPLB, and member of the National Academy of Science and Technology- Philippines. She has degrees in Statistics (BS) and Agricultural Economics (MS) from the UP; and PhD in Agricultural Economics (Major in Natural Resource Economics) from the University of Wisconsin Madison, USA. She attended the Summer Certificate on Environmental Leadership Program at the University of California-Berkeley and has more than 20 years' research experience in sustainable agriculture at the watershed level with a research focus on water governance. With colleagues, she has written and edited an award winning book, "Winning the water wars: watersheds, water policies and water institutions" (2004), whose recommendations were adopted in the Philippines' Clean Water Act. For the past four years, she led two major research programs on water in the Philippines, namely, water governance for development and water security under climate risks.

Chapter 7
Aquaculture and Water Quality Management in the Philippines

Rafael D. Guerrero III and Pepito R. Fernandez Jr.

Abstract This chapter describes the state of Philippine aquaculture resources, water quality, and other water-related concerns of the sector. In the process, the current formal and informal policy environment is discussed and evaluated. The institutional and regulatory framework described in the chapter is characterized as the product of incremental changes and developments over the years. It is also described as a reflection of the multiplicity of institutions involved, all with different hierarchical coverage, varied mandates, and representing the interest of diverse constituencies. Key problems and concerns in the aquaculture and water sectors are presented using a governance perspective. It was observed that many laws and regulations remain unenforced due to inconsistencies and lack of capacity to monitor resources and enforce laws. Also, the level of participation of diverse and multilevel partners toward sustainable aquaculture development is wanting. The challenge posed by climate change is also an urgent concern. The chapter ends with a set of general policy recommendations for the improvement of the sector.

Keywords Aquaculture • Water quality parameters • Fishpond lease • Philippine Fisheries Code • Fish kill

7.1 Introduction

The Philippines has vast water resources that are important for its development, including aquaculture, the farming of aquatic plants and animals, that provides food, livelihoods and export earnings for the country. The major aquaculture species

R.D. Guerrero III (✉)
School of Environmental Science and Management, University of the Philippines Los Baños, Los Baños, Laguna, Philippines
e-mail: aquabios@yahoo.com

P.R. Fernandez Jr.
Division of Social Sciences, College of Arts and Sciences, University of the Philippines Visayas, Miagao, Iloilo, Philippines

A.C. Rola et al. (eds.), *Water Policy in the Philippines*, Global Issues in Water Policy 8, https://doi.org/10.1007/978-3-319-70969-7_7

of the Philippines are seaweeds, milkfish, tilapia, shrimps and oysters/mussels that are grown in various aquatic environments and culture systems.

This chapter describes the state of Philippine aquaculture and other water-related concerns of the sector. In the process, the current policy environment is discussed. Key problems and concerns are presented while assessing the sector from a policy lens in terms of different governance indicators (e.g., efficiency, effectiveness, equity, sufficiency, social profitability, transparency, accountability, participation) whenever appropriate. The chapter ends with a set of policy recommendations for the improvement of the sector.

7.1.1 Philippine Water Resources for Aquaculture

The Philippines is an archipelagic country with an average annual rainfall of 2.5 m. It is endowed with vast water resources (Fig. 7.1) consisting of marine (bays), inland waters (rivers and lakes) and groundwater with a total area of 2,257,499 km^2 (Table 7.1). These resources are vital for the welfare of its people and the country's

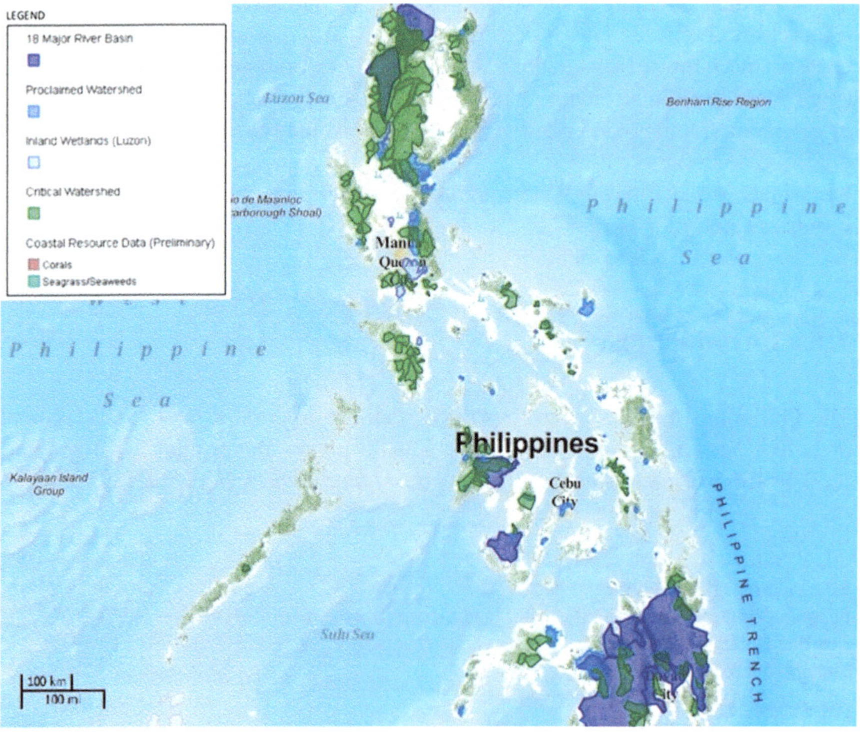

Fig. 7.1 Major river basins, watersheds, wetlands (Luzon) and coastal resources data (Source: DENR 2016)

Table 7.1 Water resources of the Philippines relative to aquaculture

Resource	Area (km^2)
Marine	
1. Coastal	266,000
2. Oceanic	1,934,000
Inland waters	
1. Fishponds	2538
2. Swamplands	2461
3. Lakes	2000
4. Rivers	310
5. Reservoirs	190
Ground water	50,000
Total	2,257,499

Source: DENR (2016)

economic development, which include fisheries, the harvesting from the wild or farming of aquatic products as well as their processing and marketing. There are 316 known fish species in Philippine rivers and lakes, some of which are endemic. Groundwater reservoirs have an estimated storage capacity of 251,100 million m^3 and a dependable supply of 126,000 million m^3 per year.

The country's surface water use is largely devoted for agriculture, with irrigation, livestock, and fisheries representing 85% of total water use. Industry and domestic sectors both comprise the rest of the total. On the other hand, groundwater use is distributed according to these uses: 63% for domestic consumption, 17% for industry, 13% for agriculture, 1% for power generation, and 6% for other sectors. Fishing is pursued for home consumption or small-scale commercial activities. There are 10 major lakes that are used for aquaculture production, while coastal or marine ranching is also present and encouraged by government (Gamolo 2008). Estimates reveal that 110,366.20 ha has been declared as alienable and disposable for aquaculture and fishpond development (Ferrer et al. 2011).

7.2 Philippine Water Policies Related to Aquaculture: Framework, Institutions, and Roles

Adopting the principle of Regalian Doctrine, the 1987 Constitution mandates that all waters and aquatic resources belong to the state. The measure and limit of water use for irrigation, water supply, fisheries, or industry is focused on beneficial use. A water permit is required, under the Water Code, for use beyond domestic purposes, including commercial fisheries and aquaculture. In general, fishing rights and rights to operate aquaculture sites are granted by local governments within their municipal waters (i.e., defined as up to 15 km from the coastline under the Fisheries Code) or agencies created by law to administer select bodies of water (e.g., Palawan Council for Sustainable Development, Laguna Lake Development Authority). Rights are granted to private individuals/firms and registered municipal fisherfolk organizations, subject to certain conditions and limitations. Fishpond licenses/permits, on the other hand, are required under the Fisheries Code. Fishpond lease agreements (FLAs) were originally set at 25 years in 1979 under Fisheries Administrative Order No.129, which is still the current duration (Yap 2010). The 25-year duration for each lease period and the 50-year maximum were affirmed in the Philippine Fisheries Code of 1998 (Republic Act 8550). Fisheries Administrative Order 125–1, s. 1991 and 197, s. 2000 set a rental fee of PhP 1000/ha leased. Such fee rate is low and fails to reflect the opportunity cost of the land and the long-term damage done on the resources. Low rental rates also encourage the conversion of mangrove forest into fishponds, more land acquisition, and extensive farming. On the other hand, the responsibility to efficiently manage pond areas is diminished (see Ferrer et al. 2011). The Fisheries Code also provides guidelines in the use and administration of lakes and rivers. Section 51 states that "not over ten percent (10%) of the suitable water surface area of all lakes and rivers shall be allotted for aquaculture purposes like fish pens, fish cages and fish traps; and the stocking density and feeding requirement which shall be controlled and determined by its carrying capacity."

The state has policy goals of aquaculture productivity and sustainability through environmentally sound management practices, especially those that relate to water quality and designated zones in which aquaculture activities are restricted. Policy guidelines related to water use for fisheries are set forth in the Philippines commitment to the 2015 Sustainable Development Goals, particularly No. 2 (i.e., end hunger, achieve food security and improved nutrition, and promote sustainable agriculture) and No. 6 (i.e., ensure availability and sustainable management of water and sanitation for all) and the "Comprehensive National Fisheries Industry Development Plan Medium Term 2016–2020" (CNFIDP 2016–2020). The CNFIDP 2016–2020 has a focused vision for increasing aquaculture production through a "sustainable and competitive fisheries industry." The following issues are to be addressed: sufficient contribution to national food security; inclusive growth within the industry; sustainable, science-based fisheries and aquatic resource management practices; compliance with international laws, policies, and standards; and enforcement of local laws and regulations; strengthened capacities in infrastructure, technology, human resource, and information sharing; and resilience to environmental hazards. The targets were developed through the combination of science-based information as presented by resource persons from academic and research institutions and actual observed situational information from industry frontliners.

The CNFIDP 2016–2020 has a target aquaculture production of the following: 4% annual increase in milkfish production; 6% annual increase in tilapia production; 10% annual increase in shrimp production; 5.4% increase in mud crab production over 5 year; 10% increase in shellfish production over 5 year; and 25% increase in seaweed production over 5 year. The Bureau of Fisheries and Aquatic Resources' (BFAR) aquaculture strategies to achieve goals include securing quality fry/seed supply through coordinated investments in propagation facilities (i.e., hatcheries, nurseries, laboratories); institutionalizing good aquaculture practices (GAqP) for key commodities and promoting sustainable aquaculture; assuring quality and traceability of aquaculture inputs and outputs; investing in species with high commercial potential; optimizing operation of mariculture parks; and ensuring climate/disaster resilience of the aquaculture sector.

To achieve these goals, BFAR got an increase in budget from PhP 6.3 billion in 2015 to PhP 6.7 billion (USD 132.9 million to USD 141.3 million) in 2016, which was applied to fund tailor-fit livelihood programs and projects for the fisheries sector. On top of the priority projects, also targeted was the completion of more than 500 community fish landing centers (CFLCs) in strategic coastal communities to reduce fishery postharvest losses from 25% to 18%. All these aim to lower and improve the socioeconomic conditions of fisher folk communities with high poverty incidence.

In the Code of Conduct for Responsible Fisheries of the Food and Agriculture Organization (FAO) of the United Nations (FAO 1995), Article 6 on General Principles (6.8) stipulates that "All critical habitats in marine and freshwater ecosystems such as wetlands, mangroves, reefs, lagoons, nursery and spawning areas should be protected and rehabilitated as far as possible and where necessary particular effort should be made to protect such habitats from destruction, degradation, pollution and other significant impacts resulting from human activities that threaten the health and viability of the fishery resources." Further, in Article 9 on Aquaculture Development (9.1.5), it is stated that "States should establish effective procedures specific to aquaculture to undertake appropriate environmental assessment and monitoring with the aim of minimizing adverse consequences resulting from water extraction, land use, discharge of effluents, use of drugs and chemicals, and other aquaculture activities." The Philippine government abides by such international stipulations to maintain the quality of its fishery exports and to promote sustainable development. The Philippine Fisheries Code of 1998, Section 47 on the Code of Practice for Aquaculture states that "The Department [of Agriculture] shall establish a code of practice for aquaculture that will outline general principles and guidelines for environmentally sound design and operation to promote the sustainable development of the industry." As such, it is the Department of Agriculture that is tasked to develop the rules on how to attain sustainable development principles and goals.

The 1998 Philippine Fisheries Code provides for the sustainable development of fishery and aquatic resources, and outlines the structure for the granting of fishing privileges. A year earlier, the 1997 Agriculture and Fisheries Modernization Act (AFMA) provided for measures to modernize the agriculture and fisheries sectors. Among the key concerns in effecting sustainable aquaculture for wetlands are obstructions to migration paths of migratory species at river mouths and estuaries, the responsibility for which is assigned to Fisheries and Aquatic Resource Management Councils (FARMCs). Section 105 of the Philippine Fisheries Code states that FARMCs determine the distance of fish pens and other contraptions from the river mouths and estuaries. Said provision also imposes a penalty of 7–12-year imprisonment or a fine from PhP 50,000.00 to PhP 100,000.00 at the court's discretion as well as cancellation of permit/license. Two other important legislations underpin the legal framework on water issues related to aquaculture. The Local Government Code (Republic Act 7160) of 1991 devolved the responsibility of basic services from the national government to local government units, including the "enforcement of environment and natural resources laws within their territories." Section 20 of the law provides each local government, through its Environment and Natural Resources Office (ENRO), the powers and functions of monitoring water quality and taking active participation in all efforts concerning water quality protection and rehabilitation.

The Clean Water Act of 2004 (Republic Act 9274) is the national policy for the protection, preservation, and revival of the quality of fresh, brackish, and marine waters. It provides for a comprehensive management program for water pollution focusing on pollution prevention. The Act aims to protect the country's water bodies from land-based pollution sources and establishes a framework for managing water quality. The Department of Environment and Natural Resources (DENR) in its Administrative Order 34 of 1990 (DAO 1990–1934) set three water quality criteria for freshwater and marine aquaculture: Class C "for the propagation and growth of fish and other aquatic resources in freshwaters (rivers, lakes, reservoirs, etc.)"; Class SA for "waters suitable for propagation, survival and harvesting of shellfish for commercial purpose" for coastal and marine waters; and Class SB for "fishery Class 1 spawning areas for *Chanos chanos* or *bangus* and other species." Water bodies designated as water quality management areas (WQMA) are regularly monitored and surveyed by the regional DENR office within an inclusive framework (i.e., governing board) that involves government agencies, local governments, and business and water utility sectors tasked to formulate policies and measures in line with the standards. The DENR Environment and Management Bureau (EMB) also conducts surveys and follow-up inspections on the requirement for industries and commercial establishments to put up and operate pollution control facilities. The DENR-EMB penalizes non-compliance with effluent standards through fines or closure.

From the aforementioned discussions, a number of Philippine institutions and agencies are involved in overseeing water governance at all levels, ranging from local, regional, to national. The DENR and the NWRB are the lead agencies charged with the protection of water resources. The NWRB acts as the principal coordinating and regulatory body. DENR's EMB Water Quality Management Section implements the Clean Water Act. Other key agencies include (1) the Department of Agriculture's Bureau of Soils and Water Management, and Bureau of Fisheries and Aquatic Resources; and (2) the Department of Science and Technology's Philippine Council for Aquatic and Marine Research and Development, which focuses on generation of new knowledge and technology; and (3) LGUs given the devolution of aquaculture services to this level. The current institutional and regulatory framework is the product of incremental developments over the years, each in response to particular challenges of the time. It is also a reflection of the multiplicity of institutions involved, all with different hierarchical coverage, varied mandates, and representing the interests of diverse constituencies. Such a framework has been described as multilevel, multisectoral, and multithematic (Malayang 2004). The result is an institutionally fragmented approach, with overlapping and fractional plans and programs (GOP 2006; Barba 2004). Because enforcement measures such as those related to obstructions and effluent practices are given to local bodies such as FARMC and the LGU, enforcement varies widely, depending on the awareness and political will of these bodies. Moreover, there are many laws that are unenforced due to inconsistencies and the lack of capacity and resources to do monitoring and enforcement activities (Luna 2005; Ferrer et al. 2011).

In spite of the presence of legal frameworks and institutions, there is evidence that points to the operation of informal systems and networks for land and water access/use. Typically, households or communities closer to water bodies with secure rights to use the water and land resources have better possibility to set up cages or ponds and generate income from aquaculture than households who do not have such rights. Several communities in the coastal areas have traditional rights to use coastal waters, which could be formalized to favor the entry of the poor into aquaculture (Cruz 1999; Sekhar and Ortiz 2006). In some cases, 86% of the respondents in an aquaculture community claim having formal and informal rights to access and use water bodies for setting up aquaculture farms. Traditional rights to use water were by virtue of their profession and/or residence, which was closer to the water bodies, or by formal permits from the local government. In case of ponds, 27% of the respondents were operating on land that was located on privately leased land, while 68% were on local government (with an agreement) land, and the rest on unauthorized areas. Researchers observe that tenants in such farms were interested in making short-term profits, which could lead to serious environmental damage in the areas. It was also difficult to assess what respondents meant by claiming that they had rights to use water because of several laws and departments that govern and have jurisdiction over water bodies in Philippines. It was observed that there was lack of clarity and respondents were not willing to discuss in detail the issue of rights for fear of being reported (Sekhar and Ortiz 2006).

7.3 The Philippine Aquaculture Sector: Performance Assessment

In 2014, the Philippine fisheries sector contributed 1.6% to the GDP of the country with a production of 4.69 million metric tons valued at PhP 242 billion. Of the production, aquaculture, the farming of aquatic organisms, contributed 49.9%, followed by municipal fisheries with 26.7% and commercial fisheries, 23.7%. The major species farmed are seaweeds (66.29%), milkfish (16.69%), *tilapia* (11.09%), shrimps (2.17%), oysters and mussels (3.76%), and others (DA-BFAR 2015).

Seaweeds (mainly the red algae, *Kappaphycus* and *Eucheuma*) are farmed in coastal marine waters using nylon monolines that are staked into the bottom or suspended from floating rafts. Milkfish (*Chanos chanos*) is cultured in brackishwater ponds, freshwater pens and marine pens/cages. The Nile tilapia (*Oreochromis niloticus*) is raised in freshwater ponds and cages in lakes and reservoirs. Shrimps

(*Penaeus monodon* and *P. vannamei*) are produced in brackishwater ponds. Oysters (e.g., *Crassostrea iredelai*) and mussels (e.g., *Perna viridis*) are grown in estuaries and coastal marine waters with bamboo poles staked into the bottom or floating rafts with suspended ropelines.

Red seaweed (*Kappahycus* sp.)

Bottom monoline seaweed culture

Floating raft seaweed culture

Milkfish (*Chanos chanos*)

Brackishwater ponds

Milkfish pen

Floating milkfish cages

Nile tilapia (*Oreochromis niloticus*)

Freshwater pond

Floating cages in a lake

Brackishwater shrimp (*Penaeus monodon*)

Brackishwater ponds for shrimp culture

A B C D

(a) Oyster (*Crassostrea iredelai*); **(b)** Fixed raft culture of oysters; **(c)** and **(d)** Green mussel (*Perna viridis*)

Studies indicate that 62 freshwater fishes have been introduced in the country from 1905 to 2013. Forty-five percent of introduced species were for aquaculture purposes. It is alarming to note, however, that 10 species (16%) are categorized as invasive, whereas four 4 species (6%) are potentially invasive (Guerrero 2014a).

Aquatic organisms require water for their growth and development and also as their medium or environment. Water quality, "the chemical, physical, biological, and radiological characteristics of water" (EMB 2008), is therefore critical in the farming of aquatic plants and animals. The important water quality parameters that affect growth, reproduction, and survival of aquatic organisms are dissolved oxygen, carbon dioxide, temperature, pH, alkalinity, hardness, nitrogen (ammonia, nitrite, nitrate), phosphate, salinity, biological oxygen demand, and total suspended solids. Other organisms (e.g., bacteria), elements (e.g., heavy metals) and compounds (e.g., insecticides) in the water can also be harmful to aquatic life.

Dissolved oxygen (DO) is the oxygen gas (O_2) "mixed between the water molecules" (H_2O). DO is derived by diffusion from the atmosphere and from photosynthesis by aquatic plants. Oxygen is required for the respiration of organisms and other metabolic processes. While cultured aquatic organisms can have varying oxygen requirements and tolerances, the desirable range of DO for most is 5–6 parts per million (ppm) or milligram per liter (mg/L). Carbon dioxide (CO_2) in the water comes from atmospheric diffusion and is produced as a byproduct of respiration by fishes and other organisms. An increase in the concentration of the gas reduces the rate of its release from the blood of fishes through the gills and causes a drop in blood pH and a reduction in the oxygen-carrying capacity of hemoglobin. Concentrations of 10 ppm and below are well-tolerated. The "acidity or alkalinity" of a solution (including water) is determined through its pH (the quantity of hydrogen ions). Measured in terms of values from 0 to 14, a neutral solution is said to have a pH of 7; that which is acidic is below 7 and that which is alkaline is above 7. Many metabolic processes in organisms and chemical reactions in the aquatic environment are influenced by pH. For most aquatic species, a pH below 4 and above 11 can slow down growth and cause death. The favored range of pH is 6.5–9.0. Pollution and ocean acidification are an ongoing concern.

Water temperature, "the intensity of heat" of water that comes from sunlight or an artificial source, is also critical in aquaculture. It affects the rate of metabolism of organisms, particularly "cold-blooded" (poikilothermic) species such as fishes and influences their reproductive activities and life cycles as well as the solubility of oxygen and other gases and solids. The favorable water temperature for growth and reproduction of fishes is 27–29 °C. Alkalinity is the "water's capacity to resist changes in pH" or a "buffering system" with carbonates (CO_3) and bicarbonates (HCO_3). Hardness, on the other hand, is the content of calcium and magnesium compounds in the water, which are essential for biological processes. For fish farming, the desired level of alkalinity is 50–150 mg/L $CaCO_3$ and 75–200 mg/L $CaCO_3$ for hardness (Wurts 2002). Increase in water temperature due to climate change is an ongoing concern.

Nitrogen in the water is in the form of dissolved gas that does not react with water and in the form of compounds such as ammonia, ammonium, nitrite, and nitrate. Ammonia (NH_3) is a waste product from wastes and feeds of cultured fishes and from bacterial decomposition of organic matter. It can damage the gills and can be tolerated by fish at 0.5–1 mg/L with toxic levels of 0.53–22 mg/L. Nitrite (NO_2) is oxidized ammonia, which binds with hemoglobin (oxygen carrier in the blood) and is tolerable to fish up to 1.5 mg/L but toxic at 10–20 mg/L. Nitrate (NO_3) is another form of oxidized ammonia, which is utilized by plants like algae in the water as a nutrient. It is not as toxic as ammonia and nitrite to fish and is harmful only at a level greater than 40 mg/L.

Phosphorus in the form of phosphate (PO_4) comes from natural sources and human activities (fertilizers, wastewater, detergents, etc.). It does not directly harm fishes. However, as a plant nutrient, it can induce algal blooms in natural waters and cause oxygen depletion with "plankton die-offs." A phosphate concentration of 0.4 ppm is recommended for fishpond fertilization.

Salinity is the concentration of salts, mainly sodium chloride (NaCl) in water. Basically, freshwater has a maximum salinity of 0.05 parts per thousand (ppt) of NaCl, while seawater has an average salinity of 35 ppt. The efficient osmoregulation (balance of salts in the body fluids and water medium) of fishes depends on their tolerances to salinity. The range of favorable salinities can be 1–5 ppt for freshwater fishes and 20–30 ppt for marine fishes.

Total suspended solids (TSS) are inorganic (clay and silt) and organic (decaying plants and animals) solids suspended in water that cause turbidity or a decrease in water clarity. Such solids can clog the gills of fish and adversely affect their growth rates and resistance to disease. A TSS concentration of 80 ppm or less is acceptable.

Bacteria in water consume oxygen for respiration. The dissolved oxygen consumption of bacteria for organic matter decomposition activities is referred to as biological oxygen demand (BOD). The acceptable levels of BOD that are tolerable for fishes are 3–6 ppm.

The contamination of waters used for aquaculture with fecal human and animal wastes that contain disease-causing bacteria is determined through fecal coliform or total coliform tests "by counting the most probable number of bacteria colonies that grow from a 100-milliliter water sample (MPN/100 ml)." The acceptable levels of fecal and total coliform for freshwater and marine waters are 5000 (MPN) for freshwater and 70 (MPN) for marine (DAO 1990–34). Any person or group of persons found violating or failing to comply with set regulations and standards are liable under Section 9 of the Pollution Control Law (PD No. 984) and/or Section 106 of the 1978 National Pollution Control Commission (NPCC) Rules and Regulations.

7.4 Philippine Water Quality Standard Related to Aquaculture

In the Philippines, the water quality criteria for freshwater and marine aquaculture is set by the DENR in its Administrative Order (AO) 34 of 1990 (DAO 1990–34). In Section 68 of the AO for Water Usage and Classification, freshwaters (rivers, lakes, reservoirs, etc.) are classified under Class C "for the propagation and growth of fish and other aquatic resources; coastal and marine waters are classified Class SA or "waters suitable for propagation, survival and harvesting of shellfish for commercial purpose," and Class SB for "fishery Class 1 spawning areas for *Chanos chanos* or *bangus* and other species." The physical, biological, and chemical parameters and criteria set for such waters are presented in Table 7.2.

From DENR Administrative Order No. 35 issued in 1990, the revised effluent standards set for Class D waters (i.e., inland waters such as lakes, reservoirs, rivers, streams, and creeks) and Class SC waters (coastal) are presented in Table 7.3.

In accordance with the FAO Code of Conduct for Responsible Fisheries and the Philippine Fisheries Code, the Philippine Department of Agriculture issued Fisheries

Table 7.2 Water quality criteria for Class C, Class SA, and Class SB waters

Parameter	Class C	Class SA	Class SB
Temperature (°C)	3° rise	3° rise	3° rise
pH	6.5–8.5	6.5–8.5	6.0–8.5
Dissolved oxygen (mg/L)	5.0	5.0	5.0
Nitrate as nitrogen (mg/L)	10	–	–
Phosphate (mg/L)	0.4	–	–
Total coliform (MPN/100 ml)	5000	70	1000
Chloride as Cl (mg/L)	350	–	–
Copper (ppb)	0.05	–	0.02

Table 7.3 Effluent standards for Class D and Class SC waters

Parameter	Class D	Class SC
Temperature (°C) (maximum rise)	3°	3°
pH (range)	5–9	6–9
COD (mg/L)	250	250
Total suspended solids (mg/L)	150	120
Total dissolved solids (mg/L)	2000	–
Total coliform (MPN/100 ml)	15,000	–

Administrative Order (FAO) 214 on the Code of Practice for Aquaculture on 2001, which recommends the following practices for water quality management:

Sec. 2 Site selection/evaluation

a. Water source in the area shall be evaluated as to its quality and quantity.

Sec. 3 Farm design and construction

d. An ideal farm shall have wastewater treatment and settling wastewater before discharging into the environment.
h. Fish cages, floating or stationary, shall be installed and kept at least 1 m between units and at 20 m apart between clusters to provide water exchange.
i. Fish pens shall be spaced 200 m apart

Sec. 4 Water usage

a. The construction and operation of deep wells for freshwater supply shall be based on a design that prevents salt intrusion into freshwater aquifers and subsidence of ground level.
b. Closed recirculating water system shall be considered in the intensive and semi- intensive farming systems.
c. Water exchange shall be minimized by maintaining good water quality through moderate stocking densities and feeding rates, using high-quality feeds and good feeding practices.

Sec. 5 Water discharge and sludge/effluent management

a. Effluents, sediments, and other wastes shall be properly disposed of through the use of waste treatment and settling ponds.
d. Discharged water shall meet water quality standards (determined qualitatively and quantitatively). Qualitative standards shall include prohibition of the release of turbid and odorous water to the receiving water, whereas quantitative standards shall include the maximum and/or minimum levels of suspended solids, measure of acidity (pH), dissolved oxygen, ammonia and other nitrogenous compounds, phosphorus, carbon dioxide and BOD.

Sec. 5 Use of drugs, chemicals, potentially toxic pesticides, and fertilizers

a. Drugs, chemicals, pesticides, and fertilizers, including lime, shall be used only when clearly justified to treat specific problems.
e. If chemicals are used, pond water shall not be discharged until they have degraded/dissipated or until the compounds have naturally decomposed to non-toxic forms.
b. Banned chemicals shall not be used for any purpose.

Worsening water quality is a serious concern for aquaculture, with experts concluding that 50 river systems are biologically dead or dying due to pollution from human trash, commercial agricultural chemicals, animal waste, and industrial waste. In Metro Manila alone, nine river sub-basins are used as dump sites. Almost 90% of wastewater being discharged in water bodies (i.e., rivers, lakes, and the sea) has little or no primary treatment. Therefore, only 10% of wastewater is treated (ADB 2012). According to the Philippine Environment Monitor of 2014 (DENR 2016), discharge of domestic and agricultural wastewater and agricultural run-off has caused extensive pollution of receiving water bodies such as rivers and lakes. Domestic wastewater contains raw sewage and detergents; agricultural wastewater has fertilizers and chemicals; and industrial wastewater can have heavy metals and oils.

An estimated 2.2 million tons of organic pollution is produced annually by the domestic (48%), agricultural (37%), and industrial (15%) sectors. Groundwater has been contaminated with coliform in 58% of the areas sampled. The adverse impact of water pollution costs the Philippine economy an estimated PhP 67 billion (more than US$1.3 billion) annually. The loss in fisheries production due to pollution has been estimated to be PhP 17 billion (WEPA 2005).

7.5 Aquaculture Systems and Water Quality Issues

Aquaculture systems are the structures, species, and methods used for the culture of aquatic organisms. An aquaculture system can be land-based or water-based. Ponds, tanks, and other structures built on land are land-based, while pens, cages, and other structures in rivers, lakes, and coastal waters are water-based. The culture methods can be extensive (i.e., with minimal or no inputs), semi-intensive (i.e., with inputs like fertilizers and supplemental feeds), and intensive (i.e., high stocking rates with feeding of commercial feeds and water management facilities). Species such as fishes, crustaceans, mollusks, and seaweeds can be grown in fresh, brackish, and marine waters.

There are issues and challenges linked to intensive aquaculture using pond, tanks, and cage systems. Fish kills occur where dissolved oxygen gets depleted from plankton die offs, themselves resulting from overstocking and overfeeding. In cage pens and floating cages, similar fish kills also happen during summer months and seasonal lake overturns, when bottom wastes or heavy organic load below rises to the surface. Indiscriminate disposal of effluent from shrimp farms into tidal and coastal waters produce disease outbreaks among shrimps. Algal bloom from pollutants affect fish flavor and their market value. Red tide outbreaks in coastal areas due also to pollution increase the danger of mussel poisoning. These problems are direct consequences of poor water quality and are exacerbated by extreme weather conditions.

7.5.1 Pond Systems

Earthen ponds that are dug-out manually or by heavy equipment in suitable sites with clayey soil and adequate supply of water from a surface (irrigation or river) or underground source are commonly used for fish and crustacean (shrimps and crabs) in the Philippines. The ponds are enclosed by dikes that are 1–2 m high, 1–2 m wide at the top (depending on the size of the pond), and 2–3 m wide at the base. Wooden and/or concrete gates for the inflow and outflow of water are provided at the opposite shorter sides of the ponds that are usually rectangular in shape. Water depths of 0.4–1.5 m (depending on the methods of culture) are maintained.

Freshwater ponds with areas of 0.1–2 ha are filled with water from an irrigation supply or pumped underground source to a depth of 0.5–1 m. Nile tilapia (*Oreochromis niloticus*) fingerlings with weights of 0.25–0.5 g are stocked at 1–5/m^2. With extensive culture, monosex male (sex-reversed) fingerlings are stocked at 1/m^2 and only fertilization (organic and/or inorganic) is applied. Fingerlings are stocked at 2–3/m^2 for semi-intensive culture with fertilization and supplemental feeding with agricultural byproducts like rice bran. For intensive culture, the fingerlings are stocked at 4–5/m^2 and fed with commercial feeds (24–28% crude protein) all throughout the growing period (4–5 mo) with pond water exchange when necessary to maintain good water quality. Yields of 1.8–8 metric t/ha per crop for the fish weighing 150–250 g each are obtained at harvest (Guerrero 2002).

The Nile tilapia is a hardy fish that can tolerate adverse environmental conditions such as low dissolved oxygen (DO) and high water temperature. It is, however, stressed when DO drops to below 2 ppm and dies when DO is less than 1 ppm. Massive mortalities ("fish kills") of the fish have occurred in the intensive culture ponds of Central Luzon (the region with the greatest area of tilapia ponds) due to DO depletion with overstocking and overfeeding of the fish that bring about nutrient enrichment (eutrophication) of the pond water and trigger "plankton die-offs." A "plankton die-off" takes place when phytoplankton blooms crash and oxygen in the water is used up by bacteria for decomposing the algal biomass.

For brackish water ponds (1–20 ha) that were developed from former mangrove areas and supplied with water from tidal rivers, a water depth of 0.3–0.4 m is maintained for growing the natural food of milkfish (*Chanos chanos*) known as "*lablab*" (benthic algae) with organic and inorganic fertilization for extensive culture at salinities of 15–25 ppt. Fingerlings of the fish (2–5 g each) are stocked at 1000–2000/ha, cultured for 3–4 mo, and harvested at sizes of 200–250 g each with yields of 250–800 kg/ha per crop. For semi-intensive culture, yields of 1.5–1.8 t/ha per crop are obtained with a water depth of 0.5–1 m, fingerling stocking of 2500–5000/ha, inorganic fertilization, and supplemental feeding. Intensive culture of the fish done with stocking rates of 10,000–20,000/ha, water depth of 1 m, use of commercial feeds, and water exchange produces yields of 2–4 t/ha per crop.

"Fish kills" have occurred in brackish water ponds with acid sulfate soils. The pond water becomes acidic with pH less than 4 when pyrite (ferrous sulfide) in the earthen dikes is oxidized upon exposure to air and converted into ferric hydroxide and sulfuric acid in water (Poernomo and Singh 1984).

According to Barg (1992), the effluent of intensive shrimp ponds during a 5-mo culture period can have 0.05–0.4 mg total phosphorus/L, 0.15–0.4 mg total nitrogen/L, 0.05–0.65 mg total ammonia/L, 20–250 mcg chlorophyll a /L, and 30–90 mg TSS/L.

The indiscriminate disposal of wastewater (effluent) from intensive shrimp culture ponds into tidal rivers and coastal waters has brought about "self-pollution" of shrimp farms and outbreaks of bacterial and viral diseases of shrimp. From the country's shrimp production of 90,000 metric t in the mid-80s, production dropped to 50,000 metric t in 2015 due to diseases. With the ban on the use of antibiotics, methods to improve water quality in the ponds for intensive shrimp culture such as closed systems for water recirculation, "green water" technology, use of settling ponds, and probiotics have been applied.

7.5.2 Cage/Pen Culture in Lakes, Rivers and Coastal Waters

The pen culture of milkfish is done in shallow lakes like Laguna de Bay, the largest freshwater lake in the country close to Metro Manila, and in estuarine rivers and coastal waters with depths of 2–7 m.

The milkfish pens of Laguna de Bay, which vary in size from 1 to 50 ha, are built of bamboo poles that are staked into the lake bottom and enclosed with polyethylene net to contain the cultured fish that are stocked at 2–3/m^2. The fish grows to market sizes of 250–300 g each in 6–8 mo of culture, thriving only on natural food in the lake without artificial feeding.

Pens in the tidal rivers and coastal waters of northern Luzon with sizes of 500 m^2 to 1 ha are built in the same way as those in Laguna de Bay with bamboo poles staked into the river/sea bottom. The cultured fish that are stocked at 2–8/m^3 and fed with commercial feeds grow to a size of more than 300 g each in 4–5 m (Dela Vega and Querijero 2006).

Laguna de Bay is a eutrophic lake that has 22 tributaries, which unload into it nutrients (nitrogen and phosphorus) from domestic and agriculture sources in its watershed that is populated by more than 20 million people. The primary productivity of the lake was determined to be 8 t carbon/ha per year. A water quality issue related to aquaculture in the lake is the occurrence of algal blooms of blue-green algae (*Microsystis* and *Anabaena*) that produce compounds like geosmin, which are absorbed by the cultured fish and give them an "off-flavor," thereby lowering their market value.

With the heavy organic loads in the bottoms of congested pens coming from fish wastes and feeds in the tidal rivers of Dagupan City, Pangasinan (northern Luzon), massive "fish kills" have taken place due to oxygen depletion, particularly in the summer months when temperature is high and water depth is low at neap tide (Inigo 2015).

7.5.3 Floating Cages in Lakes/Reservoirs and Coastal Waters

In deep freshwater lakes and reservoirs such as Taal Lake in Batangas and Magat Reservoir in Isabela, floating cages (10 × 10 × 5–10 m) made of bamboo poles or PVC pipes for frames floated by plastic drums and suspended polyethylene enclosures are stocked with Nile tilapia (*Oreochromis niloticus*) fingerlings (2–5 g each) at 500–2000/m^2. The fish is grown for 4–5 m with commercial feeds for harvestable sizes of 200–300 g each with yields of up to 10 metric t per cage (Guerrero 2014b).

With the proliferation of cages in Taal Lake and other deep lakes in the country (e.g., Lake Sampaloc in Laguna and Lake Sebu in South Cotabato), massive "fish kills" have occurred with seasonal lake overturns wherein anoxic water in the bottom, laden with fish wastes and excess feeds that are decomposed by anaerobic bacteria, rises to the surface.

In the deep coastal waters of Lingayen Gulf in western Pangasinan, circular floating cages (12–19 m in diameter and 8–12 m deep) made of galvanized iron pipes for frames floated by plastic drums and suspended polyethylene net enclosures are stocked with 5–20-g fingerlings at 50–100/m^3. The yields of such cages are 18–24 metric t per cage after 5–6 mo of culture with artificial feeding and harvestable sizes of 200–300 g each (Dela Vega and Querijero 2006).

Azanza et al. (2005) found blooms of *Prorocentrum* minimum in Lingayen Gulf where intensive aquaculture of milkfish in cages was done and related it to the "fish kills" that took place due to oxygen depletion brought about by "plankton die-off." Jacinto (2006) reported higher values of ammonia and low levels of DO in the cage and pen areas in Bolinao, Pangasinan, during neap (low) tide periods. Obliosca et al. (2003) associated feeds and fish feces that contributed to heavy organic loading in the water with proliferation of planktons.

7.5.4 Open Water Culture of Mussels in Coastal Waters

The green mussel (*Perna viridis*) is cultured using bamboo poles staked 0.5–1 m apart into the sea bottom of shallow coastal waters such as those of Manila Bay. At a depth of 1–2 m below low water in the bay, 2000–3000 spats (mussel seeds) attach per meter of the poles. The mussel thrives by filter-feeding on planktonic organisms. More than 5 t of market-size mussels (10–20 g each) are harvested per 400-m^2 of area in the bay after 4–5 mo of culture (PCARR 1977).

The high natural productivity of Manila Bay is attributed to the heavy organic load, mainly from domestic wastewater that it receives from tributaries, particularly the Pasig River that is the main drainage of Laguna de Bay. The bloom in Manila Bay of a harmful dinoflagellate, *Pyrodinium bahamense* var. *compressa*, also known as "red tide," caused the deaths of 30 people in 1988–1999 due to paralytic shellfish poisoning with the consumption of mussels having toxic levels of saxitoxin produced by the alga. In the period 1983–2002, 540 outbreaks of the "red tide" have been recorded in various bays throughout the country. The harvesting and selling of cultured mussels are prohibited in bays during a "red tide" occurrence to protect consumers.

7.6 Recommendations

The tasks of maintaining water quality for sustainable aquaculture is distributed between several national agencies where coordination, particularly at the local level, is not axiomatic. Productivity as a goal puts pressure on water quality as intensified aquaculture activity has adverse consequences. Water quality monitoring and surveillance are given to the DENR-EMB, but its regional offices do not have adequate resources and capacities. FARMCs are similarly unable to enforce rules regarding obstructions to fish migration. There are also equity issues concerning access to water resources for aquaculture. Unclear water rights for aquaculture result in extensive farming, even unto areas where no fish pond licenses and leases have been awarded. Current fish pond permits are underpriced and do not reflect the environmental costs of intensive aquaculture, particularly to the mangrove ecosystem.

Strengthening institutional mechanisms and sustaining multisectoral participation in water quality management are essential to promote water quality management. This can be achieved by developing a systematized technical and operational approach to water environment data collection, water quality or watershed monitoring, and effective flow of information. Although responsibility for monitoring can be delegated to provincial or local government agencies, support must be shared for maximum results. Outlining the roles of municipalities, educational institutions, businesses, NGOs, interest groups, and community volunteers may be useful. The best-fit monitoring framework will benefit by cooperative partnerships that bring together multilevel and diverse resources, knowledge, and commitment. Such a scheme was pursued in Iloilo Province (PENRO, UPV, CUI 2016a, b).

On water rights, there is a need to review the fees and the basis for awarding fishpond leases and permits, considering the traditional rights of fishers and fishing communities and within shorter time periods that allow for a more rational way of accessing water resources. The cost of the permit itself should be increased to include environmental costs arising from intensive aquaculture activities. Where point sources for pollution (e.g., industries and commercial establishments) and obstructions to migration paths are identifiable, a more robust system of imposing penalties should be effected.

Good aquaculture practices such as the use of settling ponds, recirculating water systems, and probiotics for preventing shrimp diseases and "self-pollution" of brackish water ponds should be applied. Other sustainable aquaculture technologies that are cost-efficient should also be developed and popularized among fishpond operators. The concept of integrated aquaculture is nothing new and has been applied in pond systems for decades in the country. The integrated multitrophic aquaculture (IMTA) is an ecosystem approach in integrated marine aquaculture that can be explored. IMTA has been proven to solve sea pollution problems associated with fish culture mainly in temperate waters (Troell 2009).

The number of fish pens/cages and feeds/feeding practices in lakes, rivers and coastal waters should be regulated within the "carrying capacity" of such water bodies to prevent "fish kills." This would entail intergovernmental coordination and multistakeholder efforts to come up with evidence-based indicators for lake/river/coastal health, as well as a monitoring system and enforcement mechanism that is efficacious. Of late, the Laguna Lake Development Authority with the DENR and the Philippine Coast Guard have joined ranks to dismantle illegal fish pens in Laguna Lake (Cordero 2017).

Finally there is an urgent need to address climate change concerns and their impacts on aquaculture, water resources, and the diverse communities that rely on them. This endeavor is a huge undertaking that will involve local, national, and international cooperation and collaboration.

References

ADB (Asian Development Bank). (2012). *Wastewater management and sanitation in Asia*. Manila: ADB.

Azanza, R. V., Fukuyob, Y., Yapa, L. G., & Tagayamac, H. (2005). *Prorocentrum* minimum bloom and its possible link to a massive fish kill in Bolinao, Pangasinan, Northern Philippines. *Harmful Algae, 3*, 519–524.

Barba, P.F. (2004). *The challenges in water resources management in the Philippines*. Paper presented at the second international conference on Hydrology and Water Resources in Asia Pacific Region, 5–8 July, Singapore. http://www.wrrc.dpri.kyotou.ac.jp/~aphw/APHW2004/proceedings/JSE/56-JSE-A519/56-JSE-A519.pdf

Barg, U.C. (1992). *Guidelines for the promotion of environmental management of coastal aquaculture development* (FAO Fisheries Technical Paper 328). Rome: Food and Agriculture Organization of the United Nations. 122 p.

Cordero, T. (2017). DENR starts dismantling Laguna Lake fish pens. *GMA News*. http://www.gmanetwork.com/news/story/597251/money/economy/denr-starts-dismantling-laguna-lake-fish-pens. Accessed 26 Jan 2017.

Cruz, H.P. (1999). Existing property regimes in the Philippines and opportunities for community title to coastal resources. In: *Power spaces and titles: Issues in community-based coastal resources management, Lundayan journal* (special edition)(pp. 34–41). Quezon City: Tambuyog Development Center.

DA-BFAR (Department of Agriculture-Bureau of Fisheries and Aquatic Resources). (2015). *Philippine fisheries profile 2014* (69 p). Quezon City: DA-BFAR.

Dela Vega, A.M. &Querijero, B.V.L. (2006). Industry perspective: Practices and needs in feeds and feeding management including financing schemes in cage and pen operations for milkfish (Bolinao Pangasinan Experience). In B.L. Querijero, C.R. Pagdilao, & S.V. Ilagan (eds.). In *Proceedings of the 1st Consultation Workshop on GUIDELINES for Establishment of Fish Cages in Lakes and Coastal Waters*, PCAMRD Book Series No. 36/2006. (pp 90–101).

DENR (Department of Environment and Natural Resources). (2016). *Water resources of the Philippines*. Quezon City: DENR.

EMB (Environmental Management Bureau). (2008). *Water quality monitoring manual* (Vol. 1, Manual on ambient quality monitoring). Quezon City: EMB, Department of Environment and Natural Resources.

Ferrer, A.J.G., Hopanda, J.C., Orquejo, M.Q., Moscoso, A.D.E., & Sadaba, R.B. (2011). Reversion of disused fishpond lease agreement areas to mangrove forests in Region VI, Philippines. Research Report No. 2011-RRN. Singapore: Economy and Environment Program for Southeast Asia. 60 p.

Food and Agriculture Organization (FAO). (1995). *Code of conduct for responsible fisheries* (p. 41). FAO: Rome.

Gamolo, N.O. (2008). Philippines' water resources explained. *The Manila Times* (Sunday Time Magazine), 3 Aug 2008.

GOP (Government of the Philippines). (2006). National water resources board. Working together to secure sustainable water for all: The integrated water resources management (IWRM) plan framework summary document. http://www.nwrb.gov.ph/Uploads/iwrmplanframework.pdf

Guerrero, R. D. (2002). Tilapia farming in the Asia-Pacific region. In R. D. Guerrero & M. R. Guerrero-del Castillo (Eds.), *Tilapia farming in the 21st century* (pp. 42–50). Los Banos, Laguna: Philippine Fisheries Association, Inc.

Guerrero, R. D. (2014a). Impacts of introduced freshwater fishes in the Philippines (1905-2013): A veview and recommendations. *Philippine Journal of Science, 143*(1), 49–59.

Guerrero, R. D. (2014b). Sustaining tilapia farming in Taal Lake. *Agriculture Monthly, 18*(8), 20–21.

Inigo, L.B. (2015). Dagupan fisherfolk warn of fish kill due to polluted rivers. *Manila Bulletin*, 17 Mar 2015.

Jacinto, G.S. (2006). Case studies: fish cage farming in coastal waters - environmental, biological, social and governance issues. In B.V.L. Querijero, C.R. Pagdilao and S.V. Ilagan (eds.) Guidelines in the Establishment of Fish Cages and Other Structures in Lakes and Coastal Waters. In *Proceedings of the 1st Consultation Workshop on Guidelines for the Establishment of Fish Cages and Other Structures in Lakes and Coastal Waters*. PCAMRD Book Series No.36/2006. (pp.48–56).

Luna, M. P. G. (2005). *Review of legislation and policies on Philippine wetlands. UNEP/GEF project: Reversing environmental degradation trends in the South China Sea and Gulf of Thailand* (p. 24). Quezon City: Department of Environment and Natural Resources - Protected Areas and Wildlife Bureau.

Malayang, B. (2004). *A model of water governance in the Philippines, Winning the water war: Watersheds, water policies, and water institutions*. Makati City: Philippine Institute for Development Studies.

Obliosca, J. M., Jacinto, G. S., & San Diego, M. L. (2003). Preliminary results on the development of marine environmental quality criteria for mariculture areas. *The Philippine Scientist, 40*, 73–87.

PCARR (Philippine Council for Agriculture and Resources Research). (1977). *The Philippines recommends for mussels and oysters*. Los Banos: PCARR. 42 p.

PENRO (Provincial Environment and Natural Resources Office of the Iloilo Provincial Government), University of the Philippines Visayas (UPV), and the Canadian Urban Institute (CUI). (2016a). *Template for preparing report cards on the health of watersheds in the Philippines*. Toronto: International Development Research Centre. 32 p.

PENRO (Provincial Environment and Natural Resources Office of the Iloilo Provincial Government), University of the Philippines Visayas (UPV), and the Canadian Urban Institute (CUI). (2016b). *A monitoring framework to communicate watershed health in Iloilo Province, Philippines*. Toronto: International Development Research Centre. 56 p.

Poernomo, A. T., & Singh, V. P. (1984). Acid sulfate soil management for brackish water fishponds. In *Advances in milkfish biology and culture*. Tigbauan: Aquaculture Department of the Southeast Asian Fisheries Development Center.

Sekhar, N. U., & Ortiz, I. (2006). *Aquaculture in the Philippines: Socio-economics, poverty and gender*. Oslo/Quezon City: Norwegian Institute for Water Resources and Philippine Bureau of Fisheries and Aquatic Resources. 22 p.

Troell, M. (2009). Integrated marine and brackish water aquaculture in tropical regions: Research, implementation and prospects. In D. Soto (Ed.), *Integrated mariculture: A global review, FAO Fisheries and Aquaculture Technical Paper. No. 529* (pp. 47–131). Rome: Food and Agriculture Organization.

WEPA (Water Environment Partnership in Asia). (2005). *Country profile – Philippines, The 1st International WEPA Forum*. Kanagawa: Institute for Global Environmental Strategies (IGES).

Wurts, W. A. (2002). Alkalinity and hardness in production ponds. *World Aquaculture, 33*(1), 16–17.

Yap, W. G. (2010). *Contribution of FLA area to Philippine fish production*. Paper presented at a seminar workshop on Fishpond Lease Agreement (FLA) Cancellation and Reversion to Mangrove Forests, 28–30 July 2010, Iloilo City, Philippines.

Dr. Rafael D. Guerrero III has a B.S. in Zoology, M.S. in Applied Zoology and Ph.D. in Fisheries Management. He was an Associate Professor and Dean of the College of Inland Fisheries of the Central Luzon State University in the Philippines and Executive Director of the Philippine Council for Aquatic and Marine Research and Development of the Department of Science and Technology. He has published numerous books on tilapia farming and vermiculture, and technical papers on aquaculture and water resources management. He is presently an Adjunct Professor of the School of Environmental Science and Management of the University of the Philippines Los Baños and Academician of the National Academy of Science and Technology, Philippines.

Pepito R. Fernandez Jr. is Full Professor in Political Science at the University of the Philippines Visayas (UPV). He has academic degrees, expertise and training in political science, human geography, and fisheries social science from the University of the Philippines, McGill University, Kagoshima University and Australia National University. He currently teaches courses in political science, liberal arts, and natural resources management. He is a recipient of various awards including Department of Agriculture-Bureau of Agricultural Research Director's Award for international refereed journal publication, and Best Research Paper Award from the Philippines Network of Educators on Environment. His most recent publications include a peer-reviewed book chapter "Creating Stable and Adaptive Governance Systems in Philippine Coasts" by the Department of Agriculture-Bureau of Fisheries Aquatic Resources (DA-BFAR) and UPV, and an online publication by the Food and Agriculture Organization (FAO) entitled "Governance Institutions and the Adaptive Capacity of Small-scale Aquaculture to Climate Change in the Philippines."

Chapter 8
Multiple and Integrated Water Resource Utilization

Guillermo Q. Tabios III

Abstract This chapter discusses multiple and integrated water resource utilization in the Philippines. Multiple use can be provided by multipurpose water systems such as reservoirs for domestic water supply, irrigation, hydropower generation, and flood control function, including water quality control function. To discuss the status, issues and problems of multiple use water system, four systems in the Philippines are showcased, namely: the Angat and San Roque Reservoirs as well as the Laguna Lake and Agusan River systems. Being multiple use, some of these water resource systems may have competing or conflicting water uses due to differing socio-economic objectives (e.g., Laguna Lake and Angat Reservoir) and even due to the physical configuration of the components of the system (e.g., hydropower plants in the Angat Reservoir). On the other hand, complementary water use occurs such as the case of hydropower generation when the same irrigation or domestic water supply releases passes through hydropower plants such as in San Roque and Angat Reservoirs. At the end of this chapter, suggestions and recommendations for harmonizing or complementing multiple water resource utilization together with land use management are described in terms of policies and management strategies to achieve effective multiple and integrated water resource utilization.

Keywords Multiple water use • Integrated water resource use • Complementary water use • Reservoir • River systems

---

G.Q. Tabios III (✉)
Institute of Civil Engineering and National Hydraulic Research Center, University of the Philippines at Diliman, Quezon City, Philippines
e-mail: gtabios@up.edu.ph

A.C. Rola et al. (eds.), *Water Policy in the Philippines*, Global Issues in Water Policy 8, https://doi.org/10.1007/978-3-319-70969-7_8

8.1 Introduction

This chapter is an assessment of water sector performance for multiple-use or multipurpose water resource systems. It discusses operational policies and practices (including programs) on multiple and integrated water resource utilization in the Philippines. In terms of multiple water use, technical, contractual, and operational problems and issues of existing multipurpose reservoirs to satisfy domestic water supply, irrigation, hydropower generation, and flood control function, including some cases with water quality control function, are discussed. In particular, certain multiple-use or multipurpose water resource systems such as the Angat and San Roque reservoirs as well as the Laguna Lake and Agusan River systems (as multiple-use river basins) are showcased to present issues and problems on competing water uses and operational protocols as well as inconsistencies or contradictions in their physical configurations and contractual obligations. Historical water use conflicts that arose from competing water uses and the remedies provided are also discussed. Also showcased are watersheds and water resource systems in the Philippines that have demonstrated or shown promise as far as harmonizing or complementing multiple water resource utilization is concerned, along with integrated water resource and land use management.

Water resource systems may either be natural or built up. The components of natural water resource systems are creeks, streams, springs, and lakes. Built water resource systems include flood channels, man-made reservoirs, and drainage ditches. For sustainable development of water resources, the latter systems must be designed and managed to fully contribute to meeting the needs of society, now and in the indefinite future, while protecting the hydrological and ecological integrity of the natural water resource systems and considering the cultural practices and socio-economic structure of the underlying communities (Vesilind and Gunn 1998).

Water resource management is the formulation and evaluation of alternative plans to accomplish water development objectives such as water supply management, water excess management, including environmental protection for sustainable use of water resources. These objectives are achieved by manipulating the hydrologic cycle through the use of hydraulic structures, water supply, water treatment, wastewater treatment and disposal, irrigation, hydropower generation, and flood control measures. Two important areas of concern in water resource management are management of water supply or use systems and management of water excess or control systems. The environmental management aspect of water management emphasizes the need for environment-based planning, design, and operation of water resource systems to ensure that these are not only technically sound but are environment-friendly. Examples of water resource systems associated with water supply and water control management are as follows:

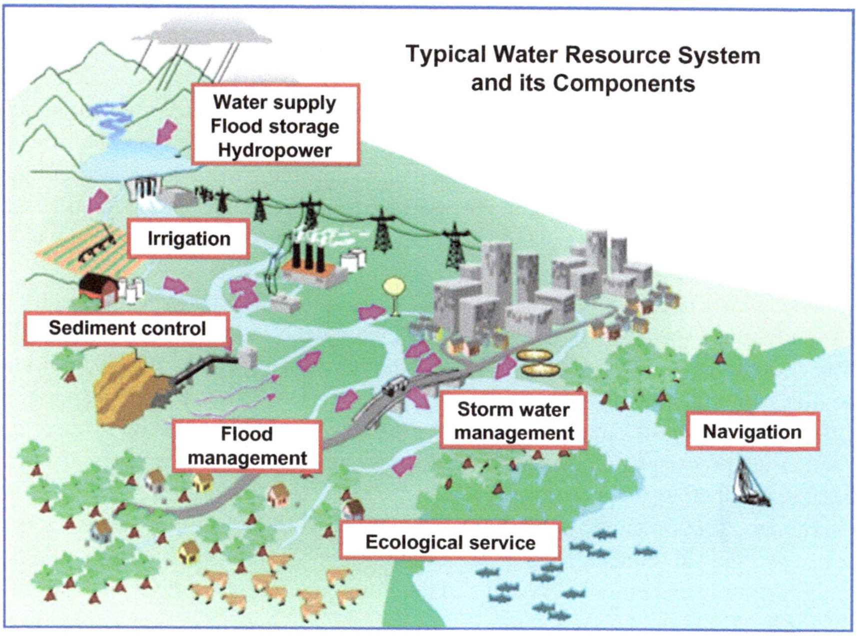

Fig. 8.1 Components of a water resource system

Water Use Systems

- Domestic and industrial supply reservoirs, towers, lines
- Wastewater treatment plants, sewers
- Irrigation aquifers, canals, ditches
- Hydropower generation reservoirs, spillways

Water Control Systems

- Drainage ditches, tiles
- Flood control embankments, levees, channels, dams
- Salinity control pumping, filters
- Sediment control fencing, reservoirs
- Pollution abatement pumping, filtering, flocculation

Figure 8.1 depicts various components of a water resource system and how they interact.

Reservoirs in particular can have multiple uses such as for domestic and irrigation water supply as well as hydropower generation while making releases for water supply purposes. Certain reservoirs can also have a flood control function by allocating through emptying a certain storage volume during flood season to contain high flows temporarily, thus attenuating the peak or spike of high flows. In certain cases, reservoirs are even used to control water quality such as when turbidity is reduced by trapping fine sediments.

Rivers can also be multipurpose for fish spawning and growth, recreation, and navigation. In particular, the flood control function of rivers with floodplains result in the spread of floodwaters into the floodplains, thus reducing water elevations and slowing water velocities due to roughness of the floodplains. Nowadays, constructed wetlands are becoming popular for water quality control, especially removal of organic wastes or pollutants.

Another major component of a water resource system is the groundwater aquifer. Groundwater appears to be single-purpose, being the major source of domestic water supply in the Philippines. However, another possible use of the groundwater aquifer is to inject urban floodwaters on a large scale, which has been proposed in cities especially with karst aquifers having high hydraulic conductivities. In certain geothermal power plants, their industrial waste or brine are injected back to some great depths below the groundwater aquifer.

The next section presents the governance framework of multiple-use water systems. In the subsequent section is a discussion of four multiple-water-use / multipurpose-water-resource systems in the Philippines. These are divided into two categories: (1) multiple use or multi-use river systems where the Laguna Lake Basin and the Agusan River Basin are presented and, (2) multipurpose reservoir systems where the Angat Reservoir and San Roque Reservoir are described. The physical features, various water uses, including competing water uses and other issues and problems on water governance, changing land use, and cultural/social trends of these systems are discussed. Policy challenges and recommendations form the last sections of this chapter.

8.2 Governance Framework for Multi-use River Systems and Reservoirs

There is no unique protocol on the governance of multi-use river systems and reservoirs. In the cases presented, different institutional arrangements and varying policies govern the systems. For instance, Laguna Lake has a dedicated agency, the Laguna Lake Development Authority (LLDA), that manages the lake waters. However, there are also other agencies such as local government units (LGUs) that have jurisdiction over its use, especially in the establishment of fish pens. Pollution is also being monitored by the LLDA, setting fees for polluters.

Other river basins that have multiple uses are sometimes declared as protected areas and are managed by the national agency, the Department of Environment and Natural Resources (DENR). The Angat Reservoir has its water allocation decided by the National Water Resources Board (NWRB). Currently, the hydropower assets of 218 MW is privately owned and operated jointly by a Korean company and SMC Power, a local company.

Conflicts arise in the use of water that may sometimes be due to absence of concrete policies for water allocation, water use, and water quality standards.

8.3 Performance Assessment of Multi-use Water Systems

8.3.1 Multi-use River Systems

Two major multi-use water resource systems in the Philippines, namely the Laguna Lake Basin and the Agusan River Basin, are presented here. The physical features, various water uses, including competing water uses, and other issues and problems on water governance, changing land use, and cultural/social trends of these systems are discussed.

8.3.1.1 Laguna Lake Basin

Laguna Lake, also known as Laguna de Bay, encompasses Metro Manila, Rizal, Cavite, Batangas, and Quezon provinces (Fig. 8.2). The land area covered by the entire basin is about 2865 km^2, whereas the lake area itself, including Talim Island, is about 900 km^2. There are 24 major watersheds/river systems that flow into the lake (Fig. 8.3). Twenty-two of these watersheds are under the jurisdiction of the LLDA, the lead implementor of the Laguna de Bay master plan (LLDA n.d.). The biggest watershed is the Marikina River basin, located north of the lake, and the second biggest is Pagsanjan River, found in the southeast part of the lake. The two watersheds outside of LLDA jurisdiction are the San Juan and Pasig river basins.

Laguna Lake is a multi-use water resource system. Among the lake's purposes and uses are flood control, fisheries production, navigation, Caliraya Reservoir-Kalayaan hydroelectric pumped-storage system, and water source for domestic water supply as well as irrigation. It is important to note that proper management of Laguna Lake should not only be confined within the lake but should extend to proper management of watersheds contributing to the lake.

Fishing is a big industry in Laguna Lake. A major portion of the lake is devoted to fisheries from commercial fish production with the use of fish pens and fish cage structures to individual, subsistence fishing. LLDA derives significant revenue from the lake by renting parts of the lake area to fish pen and cage owners. The question here is how many hectares of lake area should be allocated to commercial fish production and where should these fish pens and fish cages be located so that they do not affect the water quality dynamics and health of the lake. In 1995, the total area of the lake covered with fish pens and cages was about 15,140 ha. In 1999, the area

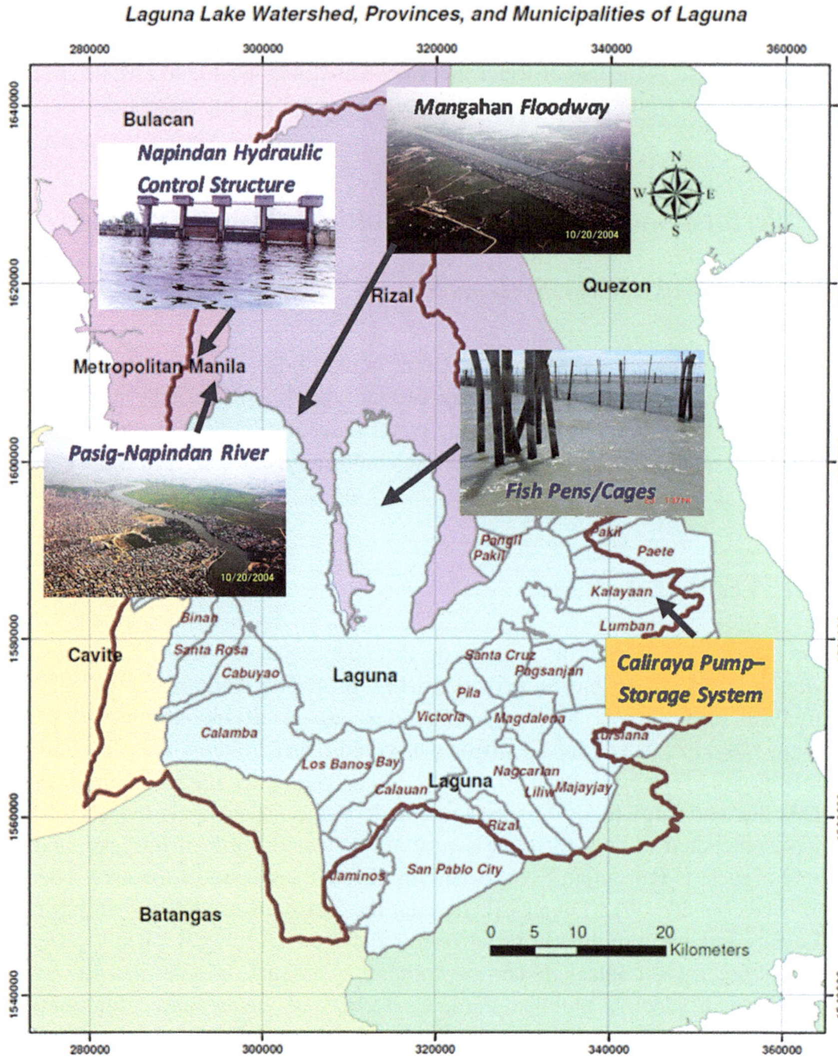

Fig. 8.2 Laguna Lake system and components

occupied by these fish structures was reduced to 11,160 ha (fish pens occupied about 6160 ha; fish cages occupied about 5000 ha) when a strong typhoon destroyed the fish structures in 1997. During these two periods, water quality in the lake in terms of biochemical oxygen demand (BOD) was about 12.5 mg/L in 1996 and about 10.4 mg/L, an improvement, in 1999 (Catalan 2000). This indicates that there may be an optimal number of hectares in Laguna Lake to be devoted to commercial fish farming.

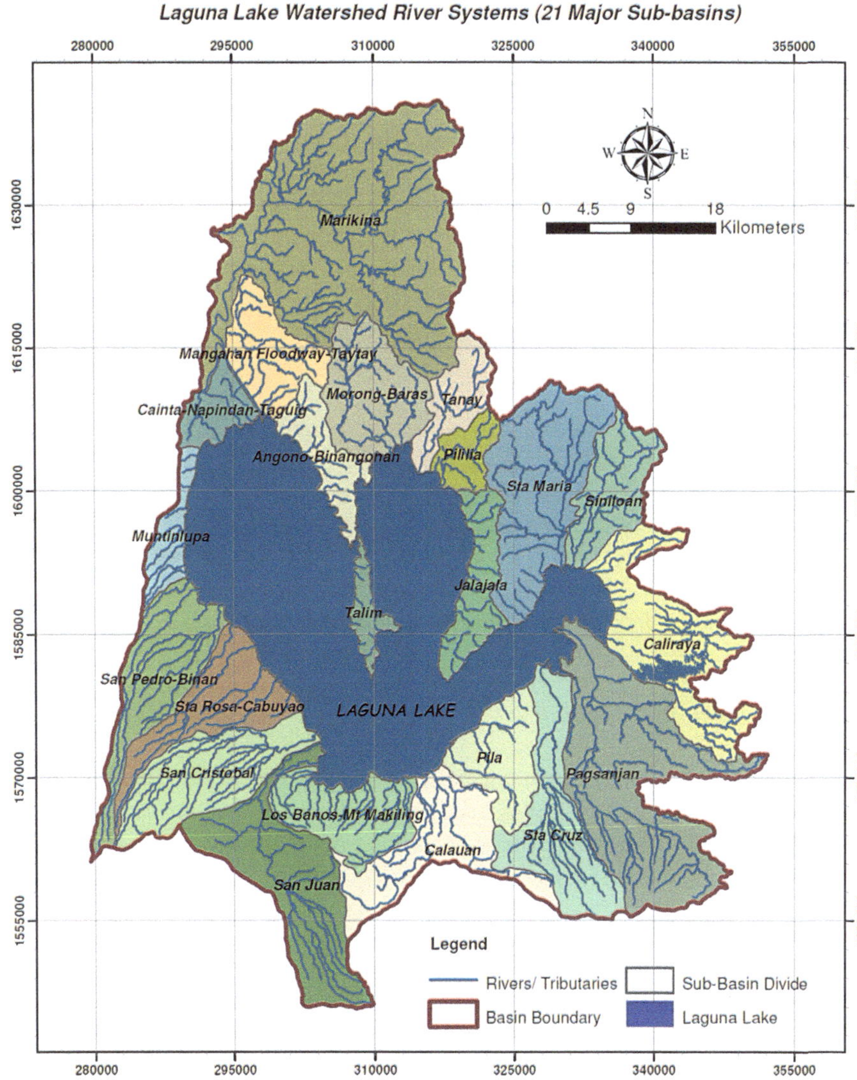

Fig. 8.3 Laguna Lake watershed and river system

Saltwater from Manila Bay through Pasig River is allowed to enter Laguna Lake, which is desirable to the fishing industry. On the other hand, lake withdrawal for domestic water supply requires a certain, minimum salinity standard that conflicts with the fisheries' salinity requirement. Thus, the issue is how to properly manage saltwater entering or exiting the lake through Pasig River, which can be controlled by the Napindan hydraulic control structure, in view of the two conflicting water quality (salinity) requirements for fisheries vs that for domestic water supply

(Tabios and David 2004). Salinity levels of 2–6 parts per thousand (ppt) observed in the lake are desirable for fish production. However, treatment of water for domestic water use could be too expensive for water with salinity greater than 0.3 ppt.

There are all kinds of pollutants dumped into the lake, but the most toxic substances are the so-called endocrine-disrupting chemicals such as heavy metals. Catalan (2000) found an alarming number of contaminated fish species in the lake and reports from fishermen and the specimens collected showed the presence of fish that are blind, with crooked spine, without tails, or with skin cancers. There are also reports of effects on people living along the shores of the lake who eat contaminated fish. They had children with reproductive deformities and certain immune deficiency and mental or physical growth deficiency symptoms.

The degree of lake pollution from industrial effluents or domestic waste around the lake varies within the year due to the seasonality of dilution effects of freshwater inflows from the watersheds around the lake, including rainfall. The dry season is the worst condition, when lake inflows and rainfall are low, thus reducing pollutant dilution. To properly manage lake pollution, monitoring of lake water quality should be taken seriously to detect sources of pollution (including timing and spatial extent) and to enable authorities (LLDA, for that matter) to control pollution. Only then can appropriate action be taken against these polluters, be it natural or man-made, through treatment, penalties, or closure of pollution sources.

8.3.1.2 Agusan River Basin

The Agusan river basin has a drainage area of about 1,193,665 ha (Fig. 8.4) (CTI-Halcrow-Woodfields 2008). The Agusan marsh is located in the midstream portion of the basin. The drainage area upstream of this marsh is about 661,200 ha and the marsh itself can inundate an area from as low as 19,000 ha to as high as 56,000 ha. The declared Agusan Marsh Wildlife Sanctuary only covers an area of about 111,500 ha.

Rainfall in the Agusan river basin has a distinct spatial variability and annual rainfall can range from 2000 mm in the plains such as the northern (downstream) part of the basin to more than 4000 mm in the mountains. On temporal basis, average monthly rainfall can be as low as 120 mm in May to as high as 350 mm in January. These rainfall variabilities definitely result in strong seasonalities and spatial variabilities of the hydrological and subsequently ecological cycles and processes in the Agusan river basin in general and the Agusan marsh in particular. For the latter, the wetland dynamics in terms of flooding frequency as well as floodplain-main channel interaction or recession rates (water stage drawdown per unit time) in the wetlands are on a seasonal basis.

CARBDP-PMO (2003) reports several proposed dam constructions in the Agusan river basin in the next few years. The proposed dam constructions will be at various locations in the upper Agusan river basin (Fig. 8.4) such as in Manat and Logum rivers; the middle basin such as in Wawa, Gibong, Simulao, Ibaoan, Umyam, and Adgaoan rivers; and one in the lower basin in Bugabus River. Likewise, in a

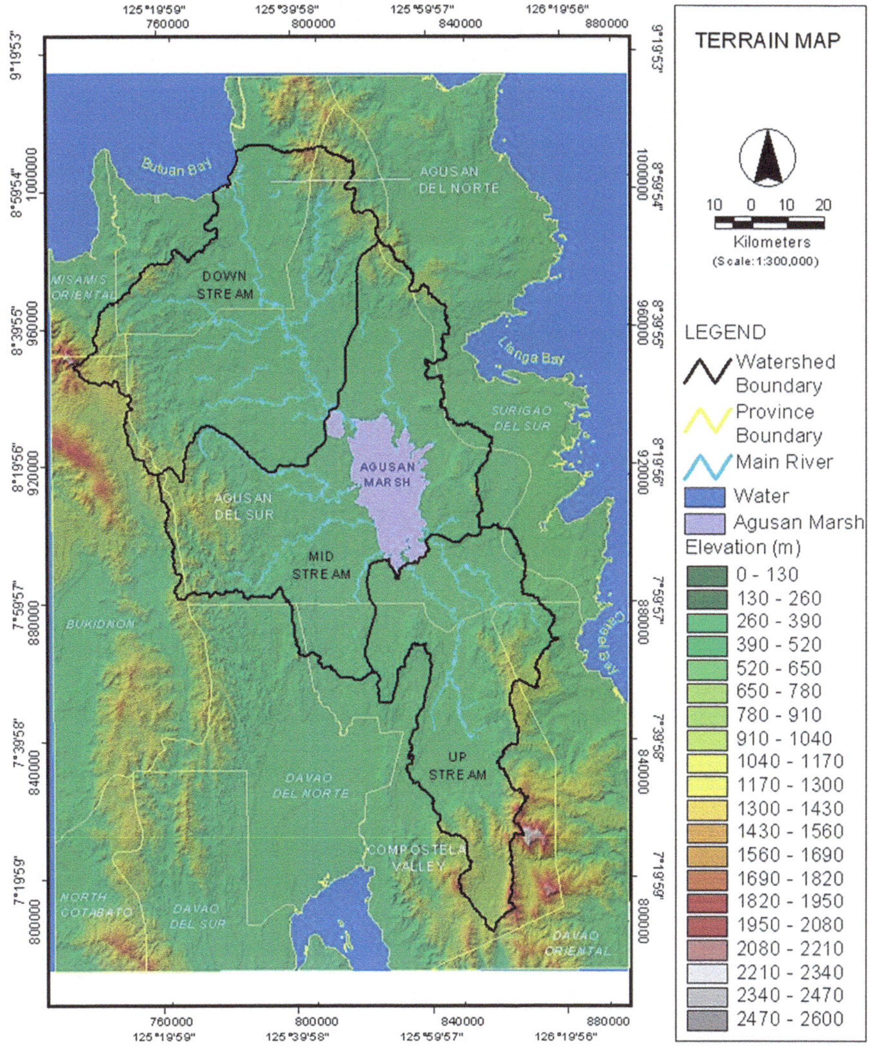

Fig. 8.4 Components of the Agusan River Basin (From CTI-Halcrow-Woodfields 2008)

report by Hassall (2003), there are planned irrigation diversion structures in Logum and Baobo rivers in Veruela, Agusan del Sur, both of which are tributaries to Agusan River upstream of the Agusan marsh.

Several issues concerning these planned reservoirs or river diversions could cut or reduce freshwater inflows to the Agusan River. For instance, the decrease in surface water inflows to Agusan marsh can modify the flooding frequency and spatial extent that essentially define the wetland dynamics of the marsh. Also, Butuan City, which heavily relies on groundwater supply, could experience drastic lowering of groundwater table elevations due to decreased recharge that comes from

Agusan River. The CARBDP-PMO (2003) report included an evaluation of the impact of proposed reservoirs which showed that there will be significant changes in the seasonal variations of the dynamics of river-floodplain interaction due to swallower river water surface elevations and also reduction of groundwater recharge and subsequently groundwater table with the proposed construction of reservoirs. The impact of the decreased groundwater flux to the coastal zone causes saltwater intrusion and hypersalinity condition, thus adversely affecting the health of the mangrove forest in the Agusan River estuary.

In the Agusan river basin and particularly in Agusan marsh, understanding the interactions of hydrology, geomorphology, and ecology in watersheds, rivers, wetlands, and estuaries is important. In Agusan marsh, studies toward a mechanistic understanding of these interactions may be lacking and are therefore needed to implement a better and scientific-based management of the marsh ecosystem.

Note that it is important to recognize the temporal and spatial scales of fluctuations of the various processes involved in the hydrology, geomorphology, and ecology interactions. For example, flood flows with accompanying high sediment loads can abruptly change the river morphology in a few hours or a few days. This is in contrast to geomorphologic changes due to bioturbation by burrowing organisms, which could take several months or years to result in significant streambed changes. Another example is the impact of the disappearance of the wetlands due to river water diversion or climatic drought that could take several years to realize and cause wetland flora and fauna to disappear. In fact, one could already be experiencing the disappearance of the wetlands and their associated flora and fauna and it is too late to realize it.

8.3.2 Multi-purpose Reservoir System

In this section, two multipurpose water resource systems or reservoir systems in particular are discussed, the Angat Reservoir and the San Roque Reservoir.

8.3.2.1 Angat Reservoir

The Angat Reservoir is a multipurpose reservoir that was constructed to supply water for irrigation to a maximum of about 31,000 ha of rice and vegetable farms in Bulacan; ensure almost 90% of Metro Manila's domestic water supply, which is regulated by the Metropolitan Waterworks and Sewerage System (MWSS) and distributed by private concessionaires Manila Water Company (MWCI) and Maynilad Water Services (MWSI); generate hydropower for the National Power Corporation (NAPOCOR) at an annual average of 500 GW-h (constituting about 5% of Luzon's power demand); and serve as flood control storage incidental to the operation of the reservoir and the afterbay re-regulation dam at Bustos (for irrigation) and at Ipo (for domestic water supply).

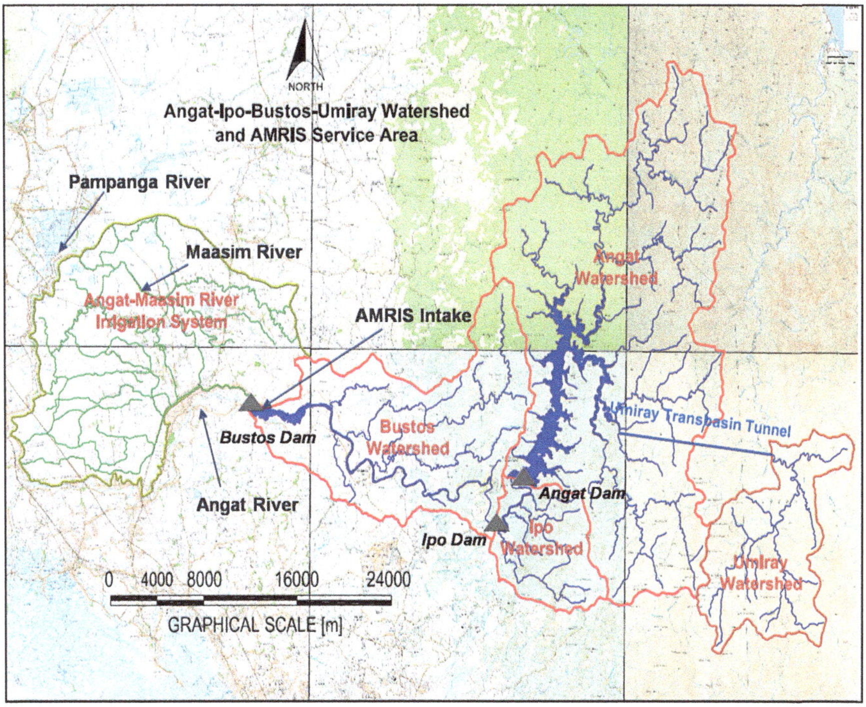

Fig. 8.5 Watershed delineation and river network of the Angat-Ipo-Bustos-Umiray water resource system (Tabios and David 2014)

The watershed of the Angat Reservoir is the Angat river basin, which has a total drainage area of 568 km^2 upstream of the Angat dam site located across a narrow gorge of the Angat River at Sitio Biniit, San Lorenzo, Bulacan (Fig. 8.5) (NHRC 2002; Jose et al. 1996). The Angat dam is a rockfill dam that is 131 m high with a road surface top at the dam crest and 630 m long that curves at a radius of 620 m. The reservoir has an active capacity of 865 million m^3 with absolute drawdown (dead storage) elevation at 158 m above mean sea level (AMSL), maximum conservation pool operating level at 212 m AMSL, design flood pool elevation at 219 m AMSL, and maximum (dam crest) elevation at 223.5 m AMSL.

Inflows to Angat Reservoir derived only from rainfall in the watersheds come at an annual average inflow of 60 m^3/s. Since 2001, however, inflow to the reservoir has been augmented by diverting water from the Umiray River watershed through the 13.1-km Umiray-Angat transbasin tunnel with carrying capacity of 25 m^3/s. The Umiray-Angat transbasin project is estimated to bring an average annual inflow of about 9 m^3/s to the Angat Reservoir. This brings a total of almost 70 m^3/s annual average inflow to this reservoir.

With regard to the physical configuration of the Angat reservoir system as shown in the schematic in Fig. 8.6, the main hydropower plant (HP) generation with a capacity of 200 MW is achieved through releases to irrigation (NIA). The auxiliary

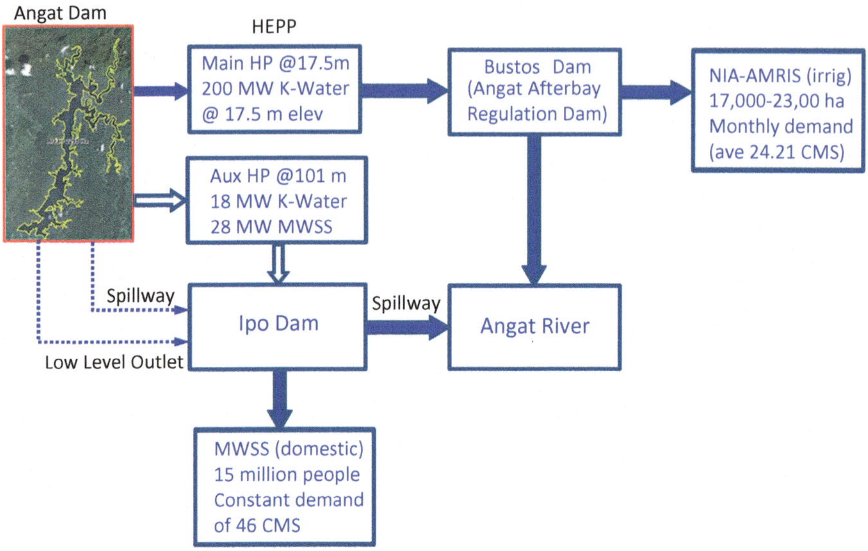

Fig. 8.6 Physical components and water demand of the multipurpose Angat Multipurpose Reservoir System for domestic water supply, irrigation water supply, hydropower generation, and flood control (Tabios 2016)

HP generation with a capacity of 46 MW is through releases to domestic water supply (MWSS). However, twice as much power can be generated at the main HP compared with the auxiliary HP for the same flow discharge since at, say, a reservoir elevation of 205 m, the head for power generation for the main HP is 187.5 m (that is, 205–17.5 m tailwater elevation) while that for the auxiliary HP is 104 m (that is, 205–101 m tailwater elevation at Ipo Dam).

With regard to NIA and MWSS water rights, originally after Angat Dam was built in 1968, NIA has 36 m^3/s, while MWSS has 22, which is a total of 58 m^3/s, the amount equal to the water rights granted to the NAPOCOR, which is to operate Angat Reservoir for their hydropower generation objectives, provided that MWSS (domestic water supply) and NIA (irrigation) water rights or allocations are satisfied on a long-term, average daily basis. The long-term average is emphasized here because NIA has seasonal water requirements in contrast to MWSS, which is fixed on a daily basis. However, since 2000, with additional water from the Umiray River transbasin transfer, water rights have evolved as follows: (1) MWSS water rights became 46 m^3/s, which is the sum of the original 22 m^3/s, 15 m^3/s referred to as conditional water rights from NIA and 9 m^3/s from Umiray River; (2) NIA-AMRIS water rights have been reduced to 21 m^3/s, which is from the original 36 minus 15 m^3/s; and (3) Bulacan Bulk water rights is 3 m^3/s, which is from the Umiray transbasin transfer. The above water rights or allocation is based on the long-term, average inflows of Angat and Umiray watersheds of 70 m^3/s.

In view of these changes in water allocation and priority, the 200-MW capacity of the main HP of Angat is seldom reached since NIA's allocation has been reduced to 21 m^3/s from its original 36 m^3/s. Also, in accordance with the Philippine Water Code, during water shortage conditions, the order of priority of water delivery for a multipurpose reservoir like Angat is to first satisfy domestic water supply, followed by irrigation demand; hydropower is incidentally generated.

8.3.2.2 San Roque Reservoir and Agno River Basin

The San Roque Reservoir is a multipurpose reservoir for hydropower generation, irrigation water supply, and flood control; it has odd water quality function, which is to trap suspended sediments to improve irrigation water quality. The reservoir is located in the Lower Agno River of Pangasinan, Philippines, and its dam construction was completed in 2002. The maximum storage capacity of the reservoir is about 840 million m^3 and its hydropower plant has a maximum operating capacity of 405 MW. Figure 8.7 shows the location of the reservoir. The San Roque Dam is a rock-fill dam structure that includes a spillway and its hydropower plant. It can generate a maximum of 405 MW of hydropower at a turbine discharge of 260 m^3/s. With regard to its flood control function, the available flood control allocation storage is about 121 million m^3, which is between its maximum surcharge water level of 290 m and 280 m.

As for irrigation, downstream of the San Roque Dam is the Agno River Irrigation System (ARIS) of the Agno River Integrated Irrigation Project (ARIIP). This regulating dam has a storage volume of 5 million m^3 and its main purpose is to store the water releases from San Roque Dam for hydropower generation (about 12 h/day operation) and released for irrigation over a 24-h period. The ARIS dam only provides 1-day carryover storage, which can be filled up within 12 h with an average inflow of 115 m^3/s, or it can be filled up in about 5.4 h with an average inflow of 260 m^3/s (maximum turbine discharge). The ARIS dam does not have a significant flood control function.

The entire Agno River, with a drainage area of 5952 km^2, is the third largest river basin in Luzon, fifth in the Philippines, next to the Cagayan and Pampanga rivers. The length of the river system from its source in the Cordillera Mountains to its mouth in the Lingayen Gulf is about 270 km, 90 km of which runs through mountainous terrain and canyons. The Agno River, after passing through mountains at an average elevation of some 2000 above MSL, forms a vast alluvial fan and delta called the Pangasinan Plain and then flows into the Lingayen Gulf. Along the upper portion of the Agno River are three major reservoirs: Ambuklao, Binga, and San Roque.

A major concern of reservoirs in general and specifically of the San Roque Reservoir is sedimentation. The Binga Reservoir was built in 1959 and its storage volume was significantly reduced by sedimentation, Its outflows into the San Roque Reservoir therefore contain significant amounts of sediments. The watersheds upstream of San Roque Reservoir likewise contribute to reservoir sedimentation, thereby necessitating the imposition of proper sediment control measures to minimize watershed sediment yields. In an NHRC study in 2007, it was concluded that

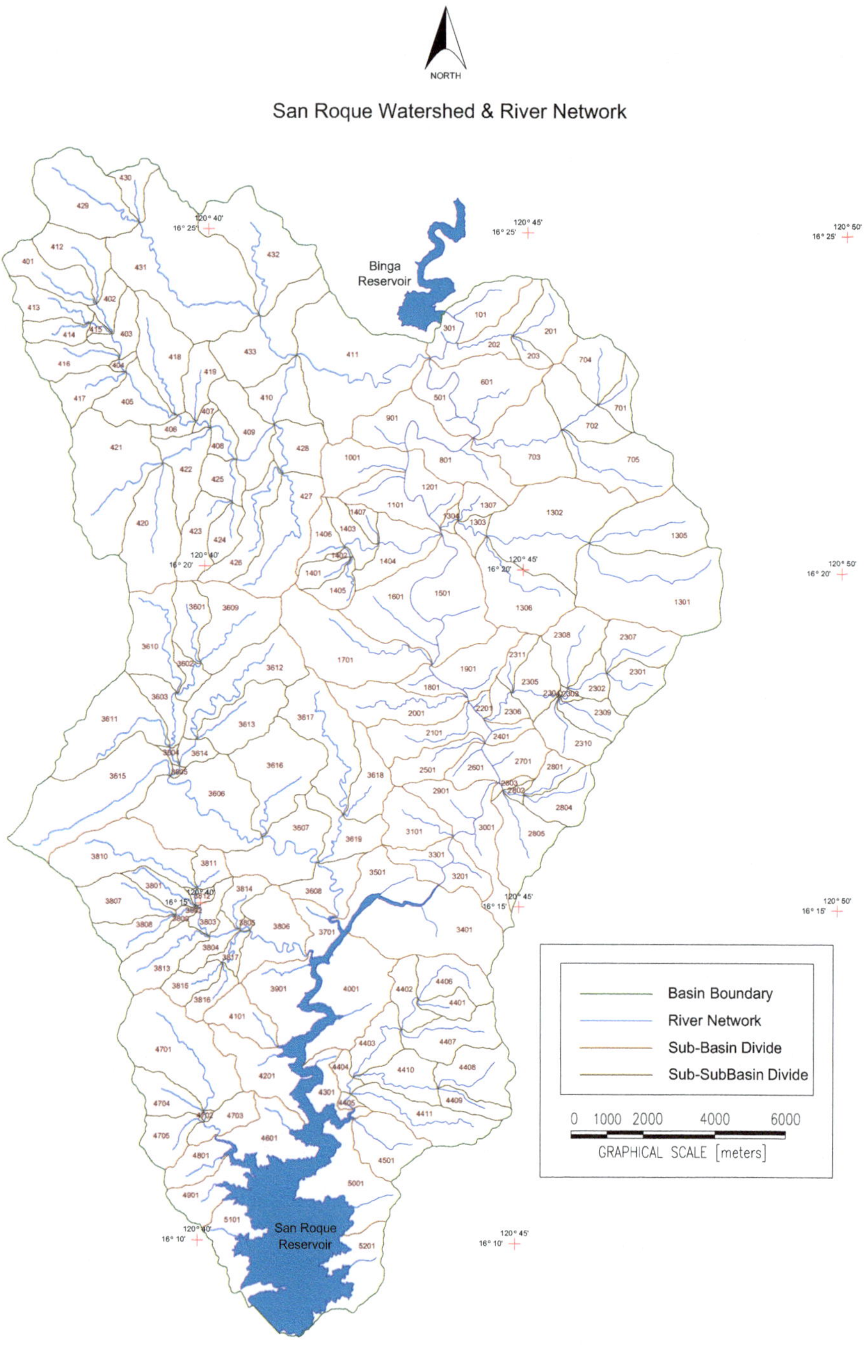

Fig. 8.7 San Roque Reservoir and the Lower Agno River Basin (NHRC 2007)

the effect of sediment accumulation in the San Roque Reservoir over the next 70 years does not have any drastic or negative impact on water supply and hydropower operations (Tabios et al. 2007). It must be emphasized, however, that sediment management control measures must be implemented to maintain at a minimum the sediment inflow scenarios considered in this study.

Downstream of San Roque Dam is the Lower Agno River flood control project. The project covers the 124-km stretch of the Agno River from San Roque Dam to Lingayen Gulf. Phase I of the project covering 54 km of Agno River from Bayambang to Lingayen Gulf is already completed; Phase II of the project covers the 23-km river stretch from Alcala to Bayambang and is not yet complete. Yet to be started is Phase III that covers the 47-km river stretch of Agno River from San Roque Dam to Alcala. This flood control project is designed to provide a level of protection to contain a 10-year return period or recurrence interval flood.

Figure 8.8 shows the design flood distribution of the various components of the Agno River flood control project of DPWH, which is designed to provide a 10-year return period or recurrence interval. In the flood frequency curve of Agno River at the San Roque dam site, the San Roque dam inflow of 2600 m^3/s corresponds to about 10-year return period flood. The flood control function of San Roque Dam was the subject of a Senate inquiry reported by Tabios et al. (2010) during Typhoon Pepeng (Typhoon Parma) in October 2009. In the said typhoon, maximum inflow (averaged over 1 h) to San Roque dam was 5547 m^3/s, which occurred around 3 AM on October 9, 2009; this corresponds to about 80-year return period flood. In fact, the actual peak inflow can even be higher than the recorded maximum inflow of 5547 m^3/s and the estimate of the actual peak flow can be above 6000 m^3/s. It may be noted that the second biggest flood that the San Roque dam experienced after its completion in 2002 was during the Central Luzon flooding in August 2009 in which the peak flow of 3029 m^3/s corresponds to a 20-year return flood.

The 20-year or 80-year return period floods can happen again and even over and over again in the Agno River basin. Thus, it is definitely worthwhile for the national and local governments to rethink if the 10-year return period design flood for the lower Agno River basin is the only level of protection they can provide the people. The bulk of the flood control or mitigation strategies in the lower Agno River basin cannot be fully relegated to the San Roque dam flood control function.

8.4 Policy Challenges

8.4.1 On Understanding the Hydrologic, Geomorphologic, and Ecologic Interactions for Effective Water Resource Management

In the planning, design, and management of water resource systems in general, understanding the interactions of hydrology, geomorphology, and ecology in watersheds, rivers, wetlands, and estuaries is important. In particular, studies toward a

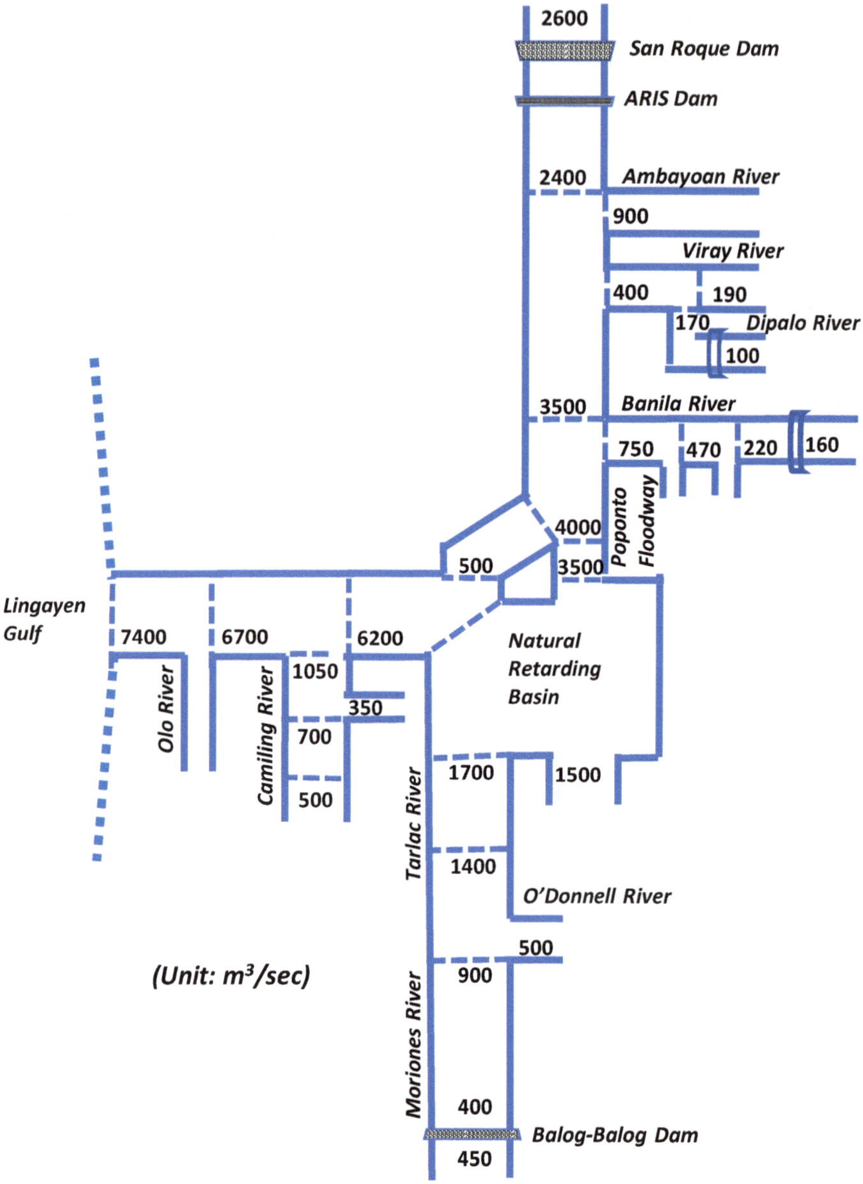

Fig. 8.8 Design flood distribution of the Agno River Flood Control Project of DPWH below San Roque Dam designed to provide a 10-year return period (recurrence interval) level of protection (Tabios et al. 2010)

mechanistic understanding of these interactions are needed to achieve a better and scientific-based management of a water resource system or an ecosystem in general, or in particular, the case of the Agusan River basin (Tabios 2008) and likewise the Laguna River basin (Tabios 2003).

The geomorphology and hydrology of a basin interact with each other since the morphology of the river system influences the shape and magnitude of the basin hydrograph, while the landform or morphology of the river system is shaped by the hydrologic or hydraulic flow regimes of the basin (Solyom and Tucker 2004). The high flow regimes are generally responsible for removing or mobilizing sediments by erosion, whereas the low flows promote deposition of sediments. During high flows, rivers can flush fines and sand in the streambed gravel matrix. The geomorphologic features of the river such as shape, slope, and sediment size dictate the type of sediment load in the river as well as stability in terms of being straight, meandering, or braided.

Hydrology including hydraulics interact with ecology in that the ecology of the river in terms of biotic structure (i.e., species, composition and communities) and life cycles of the flora and fauna highly depend on the seasonal variations of discharges, especially their dilution effects on water quality, channel velocity distribution, frequency of bankfull-discharge condition or floodplain-river interaction, floodplain recession rates, water residence times, as well as time and space frequencies of floodings (Shen et al. 1994).

The geomorphology of the river in terms of bathymetric features, sediment loads, and water-sediment balance influences ecology in terms of biogeography as well as the birth and survivability of aquatic flora and fauna. For example, the spawning behavior of certain fishes depend on the location in a river reach where they deposit and fertilize eggs in cobble bars and they rest and feed in pools (in a riffle-pool sequence) between spawning forays (Milhous 1995). Ecology can affect geomorphology in terms of bioturbation such as burrowing organisms that cause resuspension of sediments in lakebeds or streambeds.

8.4.2 On Sustainable Planning and Management of Multi-purpose Reservoir Systems

Reservoirs and dam projects in the Philippines and even other parts of the world are rarely or not at all planned, designed, and operated in the time frames of sustainable development, which is "not only for the current generation, but future generations to come." Specifically, most of these reservoir and dam projects are justified based on an economic life (i.e., period of benefit/cost [B/C] and internal rate of return [IRR] computations) of 50 years and at most 70 years, which means that it only covers the current and part of the future first generation in the context of sustainable development. Perhaps, one major reason for this practice is that reservoirs, by nature, trap sediments from upstream watersheds so reservoirs are assumed to have finite lives of at most 70 years, unless proper reservoir sedimentation management strategies are in place. Another reason may be simply because it is much more difficult to justify projects as well as for governments, including private investors, to appreciate economic planning horizons of 150–300 years.

Typically, with planning horizons of 50 years, the approach uses historical data (in the last 50–70 years if available) with regard to the hydrology and hydraulics of reservoir and dam design, water demand based on population projections and irrigation developments, and the use of a certain discount rate to calculate the B/C and IRR economic measures. However, at extended economic time horizons of 100–200 years covering four to five future generations, the planning, design, and operation studies of such reservoir and dam projects require different methodologies, different science, and consequently different policies to ensure the sustainability of these projects. Since historical data or historical past is no longer enough, there is thus the need for imagination and scenario setting for, say the next 200 years, which heavily depends on the ability of the planners and stakeholders to see into the future with regard to variabilities, changes and projections in land use, river landscape, population, political structure, and socioeconomic setup as well as climate and ecosystem changes.

In the Philippines, the familiar large-scale water resource system is the multipurpose Angat Reservoir/Dam system, which is for domestic water supply, irrigation water supply, hydropower generation, and flood control or the multipurpose San Roque Reservoir for irrigation water supply, hydropower generation, flood control, and improving irrigation water quality. In particular, its hydropower component is considered a renewable resource but, as previously mentioned, the planning and design of reservoir systems is assumed to have a finite life of 50–70 years, most likely due to reservoir sedimentation and filling up. Thus, it is not quite a renewable resource as such unless proper sedimentation controls are in place to ensure sustainable reservoir life.

Another issue with constructing reservoirs is that it could adversely impact the river downstream of the dam due to the change (or lack thereof) in sediment supply (suspended and bedload), thus altering the natural river landscape (form and alignment) and consequently the river ecology and ecological integrity. For instance, the river downstream will starve from a seasonal supply of sediments that is responsible for maintaining stable channels such as the riffle-pool sequence that prevents too much erosion or sedimentation in portions of the river downstream, thus creating flooding or bank erosion problems and even river migration to settlement areas.

Since the typical life of reservoirs is about 50–70 years (over 80% filled up with sediments), proper sedimentation management schemes should be in place to prolong the life of the reservoir, which is not only based on technical engineering measures but community involvement and stewardship to protect the watershed from forest denudation and land use practices that aggravate watershed soil erosion. On the other hand, the impact of reservoirs on modifying the river sedimentation processes (cycles of erosion and deposition) downstream of the reservoir, which begins once the reservoir is built and even several years beyond the useful life of the reservoir, requires an even bigger effort to maintain and sustain an ecologically sound river ecosystem. This would require not only a community-based stewardship but major government institutional support and financial investments.

Finally, the biggest challenge to multiuse water resource systems like the Laguna Lake and Agusan River basin is the complexity of managing these systems because of the scale and dynamics of physical systems with multiple water uses and sources, and consequently several stakeholders with varying objectives. For instance, in the Agusan River basin, the upper portion of the river system is Compostela Valley (about 350 km from Butuan City, the downstream end), which, with its mining operations is known to pollute the river. The National Irrigation Administration diverts water around Verula, which modifies the wetting and drying wetland of cycles of the Agusan marsh (150 km from downstream end). Finally, the overall effects, including water supply and especially flood control infrastructure in Butuan City, modify tidal excursion and saltwater intrusion into the mouth of Agusan River and affect the flora and fauna of mangroves of Butuan Bay. Thus managing Agusan River basin in particular must face all this multiplicity of issues and concerns and this poses a major challenge to governance and institutional synergy and coordination.

8.5 Policy Recommendations

8.5.1 On Understanding the Hydrologic, Geomorphologic, and Ecologic Interactions for Effective Water Resource Management

It is important to recognize the temporal and spatial scales of fluctuations of the various processes involved in the hydrology, geomorphology, and ecology interactions. For example, flood flows with accompanying high sediment loads can abruptly change the river morphology in a few hours or few days. This is in contrast to geomorphologic changes due to bioturbation by burrowing organisms, which could take several months or years to result in significant lakebed changes. Another example is the impact of the disappearance of wetlands due to river water diversion or climatic drought, which could take several years to realize that the wetland flora and fauna are also disappearing. In fact, one could already be experiencing the disappearance of the wetland and its associated flora and fauna and it is too late to realize it.

In this regard, the recommendation is to develop a framework and set up an institution that can continuously monitor and periodically assess strategies to plan and manage multiple-use and/or multipurpose major water resources with computerized decision-support system (DSS) that is capable of describing and capturing the hydrologic, geomorphologic, and ecologic interactions of these systems. It should be understood that while a science-based, computerized DSS is needed when dealing with complex, large-scale and dynamic water systems, it is not to replace humans (i.e., stakeholders and actors) in the ultimate planning and management decision making.

8.5.2 On Sustainable Planning and Management of Multi-purpose Reservoir Systems

There is a need to develop and implement a framework to conduct sustainable development of large-scale water resource systems that involve large-scale reservoirs and dams, including large, regulated river works so that the planning horizon is over 150–200 years. Definitely, this will involve scenario setting and evaluation of future conditions (i.e., climate, land use, socioeconomic and political structure) and physical performances of alternative configurations, dimensions, and components that will constitute a multipurpose reservoir system to ensure sustainable water resource development, including management and operations.

Furthermore, as a policy, it is definitely worthwhile to require feasibility studies related to water resource development. Construction of large dams must consider the long-term (150 years and even beyond) technical, socioeconomic, and environmental benefits and impacts of the said project. It may be noted that while the life of reservoirs can be 70–100 years (due to sediment deposition), the impact on river sedimentation (erosion and deposition process) downstream of the reservoir can begin once the reservoir is built and several years beyond its useful life.

There are possible measures or management strategies that can be adopted to minimize externalities such as the proper location of reservoir site, watershed erosion control, and reservoir sedimentation management strategies that include sediment flushing or sluicing.

8.5.3 On Strengthening Institutional Synergy and Coordination

For large-scale, complex and dynamic multi-use water resource systems like Laguna Lake and Agusan River basins, managing these systems require an understanding of the physical system (natural processes, climate, weather), social system (societal, political, economic), and human system (cultural, behavioral, lifestyles). Since there are definitely various stakeholders as well as private and government institutions that are involved in the governance of such multi-use water systems, there is the need to strengthen the synergy and coordination of these stakeholders and institutions to efficiently and effectively manage them. For this purpose, it is strongly advocated that a transdisciplinary approach be employed, one that utilizes scientific and technological tools to address the physical system; socio-political-economic studies to address the social system; psychological and cultural studies to address the human system; and, most importantly, engages stakeholders to solve problems through integrated and collaborative learning, research, and consensus building.

References

CARBDP-PMO (Cotabato-Agusan River Basin Development Project-Project Management Office). (2003). *Lower Agusan development project, stage I, phase II: Study on watershed management and FWS of the Agusan River Basin. Final Report for DPWH, Nippon Koei*, in association with TCGI Engineers and PKII Engineers. (unpubl.)

Catalan, Z. (2000). Laguna Lake–biological environments. In *Integrated Manila Bay/Laguna Lake and surround watershed environment study*. Tokyo: Japan Society for the Promotion of Science, Tokyo Institute of Technology. (unpubl.)

CTI-Halcrow-Woodfields. (2008). *Master plan for the Agusan River basin project. Final report to ADB-DENR-NWRB, technical assistance report no. 4552-PHI, Manila, Philippines.*

Hassall Associates. (2003), *Feasibility study report: Logum and Baobo subprojects. Report to the Southern Philippines Irrigation Sector Project (SPISP)*. Quezon City: National Irrigation Administration. (unpubl.)

Jose, A. M., Sosa, L. M., & Cruz, N. A. (1996). *Vulnerability assessment of Angat water reservoir to climate change, Water, air and soil pollution* (pp. 191–201). Dordrecht: Kluwer Academic Publishers.

LLDA (Laguna Lake Development Authority). (n.d.). *Primer on the Laguna de Bay master plan.* Manila: Department of Environment and Natural Resources. (Website: www.llda.gov.ph).

Milhous, R. T. (1995). Changes in sediment transport capacity in the lower Gunnison River. Colorado, USA. In *Proceedings of Man's Influence on Freshwater Ecosystems and Water Use.* IAHS Publ. No. 230 (pp. 275–280). Wallingford: IAHS Press.

NHRC (National Hydraulic Research Center). (2002). *Flow instrumentation audit and water balance study of the MWSS headworks system. Final report – phase 2.* Quezon City: UP Engineering Research and Development Foundation, Inc.

NHRC (National Hydraulic Research Center). (2007). *Reservoir sedimentation study and management plan: San Roque multi-purpose reservoir. Final report to the National Power Corporation.* Quezon City: UP Engineering Research and Development Foundation, Inc.

Shen, H. W., Tabios III G. Q., & Harder, J. A. (1994). Kissimmee River restoration study. *Journal of Water Resources Planning and Management, 120*(3), 330–349. New York: ASCE Publications.

Solyom, P. B., & Tucker, G. E. (2004). Effect of limited storm duration on landscape evolution, drainage basin geometry and hydrograph shapes. *Journal of Geophysical Research, 109*, https://doi.org/10.1029/2003JF000032; web/internet.

Tabios, G. Q. III. (2003). Modeling imperatives of Manila Bay/Laguna Lake system. In *Proceedings of the Symposium on Environmental Issues related to Infrastructure Development* (pp. 43–52). Tokyo: Japan Society for the Promotion of Science-Philippine Department of Science and Technology, Tokyo Institute of Technology.

Tabios, G. Q., III. (2008). Hydrology and related ecology-based aspects of managing the Agusan Marsh. In J. H. Primavera (Ed.), *Proceedings of the First Agusan Marsh Scientific Conference* (pp. 21–31). Jakarta: UNESCO Jakarta Office.

Tabios, G. Q., III. (2016). Competing water uses of Angat Multipurpose Reservoir with increased domestic water demand under future reservoir sedimentation and climate change. In S. J. Banta (Ed.), *Water in agriculture: Status, challenges, and opportunities* (pp. 73–90). College: The Asia Rice Foundation.

Tabios, G. Q., III, & David, C. C. (2004). Competing uses of water: Cases of Angat Reservoir, Laguna Lake and groundwater systems of Batangas City and Cebu City. In A. C. Rola, H. A. Francisco, & J. P. T. Liguton (Eds.), *Winning the water war* (pp. 105–131). Philippine Institute for Development Studies: Makati.

Tabios, G. Q., III, & David, C. C. (2014). *Appraisal of methodology in estimating irrigable areas and processes of evaluating feasibility of NIA irrigation projects. Policy Notes No. 2014–13.* Makati: Philippine Institute of Development Studies.

Tabios, G. Q. III, Delgado Jr., V. M., & Valdes, T. T.(2007). *Impact of reservoir sedimentation on water yield and hydropower generation in San Roque Reservoir, Pangasinan, Philippines*. Paper presented at the International Conference on Hydrology and Water Resources Management for Hazard Reduction and Sustainable Development UNESCO-International Hydrology Program, 19–23 Nov 2017, Manila, Philippines.

Tabios, G.Q. III, S.C. Monsod, C. Arcilla, R.S. Momo, G.J. Rabonza, C.S. Salazar, M.S. Chiu, P.D. Nilo, P.M. Ordona, M.A. Loyzaga, & W. B. Connell. (2010). San Roque Dam during Typhoon Pepeng of October 2009. Report of Investigating Committee created Department Order DO2009-10-0015 of Secretary Angelo T. Reyes, Department of Energy, Republic of the Philippines. 104 p.

Vesilind, P. A., & Gunn, A. S. (1998, July). Sustainable development and the ASCE Code of Ethics. *Journal of Professional Issues in Engineering Education and Practice, 124*(3), 72–74.

Dr. Guillermo Q. Tabios III is a Full Professor at the Institute of Civil Engineering and Research Fellow of the National Hydraulic Research Center at the University of the Philippines Diliman (UPD). He holds Bachelor's and Master's degree in Agricultural Engineering from the University of the Philippines, Los Baños, and a Ph.D. in Civil Engineering from Colorado State University. Dr. Tabios teaches and conducts researches in stochastic and computational hydrology and hydraulics, as well as water resources systems engineering. He is an Academician of the National Academy of Science and Technology, a Regular Member of the National Research Council of the Philippines and a Member of the American Geophysical Union and the International Association of Hydro-Environment Engineering and Research. He formerly held positions as Chairman of the Department (now Institute) of Civil Engineering, UP Diliman (2004–2007), Director of the National Hydraulic Research Center, UP Diliman (2008–2015) and Board of Director of the Philippine National Water Resources Board (2008–2015). Prior to joining UP Diliman in 1996, he was a Research Associate of the Hydrology and Water Resources Program at Colorado State University, Fort Collins (1985–1987) and Research Faculty of the Hydraulic and Coastal Engineering at University of California, Berkeley (1987–1996).

Chapter 9
Sustaining Water Resources with Environmental Protection

Rex Victor O. Cruz

Abstract This chapter focuses on past and current policies (including programs) on environmental protection and how these policies facilitate or constrain the implementation of policy actions to promote the sustainability of water resources in the Philippines. Alongside water policies, policies on land use management and allocation, forestry, agriculture, natural resource management, and pollution control are examined with respect to its expected outputs versus actual outputs and impacts. Cases of synergy and conflicts of environmental and water policies and what attempts were made to achieve concurrently environmental protection and sustainable water resources are presented. The chapter concludes by drawing out policy recommendations from the best practices and lessons learned from past experiences. Specifically, key policies and programs related to sustainable agriculture and food security, forest and biodiversity conservation, land use planning and management, environmental impact assessment, climate change adaptation and mitigation, and soil conservation, among others, are examined. The National Integrated Protected Areas Systems (NIPAS Law), the National Greening Program, and the Solid Waste Management Act are some of the key policies and programs that are reviewed in this chapter.

Keywords Water resources • Environmental protection • Environmental policies • Policy implementation • Integrated area-based planning

---

R.V.O. Cruz (✉)
Environmental Forestry Programme, College of Forestry and Natural Resources, University of the Philippines Los Baños, College, Los Baños, Laguna, Philippines
e-mail: rocruz@up.edu.ph

A.C. Rola et al. (eds.), *Water Policy in the Philippines*, Global Issues in Water Policy 8, https://doi.org/10.1007/978-3-319-70969-7_9

9.1 Introduction

9.1.1 State of Ecosystems and the Environment

The country was once regarded as one of the world's important repositories of pristine tropical rainforests. When the Spaniards colonized the Philippines in 1521, about 90% of the country's land area or 27 million ha were covered with forests. In 1900, after the Americans took over the Philippines from the Spanish government, the country's forest cover was estimated at about 70% or 21 million ha. Between 1934 and 1941, forest cover further dropped to 17 million ha or 57% of the land area. By 1988, the country's forest cover has dwindled to about 6.5 million ha or approximately 21.5% of the total land area of the country based on the results of a national inventory conducted in 1982–1988 by the RP-German Forest Resources Inventory Project of the DENR Forest Management Bureau (FMB) (Table 9.1). All told, the total forest cover loss between 1934 and 1990 was about 10.9 million ha, equivalent to an average annual loss of 194,000 ha (FMB 1990).

Based on land use and land cover estimates in 2010, there are only some 6.83 million ha of total forest cover remaining, of which 6.1 million ha are broadleaved forests, 0.18 are coniferous forests, and 0.27 are mangroves (FMB 2010). Of the total broadleaved forests, 4 million ha are open canopy forests and 2.6 million ha are closed canopy forests. Regionally, the largest remaining forests are found in Region 2, followed by Regions 4-B, CAR, Caraga, and 8. Among the provinces, Palawan has the largest remaining forests as of 2010, followed by Isabela, Cagayan, Agusan del Sur, Quezon, Surigao del Sur, Apayao, and Bukidon. All provinces have at least 200,000 ha of forests. More than 6 million ha of forests are located in forestland (i.e., public land), while 0.6 million ha are in alienable and disposable (A&D) or private land. As of 2012, there are 15.8 million ha of classified forestland and 14.2 million ha of A&D land.

Table 9.1 Forest cover estimates through the years (various sources)

Year	Forest cover (million ha)	% of total land area of the Philippines
1575	27.5	92.0
1863	20.9	70.0
1920	18.9	64.0
1934	17.8	57.3
1970	10.9	36.3
1980	7.4	24.7
1990	6.7	20.7
2001	5.4	18.0
2005	7.2	24.0
2010	6.8	22.8

The rapid decline of forest cover is attributed largely to mutually reinforcing direct drivers that include forest harvesting and conversion of forests to agriculture and other uses. Deforestation continues to be induced by the expansion of farming into sloping forestland, grazing, mining, and settlement development. A study by Carandang (2008) revealed that about 7.9 million ha of forests were converted into non-forest uses between 1935 and 2003. Of this, around 47% were formerly logged-over forests that were taken over by upland farmers; 44% were directly converted into agricultural, settlement, and other non-forest uses; and around 9% were permanently damaged due to the construction of logging roads, logging camps, and sawmills. The ongoing conversion of agricultural lands to agroindustrial estates and real estate adds pressures on forestland where new agricultural areas are developed to compensate for the loss of cropland through conversion and land degradation. It is estimated that some 600,000 ha of agricultural land were diverted to other uses between 1991 and 2001 (Cruz 2015).

Indirectly, population and economic growth drive deforestation. Increase in population and economic activities increase the demand for wood, food, fuel, and land, which increases the risks of forests from being harvested, converted to other uses, and encroached upon (Carandang et al. 2013). The pressure of the growing population on forests is likely to be more pronounced in areas and river basins where the current and projected population density is high. This includes the Agus, Bicol, Buayan-Malungon and Cagayan de Oro river basins that have the highest basin population density, and the Agusan, Agno, and Mindanao river basins with the highest population density in forest areas. River basins with a low percentage of forest cover either have high population density in forest areas or high basin population density.

There are around 24 million people (World Bank 2002) living in the country's uplands, which put heavy pressure on the remaining forests through farming, fuelwood collection, charcoal making, and other related livelihood activities. This upland population is likely to increase as demand for more cropland increases and available arable land is exhausted (if they are not already exhausted).

9.2 Sector Framework

9.2.1 Environment and Water Nexus

Water is an integral part of the environment. The natural and anthropogenic environment influences the quality and quantity of water. On one hand, the elements of the natural environment (i.e., climate, soil condition, land use and land cover, including forest cover, and solid wastes and other pollutants) commonly act as direct drivers of water quantity and quality that ultimately determine the availability and sustainability of water. On the other hand, the elements of the anthropogenic environment (i.e., policies, institutions, economic conditions, and demography) are the indirect

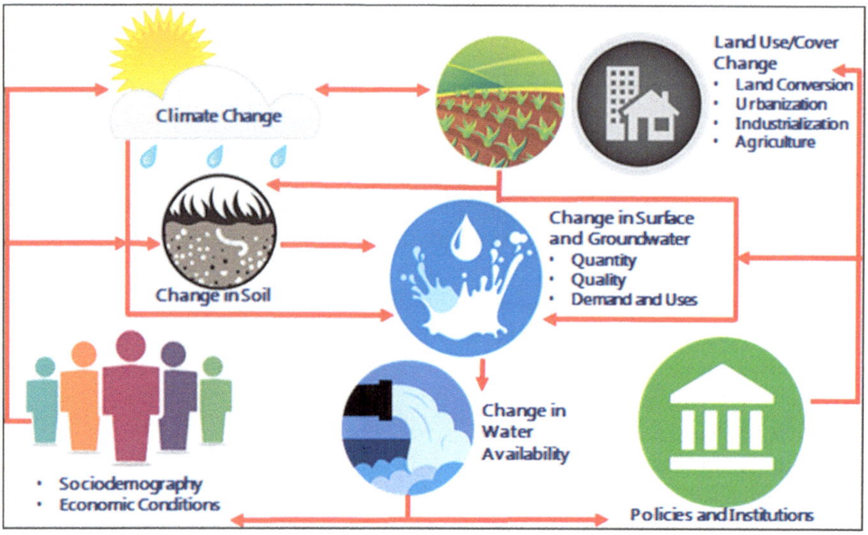

Fig. 9.1 Drivers of water availability

drivers (Fig. 9.1). Change in any of the direct and indirect drivers causes changes in surface and groundwater. Changes in the quality and quantity of water are directly attributable to changes in the direct drivers that are in turn caused by changes in indirect drivers. Among the direct drivers, land use/cover change is considered as the single most influential factor that affects the quality and quantity of water. It is even regarded to have greater impacts than climate change (IPCC 2007). Similarly, of the indirect drivers, policy is considered as the most influential since it affects the other indirect drivers (i.e., demographic, economic, and technological factors) as well as all the direct drivers, particularly land use and land cover.

The quality of policy is crucial to the protection of the individual elements of the natural and anthropogenic environment that must sum up to the protection of the diverse environmental services that are vital to human security. Sound environmental policy takes the integrated and interconnected nature of the various elements of the natural and anthropogenic environment. Policy formulation and implementation for the protection of water, land, forests, soil, biodiversity, and air, though distributed to various government agencies, must be guided by a unified framework that ensures maximum synergy and complementation and minimum conflicts and competition among various environmental policies. To achieve availability and sustainability of water, it is therefore essential that the policies specifically governing land use, forests, soil, and biodiversity are crafted with a holistic and overarching view of protecting the environment as a whole, preempting negative externalities while maximizing co-benefits on water. This will require interagency and multi stakeholder collaboration from the contemplation to the design and evaluation all the way to the implementation and monitoring of environmental policies. In this regard, appropriate institutional mechanisms must be in place to facilitate intramural and extramural policy vetting process across sectors and agencies. It is equally essential that the criteria and indicators of long-range vision for an environment commonly aspired for must be clear and consensually agreed upon by all sectors, agencies, and stakeholders. These criteria and indicators will be the limits and constraints to the pursuit of various policy goals. Likewise, proposed policies must be subject to scientifically rigorous analysis to ensure sound technical moorings and realistic assumptions, expected outputs, impacts and outcomes.

9.2.2 *Environmental Policies and Programs Affecting Water Resources*

Environmental policies and programs governing forest resource management, land use, soil conservation, conservation of biodiversity and protected areas, among many other policies, are not primarily meant to protect water resources. Nevertheless, proper design and implementation of these policies could promote the protection of existing forest cover and revegetation of deforested land; minimize soil erosion; enhance infiltration, recharge of groundwater and water retention capacity of watersheds; and reduce entrainment of solid wastes in streams that could have positive influences on water quantity and quality. Conversely, improper design and implementation of these policies could result in unmet sectorial goals and worse could have adverse impacts on water. Major environmental policies and programs that impact on water resources are considered in this chapter.

9.2.2.1 Policies on Forestry and Biodiversity Conservation

In general, policies governing forestry and biodiversity conservation (i.e., CBFM, IFMA, NIPAS Act, EO 23-2011; EO 26-11) determine the extent and quality of forest cover that in turn influence the quality of soil in relation to infiltration, surface runoff, and soil erodibility. Policies on forestry commonly aim to protect the integrity of the natural forests by regulating the use of forest resources to minimize disturbance of the natural state of the forests. Regulation is usually achieved through zoning forestland for production and protection uses, through prescription of uses of the forestland that are compatible with the designated zone, and through issuance of appropriate land tenure instruments that are consistent with timber production, agroforestry, grazing, and other revenue-generating activities.

Forestland Use Zoning Conceptually, zoning of forestland for production and protection uses is a tool commonly adopted to promote sustainability of forests and other interconnected ecosystem services including supply of freshwater within a particular landscape unit such as a watershed. Effective zonation is premised on an integrative framework wherein an ecosystem service (e.g., water supply) within a landscape unit is examined alongside other ecosystem services to ensure synergy and complementation of strategies and programs to achieve sustainability. Toward this end, the DENR adopted the Watershed Ecosystem Management Framework (WEMF) through DAO 99-01 (1999) that provides for the use of watershed as the management unit for forest management. Despite this policy however, forestland use zoning and management planning continue to be done outside the watershed framework. Consequently, land use zones are delineated with inadequate consideration of how it will impact ecosystems downstream and outside of the forestry domain and how it could contribute to the attainment of livelihood development goals and other economic development goals.

As practiced, forestland use zoning is supposed to be done through an integrative analytical process that is intrinsic to the WEMF. The determination of forestland use zones is supposed to be an integral part of a watershed wide (i.e., ridge to reef) land use-zoning process where the identification of areas for forest protection and production purposes are concurrently done with the identification of areas for agriculture, urban development, and for industrial and commercial purposes. However, the current forestland use zoning is more an intuitive than analytical process as it is based mainly on a checklist of parameters that define an area designated as production or protection zone. Most of these parameters are taken from the specifications of Presidential Decree 705 that is largely slope-based and often do not stand on solid scientific base. For instance, all forestland with a slope of at least 50% are to be designated as protected zones, even if there are areas with steeper slopes that have been sustainably used and areas with slopes less than 50% that are prone to degradation. Likewise, areas with slope between 30% and 50% are usually restricted to agriculture, when, in reality, there are areas in this slope category that are sustainably farmed. This led to forest zones that are challenging to implement as the land use zones inadequately reflect the true capability of the land as evidenced by large areas with uses that do not conform to forestland use zones. Table 9.2 shows that there are vast areas under slope categories of 30–50% and above 50% that are currently either being used for agriculture or are covered with grassland and brushland. It is reasonably possible that some of these areas that do not conform to the current zones are being used sustainably even as there are non-conforming areas that are under unsustainable uses.

The largely intuitive checklist-based forestland use zoning process is based on general knowledge that the eventual zonation of forestland will have positive impacts on the ecosystems and the communities inside a particular watershed without performing comprehensive assessment on how the natural and social systems in that watershed will likely respond to it. Knowledge on the potential changes in the natural and social systems in the watershed due to decisions based on forestland use zones is essential to enhance confidence and flexibility in deciding how a particular zone could be used for different purposes without the undesirable impacts on water and on downstream and adjoining areas.

Another weakness of the current forestland use zoning process is the inadequate if not total lack of participation of other sectors and agencies in deciding which areas are for protection and which areas are for production. As a result, land use zones are decided largely by the DENR that often fails to give adequate consider-

Table 9.2 Total area of agriculture, grassland, and brushland in various slope classes in the 18 major river basins based on 2010 land cover map of the Philippines (Cruz 2015)

	% Slope					
River basin	0–3	3–8	8–18	18–30	30–50	Above 50
Abra	28,416	66,048	97,657	70,425	26,567	1401
Abulog	13,259	20,450	27,719	20,043	6905	95
Agno	2061	5672	16,155	10,645	3003	130
Agus	500	966	1571	1436	204	9
Agusan	38,237	47,873	53,993	19,435	4506	28
Bicol	3559	4427	6022	1429	433	138
Buayan-Malungon	1527	2431	3838	2566	308	308
Cagayan De Oro	1554	8314	7095	1299	95	95
Cagayan	17,781	49,674	87,148	55,921	25,284	989
Davao	274	1742	4517	2807	773	10
Ilog Hilabangan	5566	16,088	16,862	4305	578	–
Jalaur	10,911	11,537	14,009	4438	828	–
Mindanao	7260	24,639	47,450	20,580	4950	117
Pampanga	3211	14,331	20,618	6906	1126	17
Panay	12,830	14,234	11,665	2802	438	22
Pasig-Laguna	10,291	20,198	22,826	6678	591	44
Tagoloan	690	3140	6557	3692	1341	50
Tagum-Libuganon	3106	8551	10,916	4299	834	–

ation of the relevant concerns of other sectors and agencies. In this situation, other sectors and agencies may find little motivation to adopt the forestland use zones and could encourage uses of forestland that do not conform to the designated zonation. When forestland use zoning is done concurrently with zoning for other major uses, (i.e., agriculture, urban development, and industrial and commercial uses) in a watershed landscape unit, the participation of all concerned sectors and agencies in deciding which areas of the forestland are for protection and for production uses, will be needed. Additionally, all concerned sectors and agencies that are co-owners of the forestland use zoning results are likely to accept and adopt forestland use zones as a legitimate planning consideration.

Forest Management As early as the late 70s and early 80s, the crucial roles of local communities, local government units (LGUs), and other stakeholders in forest management started to be recognized. Policies began to shift towards decentralization using various legal instruments to families, communities, LGUs, CSOs, and non-wood industry sector forest management functions previously entrusted only to wood industry sector. The intention to engage more stakeholders in forest management through devolution and the expected outcomes in terms of sustainability of forests are clear; yet, to this day, the extent and positive impacts of participation of local communities, LGUs, and other stakeholders remains low. The state, through the DENR, continues to retain jurisdiction over a much larger portion of the country's forest domain. Most of the protection areas is inside identified key biodiversity areas (KBAs) or proclaimed protected areas through the National Integrated Protected Area System Act (NIPAS). Majority of these protected areas are under the jurisdiction of the DENR while small proportions are under the stewardship of local and indigenous communities. Similarly most of the production areas is still under the jurisdiction of the DENR. As of 2012, only around 2.9 million ha of the more than 15 million ha of forestland are consigned to local communities and private sector, of which 55% are under community-based forest management agreements (CBFMA) and 34% are under integrated forest management agreements (IFMA).

The promise of enhancing the sustainability of the country's forests through democratized and participatory management involving the local communities, LGUs, and industry sector is affirmed by success stories, albeit in limited and isolated cases. Guiang et al. (2001) cited field reports showing CBFMA areas are generally better-off in terms of forest cover conditions, protection of forests against illegal users, resource use rights, and income generated from reforestation contracts. Better forest cover condition through reforestation and agroforestry means better soil condition—i.e., improved porosity, permeability, and infiltration capacity, which could enhance infiltration during storm events and increase groundwater recharge. Enhanced infiltration could also reduce siltation of rivers, lakes, and coastal and marine areas with the reduction in surface runoff that accelerates surface soil erosion. Therefore, if CBFM is successful in improving forest and vegetation cover in degraded upland farms, open areas, grassland and brushland, the quality and reliability of water resources could improve immensely.

The improvement of the condition of the forest cover and of the associated ecosystem services from the democratization of access to forest resources by local communities and industry sector is yet inappreciable due mainly to unstable policies and inadequate guaranty to capture benefits arising from its investments in forest management. Many proponents are deterred by a sudden policy change that reduces if not altogether takes away the right to harvest established tree plantations. In the aftermath of the devastating landslide and flood disasters in 2004, public outcry denouncing deforestation and illegal logging as the supposed cause of the disasters prompted the government to cancel all RUPs of existing CBFMAs and suspended the cutting of trees in IFMA areas. This change in policy happened despite the fact that there were no scientific evidences to prove public allegations.

Further, the exorbitant transaction costs incurred in navigating through the circuitous bureaucratic red tapes turn away well-meaning local community partners and investors. To illustrate, a proponent of tree plantation development in forestland will need to submit a long list of documents, including evidences of financial capability, ECC, tree plantation development plan, clearances from the LGUs, and FPIC from NCIP for areas inside ancestral domain, among many others, that could take months or, in worst cases, years to comply. Private corporations may have the capacity to persevere in complying with the requirements, but local communities are usually unable to comply unless provided with assistance such as by an NGO or an academic institution.

Forest Restoration and Protection Public support for forest restoration and protection is challenging to generate due to the intangible nature of benefits that arise from these activities. Hence, the government is usually left with most of the burden to do forest protection and restoration as more stakeholders prefer to engage in forestry projects wherein financial benefits may accrue to them in palpable measures. However, it is also a common fact that the government alone cannot successfully undertake forest protection and restoration unless there is substantial participation of key stakeholders, including the local communities, LGUs, NGAs, private sector, and civil society. There are several interrelated measures that have the potential to win the support of the stakeholders. First is the conduct of massive transformative IEC that aims to inform and educate in ways that will reform the mindsets of stakeholders and change the way they value the forests in terms of not only the financial but more so the intangible benefits derived from them. There is a need to propagate and disseminate knowledge on the total value of the forests with respect to improving the quality of life and promoting security against natural disasters associated with degraded forests. This will require growth in scientific evidences in support of the proper understanding of the roles and values of forests through increased investment in comprehensive forest monitoring and research. Relatedly, it will be crucial to institutionalize integrated area-based development planning such as using the watershed approach where forestry development planning are imbedded and concurrently prepared with the development plans of LGUs, agriculture, and other sectors. Forest protection and restoration are commonly seen as counter income generation, hence it will be needful to include in the menu of measures the provision of opportunities to earn additional income if not to compensate for the opportunities to earn income that will be lost. Through the above measures, the stakeholders are enabled to appreciate the forests as directly relevant to their respective interests, which become their incentives to make forest protection and restoration their own business.

On top of the NIPAS Law of 1992, the Philippines has three other key policies on forest protection and restoration. DAO 24-1990 and EO 23 of 2011 are on logging ban and protection against illegal logging, respectively; EO 26 of 2011 is on the National Greening Program. Results of the latest assessment of protected areas (PAs) declared under the NIPAS Law show that the limited support of LGUs is one of the constraints to the effective protection of PAs from encroachment, poaching, and other illegal activities (Guiang and Braganza 2014). A study of Cruz et al. (2013) on PAs proclaimed as ASEAN Heritage Parks (AHPs) reported that the major threats to the AHPs include illegal logging, poaching, illegal wildlife trade, illegal fishing, and illegal extraction of non- traditional forest products (NTFPs), followed by conflicts on land use and tenure, especially encroachment and expansion of settlements and agriculture. The report also cited key factors underlying the threats to effective AHP management: limited human resources (number of staff, expertise of staff, including number of park rangers) and budget, followed by lack of collaboration with other institutions, poverty among and inadequate livelihood opportunities for the growing communities within and around the AHPs.

Assessment of the total logging ban reveals that the remaining forests in all but 2 of the 32 timber license agreements (TLAs) were destroyed shortly after suspension, cancellation, or expiration of agreements (FDC 1996). This was attributed to the forest protection vacuum created upon the departure of TLA holders. This is reflective of the inadequate vetting process of the logging ban and other related policies that failed to anticipate and set in place preemptive measures to avert the threats to the forests once the TLAs pull out of their concession areas. The absence of initiative among LGUs to assist in the protection of the abandoned TLA areas contributed to quick deforestation in those areas.

The lack of support of the LGUs in enforcing the NIPAS Law and logging ban could be due to its limited appreciation of the real values of the forests to their local economy and security. It could also be attributed to the absence of a real sense of ownership among LGUs of the forest protection and development plan that is prepared and implemented independently from the local development plan. Further, the disconnection of forest protection and development plan from those of other sectors and agencies hinders the buildup of its sense of ownership of and the flow of its support to forest protection and restoration.

Based on initial reports of accomplishments, the NGP has been ahead of its planting targets, but whether this will actually translate into fully established plantations remains to be seen (DENR 2016). For this to happen, the sustained support of local communities, LGUs, and other sectors will be vital. Again, in this case, the lack of sense of ownership by and support from local communities, LGUs, and other government agencies could be a problem, with the emergence of NGP outside of the local development plans and development plans of other sectors and agencies.

9.2.2.2 Policies on Agriculture Development

The agriculture sector affects ecosystems within a watershed or other landscape units through the introduction of exotics and alien species, conversion of natural habitats, clearing of natural vegetation, use of erosive farming practices, and pollution from pesticides and fertilizer residues (Lepers et al. 2005; MEA 2005; Haines-Young 2009). Various agricultural activities could alter and degrade soil, water, and biodiversity that lead to deterioration of supply of ample water, regulation of climate, and other ecosystem services (Ramankutty 2010 as cited in UNEP 2014; Sala et al. 2009; EEA 2006). Many rivers in the country are already heavily loaded with sediments, nitrogen and phosphorus largely from runoff that originate from farmland. To reduce the impacts of agriculture on water, soil, and ecosystems, integrated area-based planning is essential where agricultural development planning is done together with the development planning for other sectors. Integrated area-based planning will facilitate proper location of areas to be cultivated, adoption of appropriate farming technologies and regulation of the use of harmful farm inputs through which economic returns from farming are optimized and adverse environmental impacts are kept within acceptable levels. This is easier said than done, considering that commodity-based planning has been the traditional mode of agricultural development planning in the country.

Commodity-based planning in agriculture is single product-focused, where more often than not, development planning for one crop is done separately from development planning for other preferred crops. As a result, the various commodity-based plans in agriculture have little synergy and often are competing for land and other production capital. Depending on the amount and kind of incentives being offered, the crops that get more financial and technical support are planted more widely than crops receiving less support. The downside of this is that, even areas that are not adequately suitable in terms of potential productivity and potential environmental impacts are planted with crops that get plenty of incentives. This is aggravated by the inadequate institutional capacity and political resolve of the LGUs to regulate the use of unsuitable areas for crop production. Additionally, the ambiguous authority of the LGUs to reclassify lands up to 15% of its total land area further accelerates the spread of crop production in areas not supposed to be farmed. The result of this is illustrated by the unregulated spread of coconut plantations even in areas with marginal production potential and worse in sloping areas from where soil could be easily eroded and that could pollute rivers, lakes, and other bodies of water (DASWCCO AMIA-NCCAG 2017). Another example is the incursion of plantation of high-value crops such as banana and pineapple into forestland in Mindanao. Crop production in unsuitable areas, especially areas that are marginally fertile, often requires large amounts of fertilizer inputs to boost production, but, in doing so, would increase the amount of nitrogen and phosphorus residues that are carried away into the rivers by surface runoff.

The passage of AFMA in 1998 trumpeted the earliest formal attempt to adopt an integrated approach to agricultural development. The AFMA provides for the delineation of SAFDZs and the preparation of integrated development plans (IDPs), together with the integration of programs and targets of various DA units and integration of DA programs and targets with those of the other agencies of government. It was also meant to ensure the balanced pursuit of production and environmental protection targets. However, the implementation of this provision of AFMA has long been held in abeyance with no significant progress toward the implementation of integrated development planning (Cruz 2016).

In 2013, the Secretary of the Department of Agriculture issued a memorandum entitled "Mainstreaming Climate Change in DA Programs, Plans and Budget." This memo calls for the shift from the traditional commodity-based planning framework to an integrated approach within the context of a watershed that calls for united efforts of various DA units and integration of DA plans and programs in synergy with those of other agencies and sectors. The DA, through the Adaptation and Mitigation Initiatives in Agriculture (AMIA) Project is pushing for the adoption of integrated development planning process using the landscape (i.e., watershed) as the planning unit to promote the resilience and sustainability of the agriculture sector as well as those of other sectors. It is expected that the committed implementation of an integrated approach to planning could improve the identification of areas where planting of certain crops will be most profitable and the use of agricultural technologies (i.e., tillage, fertilizers, and pesticides) will be least harmful to the environment, especially water resources.

Organic agriculture is also being pushed by the DA. The practice of organic agriculture is designed to promote soil conservation, raise farm productivity, reduce water pollution and other adverse environmental impacts of agriculture, minimize natural resource degradation, and protect the health of farmers and ecosystems. Evidences of economic benefits from organic agriculture though mostly positive are yet limited, while the environmental impacts particularly on water are varied (Tuomisto 2012) and need to be studied further. For now, the limited veritable proofs on the profitability and environmental benefits of organic farming will likely continue to limit the adoption of organic farming.

9.2.2.3 Policies on Land Use Management

Land use dynamically changes through time, driven by the growing demand for food, housing, and other basic commodities that need land for production. Land use change happens through land conversion, clearing of natural vegetation, soil erosion, land degradation, and urbanization. It is virtually impossible to stop land use change, but it is possible to regulate it in order to minimize its adverse impacts (Cruz 2016). In the country, land use change is largely unplanned not because of the absence of regulatory measures but because of the weaknesses in the measures that are notable in comprehensive land use planning (CLUP). The preparation and implementation of CLUPs by the LGUs is provided for in Executive Order No. 72 pursuant to the Local Government Code of 1991. The CLUP is a 10-year land use plan that is intended to guide the orderly and harmonized use of land within the boundaries of an LGU. The development of CLUPs is framed within key integrative and inclusive mechanisms, including the watershed-based ridge to reef landscape framework and participatory approach (HLURB 2013). It is expected that, through CLUP, land use in the LGUs will flourish without degrading the land and the local environment and without compromising human security against natural and human-induced disasters. Human settlements, industry and commerce, agriculture, forestry, and other major uses are supposed to be properly located through CLUP in areas where soil, water, biodiversity, and other resources will not be degraded and where there are no significant risks to life and property.

Based on a number of indicators, including the growing number of informal settlers especially in Metro Manila, the remaining pervasive poverty, the large number of people who live in areas exposed to geo-hazards, unregulated land conversion, siltation, and pollution of water bodies, CLUPs do not appear to significantly contribute to addressing these problems. Corpus (2013) cited reasons for lack of positive impacts of CLUPs on human development: inadequate capacity of LGUs to prepare robust CLUPs and implement these effectively; disjointed land use planning-investment planning and budgeting processes; failure to balance supply and demand for land; absence of mechanisms to integrate and harmonize CLUPs across LGUs; and aversion of LGU executives with short 3-year terms to commit to long-term investments required in implementing CLUPs. Likewise, PHDR (2013) cited that the reactive nature of land use planning in the country makes it difficult to regulate land use change. The LGUs have the impossible task of sorting through the many plans, yet, the reality is that, most (if not all) LGUs are inadequately endowed with expertise and financial and technological resources needed to ensure proper implementation.

9.3 Sector Performance Assessment

Environmental protection, as provided for in various policies and programs, is bound to have a positive impact on water resources, owing to the inherent connection of water with other natural resources and the environment in general. In a watershed from where freshwater supply commonly emanates, soil, climate, forest cover, land use, and land use practices largely influence the attributes of water. Normally, watersheds with undisturbed soil, unperturbed climate, intact forests, biodiversity and ecosystems, and regulated use of land and other natural resources will have regulated streamflow and quality of water. Policies and programs that are designed to maintain the healthy state of watersheds and all resources therein will naturally amount to the protection and conservation of water. This is the general expectation from the proper implementation of selected environmental policies and programs described in the previous sections. However, this expectation is largely unmet due to a number of factors that constrain the effectiveness of the environmental policies and programs.

The key issues and concerns related to policies and programs that impact on water resources are summarized in Table 9.3. Environmental policies have, thus far, limited positive impacts on water for three reasons: asynchrony and disunity of policies, ineffective policy implementation, and inadequate infusion of science in policy formulation. Briefly these reasons are discussed, along with some potential measures to address the deficiencies highlighted.

9.3.1 Asynchrony and Disunity of Environmental Policies and Programs

A watershed is made up of interconnected physical, biological, and social systems, including ecosystems. Any change in one part of the watershed such as a change in soil, forest cover, land use, and other human activities will invariably alter water and the other parts of the watershed. It is therefore appropriate that environmental policies and programs take on the interconnected character of the watershed systems for optimum synergy and maximum positive impacts. Most of the current environmental policies and programs, however, are nowhere near being properly linked to one

Table 9.3 Summary of environmental policies that impact on water resources

Policy	Desired outcomes	Impacts on water	Constraints
CBFM (Community-Based Forest Management)	Protection of OGF (Old Growth Forests)	Protection of recharge areas	Inadequate incentives (insecurity of tenure)
	Sustainable use of production forests	Improvement of water retention capacity of watersheds	Limited and insecure opportunities to capture benefits from CBFM areas
	Sustainable livelihood	Minimized siltation of water bodies, irrigation systems and reservoirs	Marginal participation of LGUs (Local Government Units) and lack of connection of CBFM to local development plans and to related development plans of other government agencies
	Progressive communities		Logging ban is a disincentive to protect forests
IFMA (Integrated Forest Management Agreement)	Protection of OGF		Inadequate incentives (insecurity of tenure, tenuous and unpredictable policy shifts);
	Sustainable use of production forests		Inadequate monitoring of compliance to regulations
	Progressive wood-based industry		Circuitous process of starting up tree plantation projects
	Sustainable livelihood		
Total logging ban	Protection of remaining natural forests		Inadequate capability for enforcement
			Inadequate measures to compensate for the lost opportunities of local communities and other stakeholders to earn income due to the ban
			Inadequate policy vetting process to anticipate vacuum in forest protection and other problems
NGP (National Greening Program)	Restoration of deforested areas and degraded forests		Inadequate assessment of forest restoration needs within the context of a watershed or landscape unit
			Absence of comprehensive monitoring
			Inadequate participation and support of LGUs and other stakeholders
			Lack of aggressive transformative IEC (Information Education Communication) to enhance correct understanding of the tangible and intangible values of the forests and to reform mindsets and values

NIPAS (National Integrated Protected Areas System)	Protection and conservation of biodiversity and habitats	Protection of recharge areas	Skewed implementation bias on biodiversity and habitat protection but little attention to water and other ecosystem services
		Minimized siltation of water bodies, irrigation systems, and reservoirs	Limited opportunities to capture other benefits from the protected areas
			Limited participation of local communities and LGUs due to low sense of ownership of PAs (Protected Areas) resulting from disconnect of PAMPs (Protected Areas Management Plan) and local development plans
			Limited budget
Organic agriculture	Reduced dependence on the use of inorganic inputs in crop production	Minimized pollution of water bodies	Low acceptability and adoption Inadequate knowledge and information on the economic and environmental benefits from organic agriculture
AFMA (Agriculture and Fisheries Modernization Act)	Sustainable crop production growth Sustainable natural resources	Minimized pollution of water bodies Efficiency in the use of agricultural water resources	Inadequate implementation of integrated approach to agricultural development planning
			Lack of effective system for monitoring impacts of agriculture on the environment, including on water
			Marginal role and participation of LGUs
			Lack of understanding on and skills to do integrated area development planning
AMIA (Adaptation and Mitigation Initiative in Agriculture)	Promote resiliency and sustainability of agriculture amid climate change through integrated landscape approach to agriculture development planning	Secured availability of irrigation water	Unpreparedness of various sectors and agencies to collaborate in a landscape approach to development planning
		Protection of water resources from pollution	Several existing sectorial policies may hinder intersector or interagency collaboration
		Protection of watersheds from land degradation	Inadequate knowledge, information and expertise
CLUP	Promote the sustainable and equitable use of land for various purposes	Minimized pollution of water bodies	Fragmentation of development plans across sectors
	Regulate land conversion	Efficient use of water resources	Too many plans that LGUs must prepare and /or implement
	Minimize adverse impacts of land uses on the environment and natural resources	Protection of watersheds from land degradation	Absence of effective monitoring of land use and its impacts on water and other natural resources
			Inadequate knowledge, information, and expertise

another. For instance, the CBFM, NIPAS, and total logging ban are inadequately linked even if their implementation overlaps in many areas. Usually, this leads to redundancy of investments, contradicting actions and outcomes, and inefficiency in the use of resources. "This requires discarding "business as usual" where interdependent interventions are carried out as disparate, parallel activities by several national line agencies, or programs within a line agency" (PHDR 2013). There is also a need for geographical convergence of plans, programs, and activities of various national government agencies and between national government agencies and LGUs to maximize effectiveness (PHDR 2013).

The perennial lack of interconnection of various environmental policies and programs is attributable primarily to the absence of institutional mechanisms to compel linkaging between the policies and programs of various government agencies, including those agencies with closely related mandates and functions. Sector-based and commodity-based planning and budgeting systems do not facilitate intersectoral and interagency policies and programs and perpetuate the silo mentality of various government agencies.

The absence of unified and integrated long-term vision and plans also makes it challenging for various government agencies to develop sectoral plans that are directed to a common outcome. To guide the trajectory of development in the country, the then MTPDP (now referred to as Philippine Development Plan (PDP) is developed for a 6-year period to coincide with the term of the president. At best, PDP provides guidance on how the short- to medium-term sectoral targets and goals could be achieved. However, PDP does not give a common and concrete long-term vision with physical indicators that will provide a stable reference for the short to medium term development planning that, regardless of who sits as president, one PDP is assured to build on the gains of the previous one and moves the country 6 year closer to a commonly desired vision. Further, it is not clear in the PDP how the various goals and targets of one sector are related to the goals and targets of other sectors; there is little guidance on how the pursuit of sectoral goals and targets could be harmonized for synergy and efficiency. The UNGSDR (2015) recognized the interconnectedness and interdependencies of the 17 Social Development Goals (SDGs) (Fig. 9.2) highlighting the fact that no single SDG can be achieved without successfully achieving the other related SDGs. In the same token, the attainment of one environmental protection goal in the country (e.g., water resource sustainability) will depend on the attainment of the other environmental goals and of other related development goals.

One of the keys to the concurrent achievement of several development goals is the adoption of the integrated area development planning approach wherein a land use framework is developed to serve as the common reference for the preparation of all sectoral development plans and other land-based development plans. A land use plan is commonly prepared using a planning unit with a defined physical boundary such as a watershed. The watershed is not only an ideal land use planning unit but is also a suitable unit for a thorough assessment of the potential and actual impacts of proposed or existing policies on water and on the other components of the watershed that interact with water. Such an assessment within a common watershed unit makes the points of contradiction and complementation of various policies lucid and guides policymakers to reach appropriate decisions.

9.3.2 Ineffective Policy Implementation

The AFMA, the Solid Waste Management Act, NIPAS, and CLUP are examples of policies with good and realizable intentions. But due to some constraints, they have had limited success in producing positive outcomes. There are three common constraints to effective implementation of these policies.

9.3.2.1 Limited Resources for Implementation

A common constraining factor is the limited and inefficient use of financial resources that are available for the proper implementation of environmental policies and programs. Associated with this constraint is the unavailability of adequate human resources and facilities that are necessary to enforce regulations and to execute programs of action. The implementation of most environmental policies, including CBFM and NIPAS, are funded through ODA, which usually is not enough and unsustainable. Funding from the national government is usually inadequate, except for the funds allocated to NGP that is perhaps the most generous and sustainable of all government investments on environmental protection. Resource limitation is no more acute than in the case of LGUs that are usually mandated to carry out substantial responsibility in the implementation of environmental policies. In AFMA and CBFM, LGUs play a significant role as extension service providers, yet they are not sufficiently endowed with the financial resources needed and usually are undermanned with properly trained extension personnel.

9.3.2.2 Absence of an Honest System for Monitoring Policy Implementation

There is usually no honest policy implementation without an honest system for monitoring. The pervasive absence of a system for effective monitoring of the performance of policy implementation and its impacts provides inadequate motivation to exercise due diligence in the formulation phase of the policy, especially in the vetting process where all the temporal and permanent ramifications of the policy once implemented are accurately assessed and where the bottom line of policy impacts is ascertained to fall within acceptable range. In addition, without monitoring, policy implementation is opaque to stakeholders and keeps implementers blind on the actual effectiveness and efficiency of the policies to achieve their objectives.

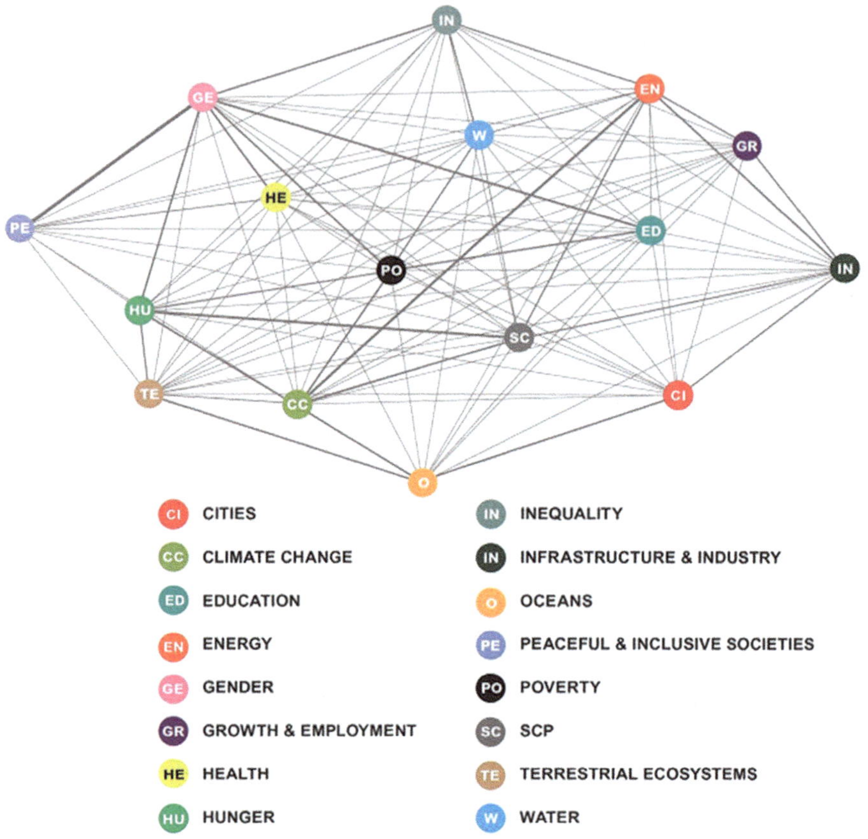

Fig. 9.2 Interconnection of the various SDGs (UNGSDR 2015)
Note: The thickness of the lines is an adjusted measure of the strength of interlinkages between goals. It denotes the number of links between two goals divided by the sum of targets under the two goals. SDG17 on "means of implementation" (which links to all other goals) was excluded from the analysis

As a result, policy weaknesses, implementation problems, and their ill effects are not corrected promptly, if at all detected. The logging ban in the late 80s and early 90s was implemented selectively by cancellation or non-renewal of agreements with erring logging concessionaires to protect the remaining natural forests from being destroyed. In the absence of a monitoring system, it was not possible to detect early on that, instead of preserving the remaining forests in the areas formerly consigned to private logging companies, the remaining forests were mostly destroyed fully due to the vacuum in forest protection created when the TLA holders abandoned their concession areas.

9.3.2.3 Marginal Role and Participation of LGUs in Policy Implementation

The LGUs are mandated, through the Local Government Code of 1991, to share with the national government in the responsibility to maintain ecological balance (RP 1991). Further, LGUs are mandated to be the comprehensive managers of natural resources within their jurisdiction. There is, however, a significant departure from these mandates of LGUs when it comes to actual implementation of national government-initiated policies on the environment. As noted in the implementation of CBFM, NIPAS, NGP, total logging ban, and AFMA, LGUs play minor roles and have minimal contribution to overall implementation efforts of the national government. This is attributable to the commonly limited financial resources that are given to LGUs to enable them to carry out their role in policy implementation. In addition, the marginal roles and responsibilities entrusted to LGUs are usually too minimal to make a significant contribution often because of their limited expertise and other institutional assets.

9.3.3 Inadequate Infusion of Science into Policy Formulation

Many environmental policies in the country are at best founded on knowledge and information that are too general and generated in areas with different circumstances. What is worse is that some of these policies are grounded on intuitions shaped by partial knowledge or truths, myths, and personal prejudices that give birth to flawed assumptions and unrealistic expectations. These are the consequences of the lack of access to correct information and knowledge that could be due to either absence of locally specific knowledge and information or absence of mechanisms to bring available knowledge and information to policymakers.

9.3.3.1 Inadequate Investment in Research

Insufficient investment in research limits the generation of basic knowledge and information, availability of innovative technologies, and building the capability of policy and decision makers, along with practitioners. The Philippines spends less than 1% of its GDP for research, way below the norm among more developed economies that usually spend at least 1–3%. This is the reason for the dearth of basic knowledge and information needed for well-informed and science-guided policies and management decisions.

Some of the critical gaps in knowledge and information are on the responses of forests and other ecosystems, biodiversity, and watershed functions and processes to changes in natural and human stressors including climate change, land use change, and changes in socioeconomic and political landscape. Systematic collection of long-term time-series data and monitoring of ecosystems, watersheds, land use, water, soil, biodiversity, and economic activities are needed to enhance the current knowledge and understanding on how water resources change amid the changes in the natural and anthropogenic environment. To this end, the establishment of a comprehensive long-term ecosystem and watershed observation network should be prioritized.

9.3.3.2 Weak Link Between Science and Policy

The large gaps in knowledge and information notwithstanding, there exist some practical opportunities to improve environmental policy formulation in the country if only the available knowledge and information, albeit limited, including the counsel of scientists and experts, could be made accessible and acceptable to policymakers. The problem is that there are inadequate efforts to translate into utilizable forms research results that mostly end up gathering dust on the shelves. Another problem is the heavy use of impersonal means of communicating research results to policymakers that, at best, is good only for raising awareness. Most research results are published in scientific journals that usually cannot prompt policymakers to take the desired action. Policy forums are also often used as venues for heralding research outputs, but these also have limited effectiveness in convincing policymakers to take the necessary steps. While impersonal means of communicating with policymakers are common, the scientific community seems to hesitate in resorting to the more personal means of communication. Personal experiences remind us that the more intimate means of communication is most effective to catch attention and to actually put across a message and eventually make an impact on the object of communication. Interpersonal communication may cause inconvenience on many researchers who are mostly apprehensive in dealing with policymakers or who are not skillful in engaging policymakers in dialogues, but it has to be integrated into the comprehensive IEC strategy so that policymaking could be enhanced through the uptake of knowledge, information, and scientific advice.

9.4 Recommendations

Sustainability of water resources is vital to and is dependent on the sustainability of soil, land, ecosystem services, human health, and livelihood. In general, environmental policies need to be integrated with one another and with related economic development policies in order to achieve the multiple but interdependent environmental protection and economic development goals, including those related to water, soil, land, biodiversity, ecosystem services, human health, and livelihood, among others. Below are the key preconditions to promote the integration of policies on water with other related policies.

9.4.1 Strategies to Promote Interagency and Multisectoral Collaboration

Interagency and multisectoral collaboration are both essential to the integration of environmental policies in order to achieve multiple outcomes efficiently and effectively. Interagency collaboration involves various national government agencies, while multisectoral collaboration involves national government agencies, LGUs, local communities, youths, CSOs, and the private sector.

Many potential measures may be undertaken to promote collaboration among different government agencies and other stakeholders, depending on prevailing circumstances. An important measure is to forge a shared vision, goals and objectives that can motivate the government agencies and other sectors to work together, to pool and share resources, to synergize, and to agree on strategic directions (Walker 2004; Mattesich et al. 2001 as cited by Majumdar 2006). This usually requires the concurrence and support of the head of state or of a superior government agency such as the Department of Budget Management or the National Economic Development Authority that may prioritize funding for interagency collaborative programs that aim to achieve interconnected sectoral outcomes. In other cases, international or local funding institutions can motivate collaboration among various sectors by giving priority grants to programs proposed by multisectoral groups.

Building high levels of trust and commitment among the various government agencies is another measure that can make collaboration effective. Other important measures that can promote successful collaboration include agreement on clear guidelines for sharing knowledge, information and other resources, distributing leadership, and sharing accountability (Dovey 2003 as cited by Majumdar 2006).

9.4.2 Institutionalized Long-Term Ecosystem and Watershed Monitoring

A comprehensive ecosystem and watershed monitoring system should be installed in all key watersheds, including the 18 major river basins. All existing monitoring systems managed by government agencies should be inventoried, mapped, assessed, and standardized. This will facilitate the quick buildup of time series of empirical databases. It could also catalyze the conduct of long-term studies that will generate information and technologies and improve current understanding on how ecosystems and watersheds change with the changing environment. An integrated knowledge management system should also be set in place to ensure that the outputs of the monitoring efforts and long-term studies are processed into forms directly utilizable to policymakers and other users. Likewise, mechanisms for data sharing and easy access should be installed. On top of it all is the need to provide secure and adequate funding for the unimpeded implementation of the activities described in this section.

9.4.3 Appropriate RDE Agenda Developed and Funded

The watershed approach to integrated development planning requires enhanced understanding of the dynamic interactions of various ecosystems and how these ecosystems are influenced by climate and human activities. Listed below are the priority thematic areas for research that are needed to improve the formulation and implementation of environmental policies. The priority thematic areas of research should include:

1. Interconnectivities and interactions of ecosystems in a watershed
2. Changes in ecosystems due to changing climate and human activities
3. Maximum and minimum thresholds for impacts of climate change and human activities
4. Accounting and valuation of natural resources and ecosystem services
5. Development of ecosystem/landscape models and simulation tools
6. Projection of future socio-ecological landscape conditions
7. Tradeoffs in changing from one land use to another

9.4.4 Accessibility of Quality Databases for Landscape-Based Planning

There is a need to enhance the management and integration of all existing databases in various government agencies and institutionalize a more liberal but secure system of access and data sharing. To the extent possible, databases should be consolidated and standardized according to themes as listed below.

1. Landform database to include all topographic, morphometric, land use, infrastructure, and land cover
2. Aquatic database to include all databases on freshwater, coastal, and marine ecosystems
3. Climate databases to include all databases collected from all types of meteorological monitoring stations maintained by various government agencies, academic and research institutions and private entities, climate scenarios from PAGASA, and international institutions
4. Geo-hazard databases to include all databases on floods, droughts, landslides, earthquakes, storm surges, tsunamis, typhoons, and others
5. Socioeconomic databases to include all demographic, economic, and political databases
6. Research databases to include all databases, information and technologies generated from all research projects funded by or through government agencies, including DOST, DA, and DENR

References

Carandang, A. P. (2008). *The forestry sector: Cost of environmental damage and net benefits of priority interventions. A contribution to the Philippines' country environmental analysis*. Manila: World Bank Philippines.

Carandang, A. P., Bugayong L. A., Dolom, P. C., Garcia, L. N., Villanueva, M. M. B., & Espiritu, N. O. (2013). *Analysis of key drivers of deforestation and forest degradation in the Philippines*. Bonn: German Cooperation (Deutsche Zusammenarbeit) and GIZ (Deutsche Gesellshaft fur Internationale Zusammenarbeit Gmbh).

Corpus, A. (2013). *Land use policy impacts on human development in the Philippines* (pp. 19). HDN Discussion Paper Series No.1. PHDR 2012-2013.

Cruz, R. V. O. (2015). *Framework formulation and study of spatial development, climate change, and the environment: Natural resources component (Consolidated Report)*. Pasig City: The Center for Sustainable Human Development, Development Academy of the Philippines.

Cruz, R. V. O. (2016). *DA – Strengthening implementation of Adaptation and Mitigation Initiative in Agriculture (AMIA Project 1) draft main report*. SEARCA-UPLBFI for the Department of Agriculture.

Cruz, R. V. O., Fernando, E. S., Bantayan, N. C., Castillo, M. L., Estoque, C. D., & Ata, J. P. (2013). *Scoping study on climate change and biodiversity of protected areas and key ecosystems in Southeast Asia*. Bonn: German Cooperation (Deutsche Zusammenarbeit), GIZ (Deutsche Gesellshaft fur Internationale Zusammenarbeit Gmbh), and ASEAN Center for Biodiversity.

DAO No. 99-01. (1999). *Adoption of the watershed and ecosystems planning framework*. Quezon City: Department of Environment and Natural Resources.

DASWCCO AMIA-NCCAG. (2017). Department of Agriculture-Systems Wide Climate Change Office. Adaptation and Mitigation Initiative in Agriculture, National Color Coded Agricultural Guide Map, Quezon City.

DENR (Department of Environment and Natural Resources) (2016). *Accomplishments of National Greening Program*. ngp.denr.gov.ph. Accessed 21 Apr 2017.

Dovey, L. (2003). *Achieving better social outcomes in New Zealand through collaboration: Perspectives from the United States*. Wellington: State Services Commission.

EEA. (European Environment Agency). (2006). *How much bioenergy can Europe produce without harming the environment?* EEA Report No. 7/2006. Copenhagen.

EO 2011-23. (2011). *Declaring a moratorium on the cutting and harvesting of timber in the National and Residual Forests and Creating the Anti-Illegal Logging Task Force*.

EO 2011-26. (2011). *Declaring an Interdepartmental convergence initiative for a National Greening Program*. Signed by President Benigno Aquino on February 24, 2011.

FDC (Forestry Development Center). (1996). *Assessment of cancelled/suspended/terminated timber license agreement (TLA) areas in the Philippines*. Los Banos: FDC, College of Forestry and Natural Resources, UP Los Banos.

FMB (Forest Management Bureau). (1990). *Master plan for forestry development in the Philippines*. Quezon City: Forest Management Bureau, Department of Environment and Natural Resources.

FMB (Forest Management Bureau). (2010). *Philippine forestry statistics*. Quezon City: Forest Management Bureau, Department of Environment and Natural Resources.

Guiang, E. S., & Braganza, G. C. (2014). *National management effectiveness and capacity assessment (NMECA) of protected areas in the Philippines* (p. 53). Bonn: Deutsche Gesellshaft fur Internationale Zusammenarbeit (GIZ) Gmbh.

Guiang, E. S., Borlagdan, S. B., & Pulhin, J. M. (2001). *Community-based forest management: Preliminary assessment* (203 pp). Quezon City: Institute of Philippine Culture, Ateneo de Manila.

Haines-Young, R. (2009). Land use and biodiversity relationships. *Land Use Policy, 26*, 178–S186.

HLURB (Housing and Land Use Regulatory Board). (2013). *A guide to comprehensive land use plan preparation 2013* (246 pp). Volume 1: The planning process. Quezon City: HLURB.

IPCC (International Panel for Climate Change). (2007). *Climate change 2007: Impacts, adaptation and vulnerability* (976 p). Contribution of Working Group II to the Fourth Assessment Report of the Intergovernmental Panel on Climate Change. Cambridge: Cambridge University Press.

Lepers, E., Lambin, E. F., Janetos, A. C., DeFries, R., Achard, F., Ramankutty, N., & Scholes, R. J. (2005). A synthesis of information on rapid land-cover change for the period 1891–2000. *Bio Science, 55*(2), 115–124.

Majumdar, D. (2006). Collaboration among Government agencies with special reference to New Zealand: A literature review. *Social Policy Journal of New Zealand Issue, 27*, 183–198.

Mattessich, P., Murray-Close, M., & Monsey, B. (2001). *Collaboration: What makes it work.* St. Paul: Amherst H. Wilder Foundation.

Millennium Ecosystem Assessment (MEA). (2005). *Ecosystems and human well-being: Synthesis.* A report of the Millennium Ecosystem Assessment. Washington, DC: Island Press.

Ramankutty, N. (2010). Agriculture and forests — Recent trends, future prospects. In *Linkages of sustainability* (pp. 11–31). Cambridge, MA: MIT Press.

RP (Republic of the Philippines). (1991). *Republic Act 7160 or the Local Government Code of 1991.* Manila: RP.

Sala, O. E., Sax, D., & Leslie, L. (2009). Biodiversity consequences of increased biofuel production. In: R. W. Howarth & S. Bringezu (Eds.), *Biofuels: Environmental consequences and interactions with changing land use.* Report of the Internatinal SCOPE Biofuels Project. Ithaca: Cornell University.

Tuomisto, H. L. (2012). "Does organic farming reduce environmental impacts?" – A meta-analysis of European research. *Journal of Environmental Management, 112*, 309–320.

UNGSDR (UN Global Sustainable Development Report). (2015). *UN Global Sustainable Development Report* (198 p). Advanced unedited version. New York: UN.

United Nations Environment Programme (UNEP). (2014). *Assessing global land use: Balancing consumption with sustainable supply* (131 pp.). A Report of the Working group on land and soils of the International Resource Panel. Bringezu S., Schütz H., Pengue W., O'Brien M., Garcia F., Sims R., Howarth R., Kauppi L., Swilling M., and Herrick J. Pari, France.

Walker, A. (2004). *Overcoming the neoliberal legacy: The importance of trust for improved interagency collaborative working in New Zealand.* Research Paper Number 11. Wellington: School of Government, Victoria University of Wellington.

World Bank. (2002). *Forestry strategy and appendices.* Washington, DC: World Bank.

Dr. Rex Victor O. Cruz is a Full Professor at the University of the Philippines Los Baños (UPLB). He obtained his bachelor and Master's degree in forestry at UPLB and his doctoral degree at the University of Arizona. His major fields of expertise are forestry, watershed management, environment and natural resources management, upland development, land use planning and climate change. The author is a member of the UN Intergovernmental Panel on Climate Change (IPCC) in 1992–1995; 1997–2000; and 2004–2007. Currently he leads three national programs, the National Research and Development Program for Watershed Management, National Conservation Farming Village Program, and the Monitoring and Detection of Changes in Ecosystems for Resiliency and Adaptation. He is a member of the Asia Pacific Forestry Network Board of Directors, and the National Pool of Technical Experts of the Philippine Climate Change Commission.

Chapter 10
National and Local Initiatives in Addressing Water Supply Sustainability

Guillermo Q. Tabios III, Rex Victor O. Cruz, Myra E. David, and Miriam R. Nguyen

Abstract This chapter discusses the national and local initiatives towards improving water management in the Philippines beginning with action plan to reform the National Water Resources Board (NWRB) in 1995, the development of the National Integrated Water Resources Management (IWRM) Plan Framework in 2006, followed by the Integrated River Basin Management and Development (IRBMD) Master Plan spearheaded by Department of Environment and Natural Resources (DENR) in 2007 to integrate past framework plans and development of strategies for sustainable river basin ecosystem management. In 2010, the National Economic Development Authority (NEDA) together with NWRB spearheaded the development of the Philippine Water Supply Sector Roadmap which recommended policy directions and institutional reforms to further strengthen NWRB and to organize a coherent, transparent and independent economic regulatory framework and as a follow up, NEDA with DOH developed the Philippine Sustainable Sanitation Roadmap in the same year. In 2011, the National Sewerage and Septage Management Program (NSSMP) was prepared by DPWH as mandated by the Clean Water Act. Despite all these framework and action plans, there is lack of strong leadership and coordination to ensure efficient and effective management of the country's water resources, since there are more than 30 agencies with water related powers causing overlaps

G.Q. Tabios III (✉)
Institute of Civil Engineering and National Hydraulic Research Center, University of the Philippines at Diliman, Quezon City, Philippines
e-mail: gtabios@up.edu.ph

R.V.O. Cruz
Environmental Forestry Programme, College of Forestry and Natural Resources, University of the Philippines Los Baños, College, Los Baños, Laguna, Philippines

M.E. David
Institute for Governance and Rural Development (IGRD), College of Public Affairs and Development, University of the Philippines Los Baños, Los Baños, Philippines

M.R. Nguyen
Community Innovations Studies Center (CISC), College of Public Affairs and Development, University of the Philippines Los Baños, Los Baños, Philippines

A.C. Rola et al. (eds.), *Water Policy in the Philippines*, Global Issues in Water Policy 8, https://doi.org/10.1007/978-3-319-70969-7_10

and fragmentation. This chapter ends with recent initiatives and recommendations on how to reform and transition from a fragmented and uncoordinated sector to one that is efficient, effective and sustainable in managing the country's water resources.

Keywords National and local initiatives · Water supply sustainability · National Water Resources Board · National Water Resources Management Body/Authority · Integrated Water Resources Management

10.1 Major Water Resource Programs and Past Initiatives

There have been numerous studies and proposals developed to help improve water resources management in the country. Some of the more important studies that have been considered are cited in Tabios and Villaluna (2011) as discussed below.

10.1.1 Action Plan for Reforms Relating to the National Water Resources Board (1995 NWRB Study Team)

This study was prepared in 1995 for the Government of the Philippines by a team of consultants financed by the Government of Japan with the World Bank acting as Executing Agency. The main recommendation of the study was to strengthen NWRB. The report has provided detailed recommendations to address fundamental needs in areas such as policy formulation, data collection and processing, national and basin planning and real-time management.

The study also proposed institutional and policy actions needed to make the technical actions effective. Recommendations fall into three categories: (i) those that can be subject of immediate Executive decisions; (ii) those that will require legislation; and (iii) those that require further study before decisions can be taken.

The study also made detailed proposals for the institutional strengthening of NWRB itself, including staffing, training, facilities and financing required to enable the same to respond effectively to the proposed reforms. Unfortunately, the recommendations of the study have not been given adequate attention and appropriate action. Seven immediate executive and administrative measures- one Presidential Executive Order, a NEDA Board Resolution and five NWRB resolutions- were proposed. However, only a few were actually operationalized. The study also proposed amendments to the laws. It also proposed to rename the NWRB to the Philippine Water Resources Authority to be headed by a Director General. It also provided some directions on the detailed studies and pilot projects that have to be done on important issues such as the full evaluation of alternatives for economic regulation. It also recommended steps, budgets, equipment and consultancy service components for strengthening NWRB.

10.1.2 JICA Master Plan Study on Water Resources Management in the Philippines (1998)

The JICA study was the first national master plan on water resources development and management focusing on the twelve (12) water resources regions (WRRs) of the Philippines. The study looked in to the water demand and supply balance of each WRR and major river basins. It also proposed measures to suffice the demand and to eliminate the water deficit. In addition to the basin level studies, it also examined the balance of water supply and demand in major cities where intensive water use is foreseeable due to high population density or economic activities.

Pending legislation on the proposed integrated planning and regulation of the water resources sector, the study supported the proposals to strengthen NWRB as an interim measure. In particular, it recommended the following:

- Attachment of NWRB to the Office of the President and later on to the DENR;
- Membership to NWRB to be limited to agencies or sectors that is responsible for policy in water resources;
- Creation of regional offices of NWRB to assume line functions delegated to deputized agencies and offices;
- Creation of a Legal Affairs Unit in NWRB;
- Provision of incremental staffing, training and equipment to NWRB; and,
- Improvement of the data collection network and establishment of a National Water Information Network (NWIN).

10.1.3 IWRM Framework Plan (2006)

In November 2006, a multi-sectoral task force developed the National Integrated Water Resources Management (IWRM) Plan Framework. The planning process was supported by the United Nations Environment Program (UNEP). The output was intended to be a directional plan to guide the different stakeholders involved in water resources management, at different levels, to either prepare their respective IWRM plans, update/enhance their existing IWRM-related plans or make IWRM an integral part of their development plans/programs.

10.1.4 Integrated River Basin Management and Development (IRBMD) Master Plan (2007)

The DENR spearheaded the preparation of the IRBMD Master plan in 2007 to integrate four principal frameworks and development strategies of sustained river basin ecosystem management, namely: integrated water resources management, integrated watershed management, wetland management and flood mitigation. These are further strengthened by three supplemental components: (i) water quality protection and monitoring framework; (ii) information and decision support systems; and iii) river protection and rehabilitation framework. The River Basin Coordinating Office (RBCO) is currently assisting 20 principal river basins by developing and implementing master plans and organizing integrated governance and management approaches and effective coordination and cooperation at all levels in the river basin.

10.1.5 Philippine Water Supply Sector Roadmap (2010)

Spearheaded by NEDA and the NWRB, the PWSSR identified the major challenges of the water supply sector and recommended some policy directions and institutional reforms to further strengthen NWRB and to organize a coherent, transparent and independent economic regulatory framework. As a result of the recommendations of the PWSSR, the SCWR was formed as an oversight body to oversee the implementation of the roadmap recommendations. Some projects emerged such as the Millennium Development Goal Fund (MDGF) 1919 Joint Programme of NEDA, DILG and NWRB with the United Nations Development Programme (UNDP) and the United Nations Children's Fund (UNICEF). This project supported a few policy studies relating to pro-poor water supply provision and capacity building efforts with the priority 36 waterless municipalities in five (5) regions.

10.1.6 Philippine Sustainable Sanitation Roadmap (2010)

As a follow-up to the PWSSR, the DOH, together with NEDA spearheaded the preparation of the Philippine Sustainable Sanitation Roadmap (PSSR). The PSSR was developed to provide an umbrella framework for the different government agencies with sanitation related mandates. Through this study, the Department of Health (DOH) affirmed its role as the lead sanitation driver and issued an Administrative Order (AO) declaring sustainable sanitation as a national policy. Sanitation for this study was limited to excreta management and urgent attention was raised to address the large number of Filipinos still practicing open defecation and mainstreaming of water, sanitation and hygiene promotion in disaster risk reduction and management.

10.1.7 National Sewerage and Septage Management Plan (2011)

The National Sewerage and Septage Management Program (NSSMP) was prepared by DPWH as mandated by the Clean Water Act. The program was recently approved at the NEDA INFRACOM Level where the most critical issue raised was the need to increase national government subsidy to motivate LGUs to invest in sewerage and septage management programs. The NSSMP proposed to support a number of highly urbanized cities (HUCs) to put in place a proper sewerage and septage management program. National Government (NG) cost share is critical to help cover capital costs and the study recommends 40% NG cost share to reduce investment cost to PhP246 M (P123 million each for LGU and WD).

10.2 Current Initiatives for Sustainable Water Supply

10.2.1 Initiatives to Promote Horizontal Unity and Integration of Water Policies, Plans and Programs Across Various Government Agencies

10.2.1.1 Proposed National Water Resources Management Body

There are pending legislations in the Congress of the Philippines relating to water which have been proposed. Since the Congress of the Philippines is a bicameral body consisting of the Senate and House of Representatives, a proposed bill must have two versions, the senate bill (SB) and the counterpart house bill (HB). Details of these proposed legislations may be referred to in the websites of Congress of the Philippines (Senate and the House of Representatives separately).

A particular proposal that is highlighted here is the National Water Resources Management Body that has been proposed in Tabios and Villaluna (2011, 2012) and also discussed by Rola et al. (2015). This has undergone several rounds of government and public consultations but unfortunately it is still pending either as an executive order as well as for legislation. The most recent legislative bill of this water resources management body is HB 00221 proposed by Rep. Linabelle Villarica in 2016.

This proposed body referred to as National Water Resources Management Authority or a water "superbody" is a new office to manage and protect the country's water resources for domestic water supply, sanitation, irrigation, hydropower, fisheries, aquaculture, flood control, navigation, and recreation including the enhancement of water quality, conservation of watersheds, control of water pollution and environmental restoration.

The organizational structure of this proposed body is the framework within which the offices arranges its lines of authority and communication and presents

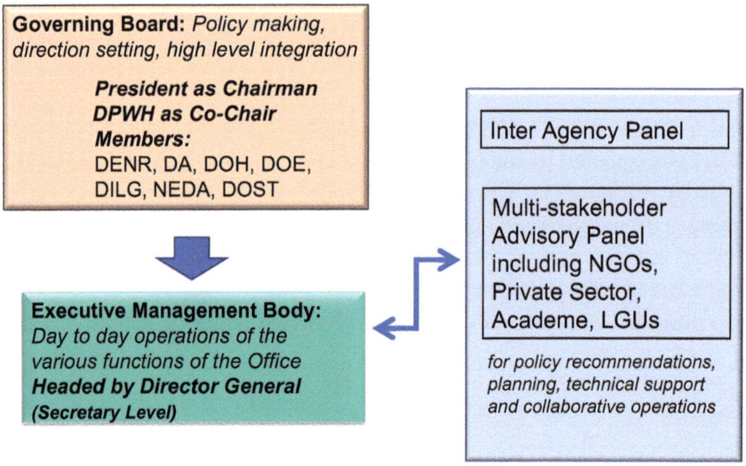

Fig. 10.1 Governing Board and Executive Management Body of the Proposed National Water Resources Management Body (From Tabios and Villaluna 2012)

detailed roles and functions of the different divisions and sections in order to achieve the objectives, mission and vision of the Office. Moreover, it determines the extent and manner to which the roles, powers and responsibilities are delegated, controlled and coordinated and more importantly tracks how information flows between levels of management, this case, from the national to the regional or local level of water governance. As shown in Fig. 10.1, the office will have Governing Board, an Executive Management Body as well as an Inter-governmental and Multi-Stakeholder Panel. The organic unit of the national office will have 10 divisions and several sections. The organizational structure follows a horizontal compartmentalization among the different divisions performing an array of technical functions including planning and policy, data collection and monitoring, scientific studies and decision support systems; infrastructure and program development, water facilities development and operations, resource regulations, economic regulations, water financing and economics, stakeholder relations and river basin organization and development (Fig. 10.2).

With the structure and function of the proposed water body, the coordination and functional relationships among the existing water-related agencies to the proposed water body are shown in Fig. 10.3.

In Fig. 10.3, the various government agencies including private organizations or companies that the water superbody will work with as partner agencies. It is recognized that there are existing agencies or organizations that have the expertise, experience and perhaps mandate to perform one or more of various functions of water management but specific to a particular water sub-sector or purpose or uses. For instance, the National Irrigation Administration (NIA) is the lead agency for irrigation while DPWH is the lead agency for flood control. In this case, the superbody will partner with these agencies to properly plan, design and operate the water

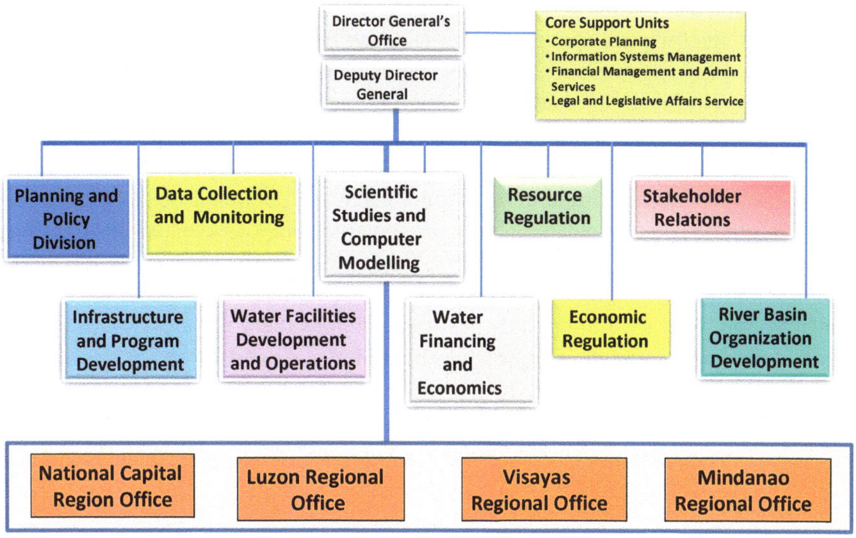

Fig. 10.2 Organization structure of the Proposed National Water Resources Management Body (from Tabios and Villaluna 2012)

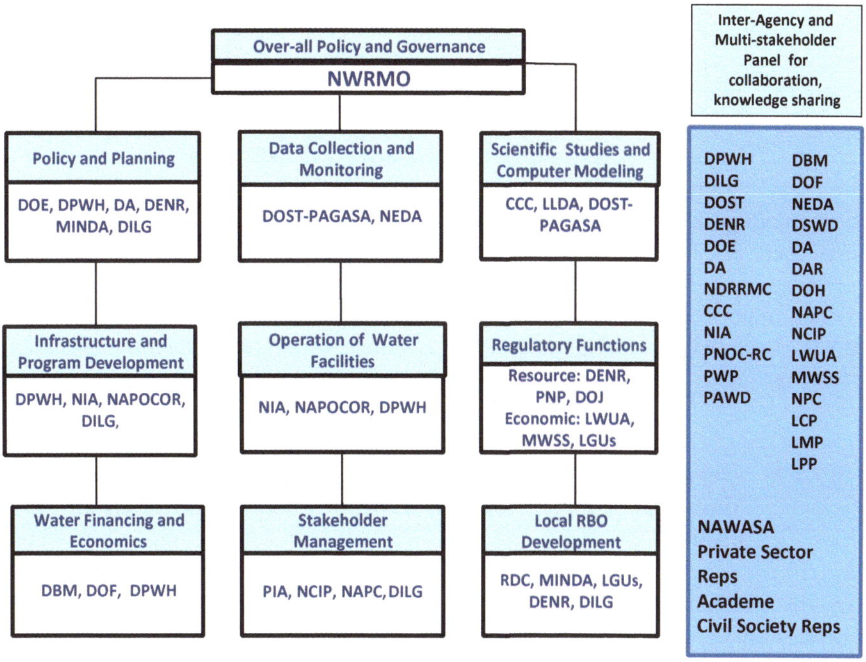

Fig. 10.3 Functional relationships of existing water-related agencies to proposed National Water Resources Management Body (from Tabios and Villaluna 2012)

system associated to these agencies. Since the partner agency may be only concerned with its own sub-sector or water use, the job of the superbody is to integrate and oversee among the various water sub-sectors to optimally plan, design and operate the natural or physical water resources recognizing the possible conflicting, competing or complementary uses of water.

To some extent, the water superbody may dictate or direct the various water partners or outsource or delegate to the partner agencies the planning, design or operation (including project implementation or construction of associated infrastructure) of the water resources system for the particular water purpose/s. Thus, it is important that the superbody must earn, gain and be recognized (through public relations and through legal instruments) as the water management authority, with a strong track record and reputation in order to be believed in and be an effective overall water manager. For the superbody to attain this stature, it must demonstrate that it is capable, able and trustworthy to perform the overall water management functions. Thus, the superbody needs a champion or champions, the right human resources (in terms of competency and manpower), machinery with associated infrastructure and significant funding to achieve and maintain its goals.

On a separate note, the water supply, sanitation and sewerage is a priority subsector that also requires a separate strong leadership support and some resources to enable the country to achieve universal coverage for safe water and adequate sanitation as soon as possible. The proposal to have DPWH as the lead agency is a rational and logical arrangement at this time since almost all the major water supply agencies are already attached agencies to this department. However, the Department should be able to set up an appropriate unit that will handle these concerns especially the strategy and action plan to achieve universal coverage in the soonest time possible. The key agencies relating to water supply, sanitation and sewerage have to be evaluated and possibly redirected to assume more pro-active, transparent and accountable roles in support of water supply, sanitation and sewerage service delivery. There is a proposal in Congress to have a separate water supply, sanitation and sewerage commission which is still in order but to clarify that its role is not envisioned to not be limited to economic regulation per se but should be linked to the goals of achieving sustainable universal coverage within the context of sustainable water resources management.

10.2.1.2 Preparation of Climate Proofed Integrated River Basin Management and Development Master Plan for the Major River Basins

By virtue of Executive Order (EO) Nos. 510i, 816ii and 50iii, a river basin control office (RBCO) was created to facilitate the development of a national master plan to minimize flooding damages and to ensure sustainable supply of water.

Pursuant to EO 510, an integrated river basin management and development (IRBMD) master plan was formulated in 2007 focusing on the eighteen (18) major river basins (with an area of not less than 100,000 ha) in the Philippines. On May 2, 2012, the Cabinet Cluster on Climate Change Adaptation and Mitigation passed Resolution No. 2012–001 declared the eighteen (18) major river basins as priority areas of concerns of the government. This is in consideration of the vital roles of these river basins as drivers of the country's economy. An important component of the IRBMD master plan is the preparation of master plan for each river basin. The river basin master plan is intended to address vital issues related to watershed conservation, river basin rehabilitation, flood control/mitigation, and water security for domestic, irrigation and industrial use, livelihood and economic development opportunities within the basin.

In compliance to the provisions of Republic Act (RA) No. 9729 or the Climate Change Act of 2009 and RA 10121 or the Disaster Risk Reduction Law of 2010 that were enacted in 2009 and 2010 respectively, the RBCO initiated the climate proofing of the existing master plans of the major river basins. The primary intention is to enhance the resiliency of the river basins by revising the master plans based on the new climate normal. Specifically, climate proofing entails updating the river basin master plans considering the implications of the new climate normals on: (1) water resources management, (2) watershed management, (3) flood control/mitigation, disaster risk reduction and hazards management, (4) wetland management (to include rivers, river deltas, marshlands and coastal areas), (5) economic development, and (6) institutional linkages and organizational structure for river basin management.

The climate proofing process consists of the evaluation of the strategies, programs and projects indicated in the master plans in terms of adequacy to address the assessed current and future climate related risks and vulnerabilities that impinge on the sustainability of water resources and other vital goods and services of the major river basins. Results of this evaluation were used to modify existing, and identify and design new strategies and projects that will enhance the adaptive capacity and resiliency of the river basins. The new strategies and projects that are commonly identified in the climate proofed river basin master plans include development of new water sources, improvement of land use and land use practices, protection of water quality through solid and wastewater management and soil conservation, forest landscape restoration, establishment of early warning systems and watershed monitoring systems, climate smart cropping systems, establishment of flood control structures and other infrastructure, enhancing the participation of stakeholders in river basin management, and enhancement of river basin governance structure among others (EIDRP-TGWGD 2016).

10.2.2 Initiatives to Promote Vertical Integration of Water Policies, Plans and Programs

10.2.2.1 The Lake Buhi Case

The case of Lake Buhi shows the unharmonized and overlapping activities/efforts of local and national institutions to protect watershed and water. Contained in the Buhi-Barit watershed, which is situated in Camarines Sur, Philippines, Lake Buhi occupies an area of 18 km^2 and with an average depth of 8 m (Caparas 2012). It is 105 masl and is surrounded on all sides by hills of more than 300 m high. Lake Buhi is the main source of water for the National Power Corporation Hydro-Electric Plant as well as irrigation of at least 100 km^2 of the downstream Rinconada towns of Camarines Sur. These towns are Buhi, Nabua, Iriga, Baao, Bula and Pili (Elazegui et al. 2016). It is the home of the world's smallest commercially harvested fish, the "*sinarapan*" (*Mistichthys luzonensis*). Moreover, the forest around the Lake is the sanctuary of more than 25 bird species.

Given the above-mentioned competing uses of the water in Lake Buhi, its integrity is however challenged due to increased fish cages, siltation, water hyacinth proliferation, among others (Elazegui et al. 2016). Binoya et al. n.d. (accessed 2016) reported of fish kill episodes, the worst occurring in 1998, with an estimated 25 million tilapia killed. Proliferation of fish cages was cited to be a major concern for the municipality of Buhi. Based on the report of Plopenio and Bimeda (2011), the fish kill in November 2010 was attributed to the fish cages covering nearly 80% of the Lake, apart from the excess feeds and oxygen depletion in the Lake water.

Elazegui et al. (2016) found out that in the case of Lake Buhi, water allocation was the primary challenge and this was aggravated by the interplay of internal and external forces and made worse by effects of climate change. There were observed 'overlaps, fragmentation, and lack of communication' in institutional arrangements to better allocate water among competing uses. The authors contend that lack of coordination among government agencies and organizations, fragmented responsibilities of multi-level government institutions in Buhi-Barit watershed, and lack of institutional capacity to manage the watershed are the major factors which must be addressed for sustained watershed governance.

Local institutions in the Lake Buhi portrayed that they were all actively engaged in protecting the watershed. However, there is a need to coordinate and harmonize all these efforts for better results. By virtue of DENR AO no. 2013–16, the Lake Buhi has been selected by the DENR as a water quality management area (WQMA) for its protection.

10.2.3 Initiatives to Promote Science-Based Policy and Management Decisions

For sustainable water resources development and management, the policies, governance structure and management strategies of water resources systems must be science-based and in particular in the context of integrated water resources management as well as sustainability science and transdisciplinary approach.

10.2.3.1 Linking Science and Policy for an Effective Integrated Water Resources Management

In the last few years, integrated water resources management (IWRM) has been advanced and advocated for effective planning, design and operations as well as sustainable development of water resources systems. It may be noted that the decision support system (DSS) is an integral part of IWRM process facilitating the use of science and technology advances in public policy. It brings together disciplines, people and institution necessary to address today's complex water resources problems. The DSS is designed so that the water resource planners, managers and other specialists who are responsible for the development and operation of our water resources systems as well as stakeholders can use the DSS to obtain an improved understanding – a shared vision – of how the water resource system functions and especially the feasible, alternative management options and their impacts or consequences. Through this shared vision, the perspectives of the agronomists, ecologists, economists, engineers, hydrologists, lawyers or political and regional scientists are all considered and respected.

10.2.3.2 Role of Sustainability Science and Transdisciplinary Approach

It is recognized that a water resources system is a complex system, dynamic thus continuously changing as well as uncertain and thus unpredictable and capable of surprise. For this reason, water resources management requires new approach which can no longer be based on traditional science but rather based on sustainability science (Komiyama and Takeuchi 2011). Sustainability science can be described according to the following elements as adapted from Yoshikawa (2011): (1) aim of study; (2) mode of change; (3) truth verification; (4) result of research; and, (5) expected outcomes.

- Sustainability science aims "to understand everything and manage the relations among the various components of the system" thus, transdisciplinary in nature in contrast to traditional science which is monodisciplinary (based on separate disciplines).

- In terms of mode of change, flood management in the context of traditional science utilizes "design flood and return period assuming stationary conditions" but sustainability science views that "water resources changes with climate, weather, social, political and economic changes both at microscopic and macroscopic levels" with the view that the earth is unstable and slowly changing although it can attain at certain periods, what is referred to as dynamic equilibrium.
- Under truth verification, traditional science had fairly relied on "experiments in the laboratory" thus a historicist and "certain or almost sure" perspective as opposed to sustainability science where nature is an "evolution in reality" and is therefore "uncertain, evolutionary and require piecewise engineering". In the context of water resources management, traditional science employs observations referred to as two-dimensional (2d) lenses or tools in line/time or three-dimensional (3d) lenses in 2-d space/time and that estimates are based on historical data, in contrast to sustainability science which employs four-dimensional (4d) lenses through computer simulations and scenario building to account for uncertainties due to socio-economic or climate and ecosystem dynamics and future changes.
- Result of research in traditional science is "knowledge for understanding" which entails analysis and can be simply a fancy or for the sake of science, referred to as dream research. This is in contrast to sustainability science which is "knowledge for action" requiring synthesis. Sustainability science would employ "integrated water resources management which covers physical, socio-economic and environment dimensions in addition to air and land management".
- With regard to expected outcomes, traditional science is for "prosperity of human beings" while sustainability science is for "sustainability of the earth".

Thus, water resources management is considered a complex problem which is an interaction of a dynamic and uncertain physical system (global and local climate variabilities, weather dynamics, river changes, and biodiversity), the social system (economic objectives, societal needs, political ambitions) and the human system (individual lifestyles and behaviours, cultural evolution). Consequently, holistic water resources flood risk management must encompass the various disciplines from physical sciences, socio-economics, political science as well as cultural and behavioural sciences. In this case, transdisciplinary approach is needed which according to van Kerkhoff (2013) "transcends disciplinary pre-conceptions, but is capable of understanding and synthesizing across a range of disciplinary and non-disciplinary ideas and theories". Transdisciplinary approach can be contrasted to monodisciplinary (isolated approach by individual experts), to multidisciplinary (additive approach bringing together a wide range of experts), and to interdisciplinary (interactive approach of several experts solving a problem together). Integral in transdisciplinary approach is stakeholder engagement and collaboration involving academics, professionals, government units, non-government organizations, communities and individuals (Tabios 2015).

10.2.3.3 Integrated Watershed Research and Development Program (INWARD)

The primary objective of INWARD a research project funded by DOST is to promote science and technology based management of watersheds in the country by facilitating the build-up of knowledge and information base on how watersheds respond to the dynamically changing environment and human activities. Through this project a network of 6 Learning Watersheds were strategically selected in various regions where different sensor-aided instruments were installed to monitor climate, hydrological processes, and soil properties together with permanent observation plots to monitor biodiversity and ecosystems as they interact with the environment. Changes in land use and land cover are also being tracked down using satellite images and complemented with field surveys. It is expected that the knowledge and information that are being generated by INWARD will help improve the robustness of policy and management decisions such as on land allocation, design of management interventions to conserve soil, water and biodiversity, and development of viable farming systems and other livelihoods that do not jeopardize the sustainability of watershed resources and ecosystem services. Additionally, it is expected that INWARD will help propagate the knowledge and skills on watershed management among students, teachers, LGU personnel and local communities that are offered opportunities to be trained in the Learning Watersheds. Lastly, management decision support tools are also being developed that will enhance the access of policy makers and managers to fresh information that will aid in improving the quality of its decisions.

10.2.3.4 Monitoring and Detection of Ecosystems Changes for Enhancing Resilience and Adaptation in the Philippines (MODECERA)

The primary objective of MODECERA a research program funded by DOST is to enhance the resilience and adaptation in agriculture, marine, and natural resources sectors through promotion of science and technology based management and policy decisions. One of the major components of MODECERA is the setting up of a long-term comprehensive monitoring system in eight (8) selected river basins strategically located in various parts of the country. The program primarily addresses the dearth of long-term time-series empirical databases on the physical, biological and socioeconomic changes in a watershed in response to changes in the natural and anthropogenic elements of the environment. One of the major outputs of the program is the establishment of Ecological and Watershed Observation Network (EWON), a network of initially 8 watersheds that were installed with a combination of sensor aided instruments and field observation plots to monitor key climate and hydrological parameters, soil properties, biodiversity, ecosystem services, human activities and land use and land cover. The EWON is designed to gather data that are needed to evaluate the changes in ecosystem and watershed functions and services in response to changes in the environment. These information are useful for making robust decisions on how the watersheds and resources therein are to be used without compromising the sustainability of ecosystems and watersheds to deliver vital services including water supply.

10.2.4 Initiatives to Promote Engagement and Participation of LGUs, Local Communities, Individuals and Other Stakeholders in Water Resources Management

The following discussions delve on some local initiatives promoting programs in managing three watersheds, namely: Sagudin-Balili watershed, Santa Cruz sub-watershed and the Tigum-Aganan watershed. This part also discusses the involved institutions in the implementation of the programs and corresponding roles they played towards effective management of watersheds. Some local initiatives that promote national level programs in one or more of the municipalities of the Santa Cruz sub-watershed include the protected area management and integrated resource management. In the case of Tigum-Aganan watershed, the local government initiated activities and instituted council/office to lead in the integrated management of the watershed. Different local agencies, both public and private were undertaking protection, conservation and rehabilitation activities.

10.2.4.1 The Case of the Balili Watershed

The Balili River is situated in Benguet, Philippines. With a length of 23.81 km, it traverses Baguio City, La Trinidad and Sablan in Benguet. The main tributary of the Balili River is the Sagudin River which has 23 tributary creeks. It is reported that the Balili River is excessively polluted and this is blamed on Baguio City's dense population contributing to most waste. Balili River is said to be the "catch basin" of Baguio City's wastewater (effluents from sewage and septic tanks) (Catajan 2015). From the Class A (fit for drinking) water classification in Balili River in 1975, quality deteriorated to Class C (fit only for agricultural and industrial purposes) in 2014.

To improve the water quality in the Balili River, the DENR designated it as Water Quality Management Area (WQMA). When an area is designated as WQMA, a governing board which serves as planning, monitoring and coordinating body, is created by DENR. This is to address water quality problems, sources of pollution, and beneficial use of receiving water body. At the same time, measures that need to be instituted for water quality improvements are determined.

Through a partnership/collaboration among watershed stakeholders (Local Government Units of Baguio City and La Trinidad, local and regional government agencies, academe, private sector, and civil society), the Balili River System Revitalization Coalition (BRSRC) was formed in August 2011 primarily for the rehabilitation of the Balili River. Consisting of almost all higher educational institutions in the area (Benguet State University, St. Louis University, University of Baguio, University of Cordillera, Pines City College, and University of the Philippines Baguio), the academe proved to be very active in its participation.

With the guidance of the University of the Philippines Baguio (UPB), the BRSRC is in the process of being institutionalized as a non-profit corporation to be registered at the Securities and Exchange Commission (SEC). A member of the Coalition itself, the UPB has been coaching the BRSRC and in fact attends quarterly and special meetings of the organization, at the same time is able to monitor the Coalition's efforts to revitalize the Balili River.

Among the major activities of the BRSRC for the rehabilitation of the Balili River are: yearly celebration of the "Balili River Day" every September and "Save the Balili River". The BRSRC launched two projects. One is called "Pollution Tracking" which will trace the source of pollution along the Balili River using GIS and an application developed by the University of Cordilleras. The other project named "Five-to-Revive" is a project by the Benguet State University aimed to raise funds for the Coalition through a bamboo coin bank which will be distributed (pia.gov.ph 2016).

With the realization of the eminent "death" of their major source of water, stakeholders began to act by organizing themselves and formed the Coalition. The Coalition is a collaboration of local institutions, including the academe, civil society and the communities surrounding the watershed.

10.2.4.2 The Case of Santa Cruz Sub-Watershed

Santa Cruz sub-watershed is one of the 24 sub-watersheds of the Laguna de Bay Basin that contributes 18–20% fresh water, the largest account of contribution to the lake (Tongson et al. 2012:13). Located on the southern part of Laguna de Bay Basin, it has a total land area of about 15,058 hectares covering several towns of the Provinces of Laguna and Quezon (Combalicer et al. 2014). The headstream of the sub-watershed is found at Mount San Cristobal, an edifice of a larger mountain – the Mount Banahaw, while the downstream or location of water discharge is at the municipality of Santa Cruz (ERDB 2015).

There are two major factors that facilitated the implementation of programs and initiatives on water management in the Santa Cruz sub-watershed. These are the: (a) recognition of a municipality-based river council, and (b) strengthening of local institutions through social processes.

Santa Cruz sub-watershed's river council is LIMAS MARINA River Rehabilitation and Protection Foundation, Inc., The name represents the six municipalities of the sub-watershed, namely: Liliw, Magdalena, Santa Cruz, Majayjay, Rizal, and Nagcarlan. Ninety per cent of the members of the river council are workers in the municipalities and who also serve as volunteers even after retirement. Some of the roles of the river council in the management of a watershed are: to formulate policies and plans, and enforce and monitor the implementation of policies affecting the Laguna de Bay Basin (Nepumuceno 2010). Since its creation in

1999, the six municipalities take turns in hosting the river council's monthly meetings. Meetings were focused on regular activities such as rehabilitating Santa Cruz's river system, giving environmental talks in schools, attending various forums on environment, and tree planting. In 2014, together with the river council, the University of the Philippines Los Banos convened several local institutions and agreed to form a technical working group to address water related problems at the watershed level. The goal of the technical working group was to develop a water resource management plan and investment plan (David et al. 2016b). The group was composed of the following:

- Department of Environment and Natural Resources-Provincial Environment and Natural Resources Office (DENR PENRO), Los Baños, Laguna
- Laguna Lake Development Authority (LLDA)
- Provincial Government-Environment and Natural Resources Office (PG-ENRO)
- LIMAS MARINA River Rehabilitation and Protection Foundation, Inc.
- University of the Philippines Los Baños (UPLB)
- Municipalities of Liliw, Magdalena, Santa Cruz, Majayjay, Rizal, and Nagcarlan

Strengthening of these institutions through social learning processes can follow after the creation of the group. Social learning is the heart of adaptive, collaborative approach in resource management through which decisions are arrived at (Lal et al. 2001; Prabhu et al. 2007:19). Consciously done collaborative engagement activities like result-based meetings can empower and strengthen the local government institutions and LIMAS MARINA river council to decide and strategize on which issue/problem to prioritize and to address through project plans. Holding of regular meetings also create camaraderie among representatives of institutions.

Social processes such as the Laguna de Bay Watershed Environmental Action Planning or LEAP (Nepumuceno 2010) and Adaptive Collaborative Water Governance (ACWG) protocol (David et al. 2016a) had been tried in the Santa Cruz sub-watershed. Both interventions directly engaged institutions in developing projects. LEAP had engaged participating institutions in implementing and monitoring the co-managed sub-projects of Laguna de Bay Institutional Strengthening and Community Participation (LISCOP) i.e., solid waste management, flood protection and ecotourism (Nepumuceno 2010). Both LEAP and ACWG protocols aimed to strengthen collaboration among institutions for improved water governance at the watershed level.

In the course of engaging the stakeholder institutions for capacity strengthening, however, constraints were experienced. The municipal representatives, in particular, may not always have the time to attend to meetings, due to competing roles. Also, the non-achievement of the goals of previous projects also affects the implementation of newly initiated projects and activities, in terms of the seeming mistrust given by the stakeholders.

On a broader issue, urban related environmental issues, such as solid waste also plague the sub-watershed. To abate this, Environment and Natural Resources Offices in Laguna organized the Association of Laguna Environment and Natural Resources Officers and Advocates (ALENROA), Inc. and had partnered with organizations to bring forth approaches where collaboration among urban municipalities may take place. The "Urban Nexus" approach has been advocated by ALENROA (ICLEI SEAS. (n.d.)). "Nexus approach on watershed management refers to complex linkages of several actors that impact one another." The approach breaks the 'silo' thinking within local governments – a sectoral thinking. Instead, local governments should collaborate and cooperate for a common cause in managing watersheds. Activities of ALENROA did not only involve LGUs of the Santa Cruz sub-watershed but other neighboring municipal/city executives within the region of CALABARZON. Such partnership is believed to be a venue for institutional strengthening and building up of 'champions' required in addressing urban-related issues.

10.2.4.3 The Case of Tigum-Aganan Watershed

The Tigum-Aganan Watershed (TAW) in the Province of Iloilo, Philippines, spans 29,700 hectares of which 19,300 hectares covers Tigum Watershed while Aganan Watershed embraces the remaining 10,400 hectares. The Tigum River passes the municipalities of Maasin, Cabatuan and Sta. Barbara while the Aganan River crosses the municipalities of Alimodian, Leon, San Miguel, Oton and Pavia. These two rivers converge in the municipality of Pavia to form the Jaro River which drains to the Iloilo Strait (EIDRP-TGWGD 2016).

Designated as a reserve in 1923, Tigum-Aganan watershed is the main source of water for Iloilo City. The Maasin watershed is the headwater of the Tigum River which supplies the water requirements of residents in Iloilo City as well as neighboring towns (Francisco and Salas 2004). On the other hand, the Aganan River is the major source of irrigation water to farms in this watershed (Vogel et al. 2013).

In 2000, the Iloilo Watershed Management Council (IWMC) was established by the provincial government of Iloilo to manage all watersheds in Iloilo province with the governor as the chair (Francisco and Salas 2004). This was seen as a more sustainable institution in the protection of the watershed. Under the IWMC, watershed management units were created, and one these is the Tigum-Aganan Watershed Management Board (TAWMB). Salas (2015) reported that the major challenges faced by the communities in the watershed are drying of the Tigum River, increasing soil erosion and sedimentation.

Various local institutions in the province are doing watershed level programs to protect and preserve water resources. For one, the Metro Iloilo Water District (MIWD) initiated awareness campaigns and IECs about possible water scarcity for drinking water by 2020, thus the need to protect Maasin watershed. Consisting of LGUs, DENR, NGOs, and MIWD, the Maasin Task Force was formed and took over management of the watershed reservation. NGOs, on their part undertook campaigns which paved the way to the creation of the "Save the Maasin Movement." The LGU spearheaded the rehabilitation of the watershed through reforestation of the upper portion of the watershed. The DENR set up tree plantations in areas cultivated by communities in the watershed. Residents participated in tree planting and fund raising activities in support of these initiatives. These efforts resulted to the reforestation of 500 hectares forestlands in the watershed (Francisco and Salas 2004).

10.2.4.4 The Conservation Farming Village Program

The Conservation Farming Village (CFV) Programme was designed to transform erosive farming in sloping lands (uplands) into a transformative development strategy for improving the income and quality of life of farmers, and for promoting the resilience and sustainability of upland areas in particular and of watersheds in general. It involves the active collaboration and shared learning of upland communities, local government units (LGUs), and academia in enhancing the capability of upland communities in practicing conservation farming and other related technologies. Fusion of science and local knowledge is the key strategy adopted by CFV in transforming the traditionally erosive farming systems into more conservation-biased systems. Operationally the upland farmers are being made aware and trained on conservation oriented techniques from planning to implementation but were given the freedom and calibrated technical guidance to reengineer their farming systems as they see proper. This strategy builds up a sense of ownership and facilitated the voluntary adoption of changes in farming practices of the farmer participants.

The first phase of this programme involved the establishment of 15 barangays as CFVs with 5 farmer volunteers that were capacitated to serve as model farmers practicing S&T based upland conservation technologies. Eventually the 15 CFVs increased to 75 CFVs in five different towns from 5 different provinces. In addition, there were more than 500 adoptors by the end of the first phase covering at least 400 ha of upland farms. The second phase involved the upscaling of phase 1 by building the capacities of upland farmers and communities, LGU executives, and technical personnel to implement the CFV programme in other upland areas of the country outside of the original sites. The target of Phase 2 was to facilitate the establishment of at least one new CFV model farm in each of the 15 new provinces. Enhanced capability of the farmers and LGUs were vital to the success of CFV in these sites. Five training courses were conducted with 272 participants consisting of farmers, LGU officials and academia. Participants visited existing CFV sites. Eighteen farms were established in the 11 provinces where CFV capacity building activities were implemented. Phase 3 involved further replication of CFVs in additional sites and the assessment of the impacts of CFVs on site quality and on the household economy in various CFV sites. To date CFV has transformed the mindset and practices of more than 1000 farmer-adopters. It has also capacitated almost 6000 farmers through more than 100 training courses that it has implemented. At present preparations are being made to explore the nationwide scaling up of CFV with the participation of DENR, DA, DAR, and DSWD.

10.3 Summary and Conclusions

One of the major issues and challenges of water resources management in the Philippines is water governance aside from limited water availability when and where needed, inadequate access to safe water and sanitation, need for improved irrigation services, flooding issues and concern and environmental degradation. The problem in water governance in particular is due to existence of multiple and fragmented water-related agencies in the Philippines and there is no clear and consequently weak leadership water resources management agency in the country. Thus, the plan to have a strong water resources management agency is long overdue and it is a welcome development especially in the light of the complex problems that the country is facing in terms of extreme climate variability and growing water-related hazards.

It may be noted that currently, all the water sub-sectors in the country like water supply for domestic use, sanitation, sewerage, energy, irrigation and flood management have their own separate roadmaps. The water sector master plan should take into consideration all of these sub-sector roadmaps and orchestrate it so that there is harmony and cooperation among all these sub-sectors. Thus, a strong high level committed leadership with appropriate support of resources (manpower, equipment, budgets) is necessary to make the proposed master framework plan.

The National Water Resources Board (NWRB), the lead national government agency mandated by law to coordinate and regulate water resources management in the Philippines is unfortunately ill-equipped to perform all its functions. With an annual budget (circa 2015) of about P90 Million and a staff complement of 100 filled up positions, it is unable to perform much of the functions required to develop and manage the country's water resources. The NWRB's current focus is mainly on policy and program coordination, resource regulation and economic regulation. On policy and program coordination, the agency has difficulties orchestrating all the agencies engaged in water-related programs. The NWRB's data on water resource assessment is very much outdated with very limited updating since 1980. They currently do not have any capacities on scientific modelling and computerized decision support systems that will aid decision makers and planners in the optimal management of the country's water resources.

In view of this, a major recommendation in this chapter is to reconstitute, elevate and strengthen the current National Water Resources Board (NWRB) into a water superbody called the National Water Resources Management Office or Authority with a mission to be an efficient and effective agency that manage and protect the country's water resources for domestic water supply, sanitation, irrigation, hydropower, fisheries, aquaculture, flood control, navigation, and recreation including the enhancement and maintenance of water quality, conservation of watersheds, control of water pollution and environmental restoration without compromising the natural ecosystem functions and services. For this purpose, the organizational structure of this office follows horizontal compartmentalization among 10 divisions performing an array of technical functions namely: (1) planning and policy; (2) data collection and monitoring; (3) scientific studies and decision support systems; (4) infrastructure and program development; (5) water facilities development and operations; (6) resource regulation; (7) economic regulation; (8) water financing and economics; (9) stakeholder relations; and (10) river basin organization and development.

Some major features of this super waterbody by virtue of the design of the various divisions in its organizational structure are as follows:

- The water policy and operation studies should be based on full-time science and technology support institutions rather than dependent on episodic foreign TA or local funds.
- With permanent and fully supported S&T water institutions, continuous updating and retrofitting of water policies, plans and infrastructures can be made to address evolutionary changes in climate, land use, socio-economic and political changes. In particular, comprehensive water resources plans for various water uses integrated with land use plans for each WRR which will be updated annually due to the unfolding or evolution of climate change, climate variability, land use change and socio-economic-political changes.
- Establishment of a centralized water resources data center and with real-time data acquisition system from field/gauging stations through satellite or cellular communications and the basic processing of raw water data for almost real-time, medium-term and long-term uses or purposes.
- Policies and investments towards providing water supply needs and access to safe water and sanitation must be reassessed and prioritized. Consequently, this requires reassessment and prioritization of water investments which includes human resources (to study, develop policy and plan) and financial investment (to build infrastructures or to monitor and optimize operations).
- It includes a strategic regional presence, otherwise the criticism of having a central agency that is not accessible to other stakeholders will continue to haunt the system which will undermine its capacity and credibility to be the Authority that all stakeholders must link up with in terms of water resources management.
- Provide capacity building and training on RBO development considering the appropriate mix of public and private provision and management of water resources as well as greater user participation in decision making and actual management.
- Initiate scaling up of promising local initiatives in areas found applicable by involving stakeholder institutions even during the start of the planning phase of any project undertaking until its desired outcome is achieved.

References

Binoya, C., et al. (n.d.). *Lake Buhi*. Society for the Conservation of Philippine Wetlands (SCPW). http://www.psdn.org.ph/wetlands/wetland_buhi.htm. Accessed 12 Dec 2016.

Caparas, K. (2012). "Lake Buhi: Home of the smallest fish in the world." https://www.vigattintourism.com/index.php?/tourism/articles/Lake-Buhi-Home-of-the-Smallest-Fish-in-the-World. Accessed 12 Dec 2016.

Catajan, M. E. (2015). "Balili River clean-up seen as an uphill battle". *Sun Star Baguio*. http://www.sunstar.com.ph/baguio/local-news/2015/06/11/balili-river-clean-seen-uphill-battle-412680. Accessed 7 Dec 2016.

Combalicer, E. A., David, M. E., Pulhin, J. M., & Rola, A. C. (2014). Biophysical characteristics of the Sta. Cruz River watershed, Laguna: Preliminary results. (WGD Working Paper No. 2014–14).

David, M. E., Rola, A. C., & Pulhin, J. M.. (2016a). *Adaptive collaborative water governance: The case of Santa Cruz sub-watershed, Laguna, Philippines*. Presented during the 8th National Social Science Congress, Lyceum of the Philippines, Batangas City on June 15–17, 2016.

David, M. E., Rola, A. C., & Pulhin, J. M. (2016b). Development of a protocol on adaptive collaborative water governance for improved Santa Cruz watershed Management in the Philippines. *Ecosystems & Development Journal, 6*(2), 35–51.

EIDRP-TGWGD. (2016). *Emerging interdisciplinary research program-towards good water governance for development: A multi-case analysis* (Year 4 Annual Report (July 1, 2015–August 31, 2016)).

Elazegui, D. D., Rola, A. C., & Allis, E. (2016). Enhancing institutional dynamics for multiple uses of water amidst climate-related risks: The case of Lake Buhi, Philippines. *Lakes and Reservoirs: Research and Management, 21*(3), 224–234.

ERDB. (2015). *Vulnerability assessment of Sta Cruz watershed to climate change*. College: ERDB.

Francisco, H. A., & Salas, J. C. (2004). "Realities of watershed management in the Philippines: The case of the Iloilo-Maasin watershed" (Discussion Paper Series No. 2004–22). Makati City: *Philippine Institute for Development Studies (PIDS)*. 25 pages.

ICLEI SEAS. (n.d.). http://seas.iclei.org/logos/logos-6/iclei-seas-presents-urban-nexus-at-laguna-environment-summit.html. Accessed 4 Nov 2016.

Komiyama, H., & Takeuchi, K. (2011). Sustainability science: Building a new academic discipline. In H. Komiyama, K. Takeuchi, H. Shiroyama, & T. Mino (Eds.), *Sustainability science: A multidisciplinary approach* (pp. 2–19). Tokyo: United Nations University Press.

Lal, P., Lim-Applegate, H., & Scoccimarro, M. (2001). The adaptive decision-making process as a tool for integrated natural resource management: Focus, attitudes, and approach. *Conservation Ecology, 5*(2), 11.

Nepumuceno, D. (2010). *Improving water security for the future: Learning from the IWRM experience in Laguna de Bay*. A discussion paper presented at the River Basin Study Visit in Spain, May 26, 2010.

Philippine Information Agency (PIA). (2016). Stakeholders renew commitment to save Balili River. http://news.pia.gov.ph/article/view/41474356304/stakeholders-renew-commitment-to-save-balili-river. Accessed 8 Dec 2016.

Plopenio, J. C., & Bimeda, S. B. (2011). *Water quality monitoring in Lake Buhi, Buhi, Camarines Sur*. Naga City: Institute for Environmental Conservation and Research, Ateneo de Naga University.

Prabhu, R., McDougall, C., & Fisher, R. (2007). Adaptive collaborative management: A conceptual model. In *Adaptive collaborative management of community forests in Asia. Experiences from Nepal, Indonesia and the Philippines*. Bogor: Center for International Forestry Research.

Presidential Decree No. 1152. Philippine environmental code.

Rola, A. C., Pulhin, J. M., Tabios, G. Q., III, Lizada, J. C., & Dayo, M. H. F. (2015). Challenges of water governance in the Philippines. *Philippine Journal of Science, DOST, 144*(2), 197–208.

Salas, J. C. (2015). Tigum Aganan watershed. *A Case Study*. http://www.philwatershed.ph/wp-content/uploads/2015/07/Tigum-Aganan-Case-Study.pdf. Accessed 6 Oct 2016.

Tabios, G. Q. III. (2015). *Need for holistic flood risk management: Case of Pasig-Marikina River Basin of Metro Manila*. 11th APRU Research Symposium on Multi-Hazards around the Pacific Rim (APRU MH2015 International Symposium), UP Diliman, October 8.

Tabios, G. Q. III., & Villaluna, R. A. L. (2011). *Status, challenges and proposed National Water Management Superbody for Philippine water resources sector development*. A Policy Brief Submitted to National Economic Development Authority, Philippines, October.

Tabios, G. Q. III., & Villaluna, R. A. L. (2012). *Development of the implementation and operational plan for the National Water Resources Management Office*. Submitted to the Interagency Committee on Water and National Economic Development Authority. Quezon City: NEDA. 96 p.

Tongson, E. E., Hernandez, E. C., & Faraon, A. A. (2012). *Hydrologic atlas of Laguna de Bay 2012*. Quezon City: Laguna Lake Development Authority and WWF-Philippines.

van Kerkhoff, L. (2013). Developing integrative research for sustainability science through complexity principles-based approach. *Sustainability Science,* (2014) 9, 143–155.

Vogel, J. M., Smith, J. B., Brown, P. E., Troell, J., & Ray, A. (2013). *An assessment of water security, development, and climate change in Iloilo, Philippines and the Tigum-Aganan Watershed. Technical report*. USAID. http://pdf.usaid.gov/pdf_docs/PA00KJG8.pdf. Accessed 6 Oct 2016.

Yoshikawa, H. (2011). Science and technology for society. In H. Komiyama, K. Takeuchi, H. Shiroyama, & T. Mino (Eds.), *Sustainability science: A multidisciplinary approach* (pp. 256–271). Tokyo: United Nations University Press.

Dr. Guillermo Q. Tabios III is a Full Professor at the Institute of Civil Engineering and Research Fellow of the National Hydraulic Research Center at the University of the Philippines Diliman (UPD). He holds Bachelor's and Master's degree in Agricultural Engineering from the University of the Philippines, Los Baños, and a Ph.D. in Civil Engineering from Colorado State University. Dr. Tabios teaches and conducts researches in stochastic and computational hydrology and hydraulics, as well as water resources systems engineering. He is an Academician of the National Academy of Science and Technology, a Regular Member of the National Research Council of the Philippines and a Member of the American Geophysical Union and the International Association of Hydro-Environment Engineering and Research. He formerly held positions as Chairman of the Department (now Institute) of Civil Engineering, UP Diliman (2004–2007), Director of the National Hydraulic Research Center, UP Diliman (2008–2015) and Board of Director of the Philippine National Water Resources Board (2008–2015). Prior to joining UP Diliman in 1996, he was a Research Associate of the Hydrology and Water Resources Program at Colorado State University, Fort Collins (1985–1987) and Research Faculty of the Hydraulic and Coastal Engineering at University of California, Berkeley (1987–1996).

Dr. Rex Victor O. Cruz is a Full Professor at the University of the Philippines Los Baños (UPLB). He obtained his bachelor and Master's degree in forestry at UPLB and his doctoral degree at the University of Arizona. His major fields of expertise are forestry, watershed management, environment and natural resources management, upland development, land use planning and climate change. The author is a member of the UN Intergovernmental Panel on Climate Change (IPCC) in 1992–1995; 1997–2000; and 2004–2007. Currently he leads 3 national programs, the National Research and Development Program for Watershed Management, National Conservation Farming Village Program, and the Monitoring and Detection of Changes in Ecosystems for Resiliency and Adaptation. He is a member of the Asia Pacific Forestry Network Board of Directors, and the National Pool of Technical Experts of the Philippine Climate Change Commission.

Dr. Myra E. David is Assistant Professor of the Institute for Governance and Rural Development, College of Public Affairs and Development, University of the Philippines Los Baños (UPLB), Laguna, Philippines. She teaches extension and education courses for both graduate and undergraduate students. Among her research interests are developing a model for adaptive collaborative water governance at the watershed level, and community based solid waste management. She holds a PhD in Extension Education and MS in Social Forestry from UP Los Baños.

Dr. Miriam R. Nguyen is a University Researcher at the Community Innovations Studies Center (CISC), College of Public Affairs and Development (CPAf), University of the Philippine Los Baños (UPLB). She completed Bachelor's degree in Economics and Master's degree in Agricultural Marketing. She earned her Doctor of Philosophy in Community Development with Environmental Management as cognate at UPLB in 2006. Her research work generated knowledge on community and agrarian studies, enterprise development, institutional development, policy and socio-economic studies, resource and environmental management as well as governance. Her recent involvement on environment-related research works center on governance of irrigation water and watershed; food security and resilience of different types of communities towards climate change.

Chapter 11
Water Demand Management and Improving Access to Water

Corazon L. Abansi, Rosalie Arcala Hall, and Ida M.L. Siason

Abstract The chapter reviews and examines water-related program interventions–social, economic, institutional–that have directly and indirectly influenced water demand management in the Philippines since 2000. Demand focuses on water users and the human dimensions of water use, including degradation of water quality, excessive drawing from aquifers, non-consumptive uses of water, and the organizations that have evolved to represent the various stakeholders. The chapter then examines the feasibility of the emergent emphasis on policies that stress making better use of existing supplies in combination with decentralization and participation of water users, even as new sources are explored. Initiatives by the Department of Interior and Local Government toward expanding water access in areas not served by municipal water districts, participatory management schemes for irrigation associations by the National Irrigation Administration, and intersectoral formation for surface water/river quality management by the Department of Environment and Natural Resources are probed. The chapter describes initiatives such as inter-municipal water transfers, independent community-based collective arrangements for domestic and irrigation provisioning (through cooperatives) and riparian upstream-downstream coalitions for water quality in select locality cases. Such policies and interventions aim to influence demand along principles of efficiency, equity, and sustainability. This chapter explores the application of water-demand programs by select national government agencies to their respective client-groups. The institutional arrangements thus created by these program applications are "grey area" because they are not as yet grounded on water rights or adequately covered in the existing Philippine Water Code. But they portend to better/improved ways by which water can be more equitably accessed.

C.L. Abansi (✉)
Institute of Management, University of the Philippines Baguio, Baguio City, Philippines
e-mail: corazon.abansi@gmail.com

R.A. Hall
Division of Social Sciences, College of Arts and Sciences,
University of the Philippines Visayas, Miagao, Iloilo, Philippines

I.M.L. Siason
University of the Philippines Visayas, Iloilo, Philippines

A.C. Rola et al. (eds.), *Water Policy in the Philippines*, Global Issues in Water Policy 8, https://doi.org/10.1007/978-3-319-70969-7_11

Keywords Water demand management • Water recycling • Water saving-devices • Community participation in water management • Non-price water conservation program

11.1 Understanding Water Demand Management

Rapid population growth, economic development, urbanization, and industrialization have taken their toll on the country's water services and resource base (Araral and Wang 2013, NEDA 2010). The sustainability of the country's water resources is threatened both in terms of quantity and quality. Unless current water usage patterns are changed, future water demand will significantly exceed existing available fresh water resources. There is a need to look at water demand management (WDM) as an alternative management mechanism to address the water problem in the country.

In its simplest, WDM is defined as the management of the total quantity of water abstracted from a source of supply using measures to control waste and undue consumption (Herbertson and Tate 2001). A more elaborate definition states that WDM is the adaptation and implementation of a strategy by a water institution or consumer to influence the water demand and usage of water in order to meet any of the following objectives: economic efficiency, social development, social equity, environmental protection, sustainability of water supply and services, and political acceptability (Jalil and Njiru 2006).

The traditional approach of hydrologists and water resource engineers has been to focus on the supply side and the assessment of available water resources. Forecasts of water demands have often been provided by different institutions with a wide range of uncertainty because of the (i) limited data on historic actual water use; (ii) high levels of uncertainty in establishing efficiency of water use, with significant losses likely in irrigation, urban, and industrial water use; and (iii) uncertainties in the basic economic, social, and demographic assumptions required for water demand forecasts (Herbertson and Tate 2001). As a result, there is a high degree of uncertainty in current forecasts of the supply-demand balance. There are large variations in local availability both in space and time. To ensure long-term sustainability of water resources, the focus of water management needs to shift from a traditional supply side management to demand side management. It is a relatively new branch of water resource science but it offers a promising alternative for sustaining the world's freshwater supplies in the next century and beyond.

In the Philippines, WDM can improve the existing supply-demand balance, especially in water-stressed areas. Though augmentation of supplies would be required to meet the growing demand in urban centers, our future lies in effectively controlling our demand for water resources and efficiently managing and using available resources. The approach calls for the development of a WDM strategy that would aim at reducing losses in the system, improving operational efficiencies, promoting rational use of water resources, ensuring equitable distribution of resources, and exploring alternative sources such as recycling of wastewater for non-potable uses. Implementation of such an approach would require reforms in terms of providing adequate regulatory, institutional, and legal frameworks for delivery of services; tariff reforms that ensure financial sustainability of operations of the utility; and internal reforms within the utility that aim at improving operational efficiencies and reducing losses to acceptable levels (Dziegielewski 2003).

Water demand management therefore requires the implementation of policies and/or measures, which serve to control or influence the amount of water used and thereby lead to improved efficiency in production, transmission, distribution, and use of water. However, a cursory look at existing national policies shows that addressing water tends to be punitive in its provisions rather than incentive-based. Failure of implementation is not new due to the weakness of existing administrative mechanism. Therefore, policies must emphasize on improved management of existing systems rather than augmentation of supply and promoting optimum utilization of available water resources through increased consumer awareness and better operational efficiency at the level of service providers. WDM should be viewed as complementing the supply management efforts and not replacing them, hence, demand-side management is better used in conjunction with conventional approaches (Inman and Jeffrey 2006).

11.2 Goals of Water Demand Policies and Interventions

In practice, WDM comes down to three key goals: efficiency, equity, and sustainability (Turton 1999).

11.2.1 Economic Efficiency

Economic efficiency aims to minimize the amount of water required for a particular purpose. Improving water efficiency can come in the form of reducing losses in the distribution system, reducing the amount or the quality of water needed to perform a task, and changing the timing of water use. Consumers can use water efficiently by reducing water wastage and by choosing more water-efficient products. Reducing the quality of water used is also an alternative since lower quality, less costly water can be safely used for many purposes such as flushing toilets, landscape irrigation, and industrial uses. In terms of timing, using water at off-peak times reduces strain on the water delivery system while irrigating crops at night reduces losses from evaporation. Efficiency also requires that the system is able to cope with water shortages.

11.2.2 Social Equity

WDM strategy must take into account the goal of social equity. Water pricing must not be a barrier to reasonable access to clean water by the poor in both urban and rural areas. The situation of poor women and children requires that policymakers find ways to provide for at least a minimum supply of clean water. Women are frequently the "water managers" in the household, and difficulties in their efforts to ensure the family's water supply can become a barrier to opportunities for education and employment. WDM strategies must also be sensitive to local customs and traditional water rights. Community participation in water management decision making ensures that the benefits of WDM are understood and widely accepted and will greatly increase the likelihood of success.

11.2.3 Environmental Sustainability

There is rapidly growing evidence of the effects of climate change on water availability. The Philippines, which is particularly vulnerable, can expect increased temperatures and reduced rainfall, according to the 2007 Report of the Intergovernmental Panel on Climate Change (IPCC 2007). In the past, water management has generally been based on the assumption that climatic conditions would remain more or less the same. Given the IPCC's predictions, governments need to be prepared to re-evaluate their policies and institutions for the management of water resources. WDM is an effective strategy for adaptation to the current challenge of water scarcity and will become more so as climate change further reduces the availability of water and land. Policies based on WDM strategies contribute to preparedness and social resilience in the face of the challenges that lie ahead for the Philippines.

11.3 Water Demand Management Strategies in the Philippines

This section discusses the different strategies in managing water demand in the Philippines. Broadly, these strategies are categorized as economic, technical, and socio-political.

11.3.1 Economic Strategies/Instruments

Markets and water prices could become highly efficient instruments of appropriate water allocation, if successfully applied. Pricing strategies, such as increasing block tariffs, are commonly applied in water demand management, especially in the urban areas (Araral and Wang 2013). Marginal-cost pricing is viewed as an effective instrument of demand management because consumers will respond to marginal prices by adjusting their consumption (Olmstead et al. 2007). In the Philippines, increasing block tariff is used by water districts. The underlying principles essentially rely on economic valuation while also balancing affordability, conservation, and sustainability.

There are approximately 5400 water service providers in the Philippines (Llanto 2013; NEDA 2010). This figure is not definite because the government does not have accurate information on the water supply sector. There are 879 water districts in the Philippines, 476 of these are registered with the National Water Resources Board (NWRB) and only 511 are operational (World Bank 2005). NWRB is the leading regulatory government agency responsible for ensuring the optimum exploitation, utilization, development, conservation, and protection of the country's water resources. Two private concessionaires, Maynilad Water Services, Inc. and Manila Water Company, are the contracted water providers in Metro Manila. Outside Metro Manila, the major water service providers are the local water districts regulated by the Local Water Utilities Authority (LWUA). A water district is a utility serving a city or municipality but is legally and financially separate from this local government. Local government units (LGUs) and community-based organizations (CBOs) are the biggest domestic water providers serving 55% of those with access to water (NEDA 2010). This is followed by the water districts at 20% and private operators at 5%. The remaining 20% of the population rely on informal sources (self-provisioning, vending) to satisfy their water needs.

11.3.1.1 Economic Strategies of Water Districts

Water districts manage demand by charging for water in proportion to the amount and characteristics of use. Water use is regulated by check valves and meters, signaling the message of wise use of water. Metering of service connections is seen as the economical and equitable procedure for arriving at the appropriate water charges. Metered rates are imposed on the basis of the actual volume of water consumed for a billing period (LWUA 2014a).

The LWUA sets the rules on water tariffs of the WD (LWUA 2014b). The prescribed rate structure consists of two components, the minimum charge (MC) and the commodity charge (CC). MC is also known as the service charge or the demand charge. Minimum charge is fixed, depending on the size and classification of service connection (customer class) and should cover the first 10 m^3 of water consumption. MC covers the fixed cost required to carry on vital water supply functions not directly connected with production and distribution. MC is set using two criteria:

1. MC should be within the ability to pay of the low-income users. It is established that 10 m^3 is enough to supply the domestic water need of low-income users. Hence, the MC for residential water should not exceed 5% of the average monthly family income of the low-income group.
2. MC varies with the size of the meter; the bigger the meter, the higher the MC.

The CC is the amount per cubic meter in excess of MC and covers expenses related to production, distribution, and all costs not covered from MC. This marginal component of the pricing scheme is expected to affect water demand because successively increasing consumption is charged at higher rates so that the basic requirements of consumers are met and more luxurious uses are charged at higher rates (LWUA 2014b).

This tiered pricing scheme is used as a fundamental tool to ensure the sustainability of water systems and the efficiency of water allocation. Massarutto (2007) argues that water pricing aimed at water allocation and cost recovery requires different approaches to pricing as recovery of costs does not necessarily match the cost for allocating scarce water resources. Various authors stressed the economic aspect of determining water rates, stating that, in computing how much water should be charged, it is important to consider generating revenue so that the water provider can include recovering production costs and providing further investment (Hanemann 2005, Glennon 2004, Min 2007).

Recognizing these, the Philippine water districts have slowly adapted market benchmarks for their performance. The water supply industry maintains that the public can be served best by self-sustained enterprises adequately financed with rates based on sound engineering, social, and economic principles. A properly operated and managed water utility should be a self-sufficient enterprise. The challenge for the provider is to identify tariff structures that ensure cost recovery while increasing take-up of basic services among the poor. A modeling of the demand function for the entire market of potential users of the service using survival curve estimation techniques and knowing the costs faced by the provider was done using the Metropolitan Cebu Water District as example (Box 11.1).

Box 11.1: Water Tariff and Demand, the Case of the Metropolitan Cebu Water District

Implementing tariff structures with differential pricing makes it a potentially powerful tool for increasing take-up of important services among the poor, enabling providers of basic services to aid in more sustainable development while still allowing the provision of the service to be viable over the long term. However, altering tariff structures may not work for all types of services where the differences in willingness to pay (WTP) are less determined by level of income. This approach was analyzed using a contingent valuation survey that captured household and business WTP for access to improved water and new sanitation services provided by the Metropolitan Cebu Water District (MCWD) in the Philippines. Results showed that up-front connection charges are a major deterrent to more units opting to connect and access MCWD's water services. Charging a much smaller up-front connection charge and amortizing the remaining costs of the connection charges into the monthly fee results in a substantial rise in demand from 31% of households to 50%. Moreover, this tariff structure results in 30.4% of non-connected households connecting to water services. A tariff structure that price differentiates to equalize demand across different subgroups increases take-up of water services by households to 56.8% based on household income subgroups or 62.2% using geographic subgroups. Moreover, it increases take-up of water services by poor households from 12.2% to 61.9% of the population. This is nearly a 500% increase over the current pricing scenario. Price differentiation, however, is found to be a less effective tool for increasing total household take-up of potential new septage and sewage services. It is shown to only increase overall household demand by 0.8% in the case of monthly septage service fees and 2.4 percentage points in the case of sewage services. However, it is effective in increasing take-up of services by 4.2 percentage

Source: http://www.lwua.gov.ph accessed on April 28, 2016.

11.3.1.2 Economic Strategies of LGU Water Systems and CBOs

A water system operated by the LGU directly provides water through the city or municipal engineering department. This water system delivers domestic water supply to residents in the areas of high population density. The LGU water system is usually established through loans by the LGU or grants from national politicians and international donors. Members of the governance board are selected by the local government officials while the water manager is usually the municipal engineer.

LGU water systems manage demand by collecting fees from users and using the revenue from these fees to cover variable costs and other maintenance costs. In most cases, revenues generated from fees may not be adequate to cover maintenance costs and the LGU subsidizes the rest.

Community-based organizations operate water supply systems through the support of national government or non-government organizations (NGOs) or their own village government. Some are privately organized by a group of families, while others are organized by individual tribal leaders to cater to member-households. These are loose organizations, mostly without legal basis, but guided by customary rules. Some supply both domestic water needs and irrigation. Some of these organizations charge very minimal water fees, but in most instances, water is free.

11.3.1.3 Comparison of Prices of Water Districts, LGU Water Systems and Community-Based Organizations

Results of a survey of managers of water organizations conducted in 2013 (Rola et al. 2016) show that pricing as a strategy to manage demand depends on the pricing policies adopted by the organization. Water prices of the LGU-based water system are based on the cost of repair and maintenance whereas for CBOs, pricing is based on the approval of the members of the cooperative and expenses for maintaining and operating their water system. Water charges are computed based mostly on partial-cost recovery. The complexity of water pricing as shown by these two providers arises from the belief that water is partially a public good so government should be involved in the pricing process (Min 2007). There is also this perception of water as an integral part of the ecosystem, a natural resource, and a social and economic good, hence managers are faced with the issue on how, practically, to achieve the right balance between managing water as an economic and a social good (Gleick et al. 2002, Linton 2010, Feitelson 2012).

Comparing prices across water providers, water districts charge the highest for water use (Table 11.1). Commercial users are charged higher at an average of P293.43 per month for the first 10 m^3 of water. Residential water users, on the other hand, are charged an average of P185.55 for the first 10 m^3 per month. Charges for residential use range from P100.00 to P370.00, while commercial water charges

Table 11.1 Water charges (in pesos/10 m^3 per month), by water organization

Organization		Average	Minimum	Maximum
Water district	Residential	185.55	100.00	370.00
	Commercial	293.43	125.80	614.00
Local government	Residential	51.51	0.00	204.70
	Commercial	117.50	100.00	135.00
Community based	Residential	62.14	0.00	838.00
	Commercial	320.00	65.00	900.00

Source: Rola et al. (2016)

range from P125.80 to P614.00. Community-based organizations have rates averaging P62.14 for residential users. Most (41%) of the CBOs provide the water for free. Hall et al. (2015) argued that water pricing remains heavily subsidized by the state with many politicians refusing to impose water prices at cost recovery.

Water charges for irrigation vary by season, with an average of PhP 944 per hectare during the wet season and PhP 1184 during the dry season.

A common problem encountered by LGU-based organizations and CBOs is the failure to pay the water fee on the part of some users. Where local governments control the water system, communities in general resist investments or abstraction limits that increase fees or impede access (Hall et al. 2015). Agrawal and Goyal (1999) argued that variations and changes in demographic pressures influence the ability of water providers in rural communities to create enforceable rules and resolve conflicts especially for common pool resources. Van Koeppen et al. (2007) agreed that rule setting and enforcement is the Achilles heel of any water policy, and community-based water policies have both strengths and weaknesses in this regard. It is morally more difficult for rural communities to hold other water users, relatives, and neighbors accountable to restricting water use for livelihoods or to use the sanction of cutting water delivery to enforce agreed obligations such as payment of fees on time.

Young (2002) and Herrfahrdt-Pahle (2010) elucidate the importance of ensuring a close fit between social and ecological systems through the existent institutional interface. There is growing recognition that pure market modes are quite rare and that market-based mechanisms require effective regulation to ensure that social and environmental needs are met (Bakker 2003, Freeman and Kolstad 2007). In areas dominated by informal water economies, Abansi et al. (2016) proposed a model that combines and effectively integrates both economic and cultural dimensions to manage water demand and use pattern.

11.3.2 Technical Strategies

Any activity, practice, technological device, law, or policy that can potentially reduce water use may be considered a demand management measure. Hundreds of different measures can be found in the literature (Dziegielewski et al. 1993). Technical interventions aim to meet the existing needs of individual users and uses with less water. These can come in the form of (1) reducing the quantity or quality of water required to accomplish a specific task, (2) adjusting the nature of the task so it can be accomplished with less water or lower quality water, (3) reducing losses in movement from source through use to disposal, (4) shifting time of use to off-peak periods, and (5) increasing the ability of the system to operate during droughts. These five components mostly point to the efficiency by which water is used; improvements can free up significant quantities of water.

11.3.2.1 Introduction of Water-Saving Products and Technologies in Domestic and Industrial Use

Technologies and efficient water use practices can be employed to reduce water use for domestic, commercial, and industrial users. For example, one shower head is considered more efficient than another if it can accomplish the same purpose by using less water or other inputs. For domestic water use, some of the available devices and products are low-flow shower heads, shower flow restrictors, toilet-tank inserts, faucet aerators, low-flush toilets, dual-flush toilets, insulation of hot-water pipes, horizontal axis washing machines, low-pressure supply connections, pressure-reducing valves, water-efficient landscape designs, and irrigation practices. The water efficiency gains from these devices over traditional counterparts can be substantial without diminishing the fulfillment of the original purpose for which water is used. For example, ultra-low flow toilets consume as little as 1/5th to 1/7th of what traditional toilets consume. Toilet dams or other water displacement devices block part of the tank so that less water is required to fill the toilet following each flush. Some problems may occur with the need to double flush, but water savings from these devices are estimated at more than 10%.

In the industrial sector, technological devices include counter-flow washing and rinse systems, reuse of processed water, recirculation of cooling water, ozone treatment for cooling towers, treatment and reuse of blow-down, and water recycling.

The extent of use of these water saving devices and technologies depends largely on the market penetration of these products, the types of industries linked to the system, and the technologies available for the domestic market. As a more aggressive strategy, a municipality can enact standards for water-using appliances, at least for upcoming localities with new construction. A review of the Water Code of the Philippines and Water Crisis Act of 1995 did not show provisions on incentives and rewards for the use of water-saving devices. Only the Clean Air Act of 2004 has a provision (Section 26) for incentive schemes, but these are targeted to technologies and devices for pollution control and clean production. The 2014 Investment Priority Plan of the Bureau of Investment includes incentives for water pollution control and energy-saving technologies but none on water-saving technologies.

11.3.2.2 Introduction of Water-Saving Products and Technologies for Irrigation Water

In agricultural irrigation, the water efficiency gain of drip irrigation over furrow irrigation has been documented by many studies (Tagar et al. 2012, Payero et al. 2005, Yonts 2008). A water-saving technology developed by the International Rice Research Institute called alternate wetting and drying (AWD) can help farmers reduce the amount of water use by eliminating the traditional method of keeping their rice fields continuously flooded (Rosellon 2016). The water level is allowed to drop 1–15 cm below the soil surface before reflooding the field. A perforated water tube, which could be made from PVC pipes or bamboo, is used by farmers to monitor the water level below the soil. According to the Department of Agriculture, AWD reduces irrigation significantly compared with farmers' practice and, in most cases, lessens fuel consumption for pumping water by about 30 L/ha (Fernandez 2015). A field study by the Philippine Rice Research Institute on AWD showed water savings ranging from 16 to35% without decreasing grain yield.

In rice production, a water-saving technology called "aerobic rice culture" requires aerobic rice varieties to be grown in well-drained, non-saturated, and non-puddled soils. Growing rice aerobically saves water by eliminating continuous seepage and percolation, reducing evapotranspiration, and eliminating wetland preparation (Fernandez 2015).

In landscaping and gardening, native species that have high survival under local rain and climate conditions can save large amounts of water. Even though the water consumed for gardening activities in the Philippines may be relatively less, this is also important as the government maintains most of public gardens.

The widespread adoption of AWD, aerobic rice culture, and other alternative technologies depends on incentives for irrigation. Most irrigation systems in the Philippines currently charge farmers a flat rate, based on hectarage, regardless of the actual amount of water they use. A re-visit of the irrigation service fees as well as the type of irrigation system (national irrigation system or communal irrigation system) will have policy implications. Likewise, the manageability of technology can contribute to an enabling environment for farmer adoption. It is often argued that if water is available and easy to apply, then farmers will use more than necessary. Therefore, to promote farmer adoption, technology should be effective, easy to apply, in the desired amount, easy to operate and maintain with local resources, and affordable.

11.3.2.3 Grey Water Reuse

Grey water is wastewater from baths, sinks and washing machines, accounting for about 60% of the outflow from homes. It contains little pathogens and 90% less nitrogen than toilet water, so does not require the same treatment process. Grey water use reduces irrigation water needs, increasing availability of freshwater for other primary uses. The benefits of grey water use especially for agriculture have been demonstrated in many countries such as Africa, Australia, Egypt, Lebanon, Mexico, and Palestine, among others. In the Philippines, wastewater from baths, sinks and washing machines usually go directly to the storm drain, without the benefit of reuse for gardens and farms. Wastewater treatment plants are not available in cities and municipalities, with some exception. Baguio City has a water treatment plant but has been operating by 4000 m^3 more beyond its capacity of 8000 m^3. This urged the local government of Baguio to prepare a site for a decentralized wastewater treatment facility for those who are not serviced by the Baguio Sewage

Treatment Plant. The new facility will prevent wastewater from draining directly into the Balili River System. The Baguio LGU also requires that all high-rise buildings have their own sewage treatment facilities (Refuerzo 2015).

11.3.2.4 Minimizing Non-revenue Water

Minimizing non-revenue water (NRW) is a supply-side water demand management policy. NRW is the difference between system input volume and billed authorized consumption. Accurate information on the quantity of water in and out of the system is paramount to track and control operational efficiencies by helping identify and prioritize actions to reduce losses.

In the Philippines, water districts and concessionaires aimed for metering of water flows at the transmission, distribution, and consumer end to reduce NRW, thus providing information on the amount of water entering the system and consumed. The case of Maynilad's project extending water services to informal settlers to the west zone of Manila significantly contributed to minimizing their NRW. More details about this project are provided in the succeeding sections of this chapter.

Pressure management and leak detection and repair are strategies to reduce NRW. A water supply system is generally designed to operate at pressures, which lead to minimum head loss, while also ensuring adequate pressure to consumers located at the tail end of the network or on levels above the ground floor. High pressure can lead to high system leakage and increased water usage by consumers.

Water supply in many Philippine cities generally faces problems related to supplies at low water pressure. Consumers in many cities such as Baguio, Bulacan, and Iloilo complain of low water pressure and resort to the use of online booster pumps. In a community located near a university in Baguio City, water provided by the Baguio Water District and private water delivery services is not enough for boarding houses and dormitories so owners have drilled wells as deep as 300 feet. These wells are so close together that community leaders fear that empty water tables may already be unsafe for the people living in the area.

11.3.2.5 Peak Season Interventions

Section 18 of the Water Code of the Philippines recognizes the possibility of recurrent water shortage from either peak demand periods and/or times of crises and emergencies. Recurrent water shortage refers to the natural periodic diminution of water in a source of supply to a volume of rate flow insufficient to meet water requirements. Peak water use describes the time of year when residential water use is at its highest. A survey conducted by the NWRB with World Bank in 2014–2015 puts consumption pattern per person per day as varying, largely depending on the water provider. Private sector provides the highest at 120 liters per capita per day (lpcd), followed by water districts at 108 lpcd and the least by LGU at 99 lpcd and CBOs at 86 lpcd. While average per capita consumption is about 96–100 liters of water per day, during peak season (hot and dry summer months), the same household can use about 129–144 liters of water per capita in a day (DILG 2008). Peak-season water demand management interventions take the form of public announcements through websites and news articles urging the public to use water prudently. Practical tips on monitoring daily water use, limiting water usage, and reusing water stretches the precious water supply to provide everyone's water needs during the sweltering months (PCDSPO 2014). Watering of plants is done only when needed, laundry is washed only on full load, and plumbing fixtures are efficiently set up. Peak hours can be avoided when taking showers, cooking, and doing laundry.

Section 8 of the National Water Crisis Act of 1995 provides for anti-pilferage in whatever form, thus presenting a strong policy support to prevent and minimize water loss during peak season.

11.3.3 Social/Political Strategies

The human dimension of water use includes understanding how people use water, their consciousness and knowledge about water, and the behavioral determinants of water conservation. An improved understanding of these elements will help the government develop policies and strategies that can facilitate appropriate behavior change.

Water providers are also implementing non-price conservation measures such as public awareness and information; education, and recycling/reuse with the expectation that these programs will reduce residential water demand (Michelsen et al. 1999). Changing behaviors associated with high water use and encouraging the use of water-saving devices and practices often are seen as effective means to achieve a reduction in water use.

11.3.3.1 Policies and Interventions that Promote Public Awareness and Information

To understand how citizens use water, there is a need to examine the regulatory framework within which water-related behavior takes place, specifically the extent to which individual consumers are engaged. This is basically found in enacted laws pertinent to water. To what extent do these laws directly specify the involvement of water consumers? The Water Code (1976) is explicit in recognizing the need for undertaking public consultation and hearings such as for appropriating water use, adjusting supply during water shortage, and revocation of permits. The National Water Crisis Act (1995) directs nationwide consultations on the water crisis, although the rest of the law is largely devoted to forms of water pilferage and their corresponding penalties.

Only in Letter of Instruction (LOI) 744 (1978) is public education program mandated of water districts on aspects of their operation, including conservation. Likewise, it requires districts to prepare a comprehensive program and system of both formal and informal public consultation when considering increases in water rates. The LOI further mandates the review of the composition of water district boards to ensure that consumers are "properly and fully" represented, although there is no mention of how board members are to relate with the constituency they represent.

The Philippine Clean Water Act of 2004 communicates a greater awareness of the consuming public, initially in its declaration of policy–the pursuit of public health and quality of life; encouragement of "civil society and other sectors in their efforts to organize, educate and motivate the people in addressing pertinent environmental issues and problems at the local and national levels." Two provisions substantiate these policies with the mention of "due public consultation" and the inclusion of individuals and civil society among possible recipients of incentives and rewards for outstanding participation in quality water management initiatives and programs.

Public information and education campaigns continue to be popular means of encouraging users to adopt and maintain long-term water conservation behavior. In general, there has not been a systematic water educational program that has been replicated across the country. There may be occasional campaigns at the local (e.g., city, municipal) levels, which are largely prompted by alarmingly low water supply levels. Such campaigns usually attempt to persuade water users to conserve water, and they also provide users with information on how to do so, such as taking shorter showers, scaling back on lawn watering or car washing, or installing water-saving fixtures. Simple techniques for indoor residential are illustrated which include checking running toilets, dripping faucets, and other household leaks. Most leaks within a home can be easily fixed without investing in new equipment and can be performed with the knowledge and guidance of the local water utility.

During drought, a nationwide public education campaign is conducted, encouraging people to conserve water. In the summer of 2015, Angat Dam, which supplies 97% of Metro Manila's water need, had dangerously low levels. Water officials called for reduction of water use, giving suggestions on how to stop wasteful practices. Newspapers, radio, and television carried stories on the drought, usually accompanied by an exhortation to conserve water.

The above notwithstanding, two water projects, which can be models for increasing public awareness and involvement within communities are described in Box 11.2.

Box 11.2: Model Projects for Public Awareness and Community Involvement

Case 1: Manila Water

Manila Water, the private concessionaire providing water in the east zone of Metro Manila, set up in 1998 the "Tubig Para Sa Barangay" (TPSB, Water For The Community) to serve the water needs of poor marginalized households to easily connect to a piped-in water supply. Of interest here is the community aspect where Manila Water forged a strong partnership with the community and other stakeholders such as local government and non-government organizations. The project divides the service area into smaller territories each handled by a territory manager, whose responsibilities include calling meetings to ensure that all concerns are addressed. Community sense of ownership and responsibility over the project is fostered by involving residents in every stage, holding several public consultations and dialogues, and allowing community members to play crucial roles in the management, billing, collection, maintenance, and monitoring of each water connection. This ongoing program has uplifted the lives of poor people beyond water access, enabling them to pursue livelihood programs to augment limited incomes and raising their quality of life.

Case 2: Maynilad Water Services

The Maynilad Water Services Inc., the second concessionaire providing water to the west zone of Metro Manila, launched its Samahang Tubig Maynilad (STM; Maynilad Water Associations) in 2009. STM extended water services to informal settler families, as these poor households would otherwise not have legal rights and financial resources to apply for water connections from a water district. The approach uses a community-based water management system in which residents are directly involved in the day-to-day operations and administration of the water supply, largely similar to TPSB of Manila Water. Through community organizing, residents are grouped into associations and undergo training for value education and water management to enable them to manage their community water system. Once organized, members of the STM sign a Memorandum of Agreement with Maynilad Inc., committing to their duties and responsibilities as water system administrators. Maynilad installs a bulk water wherein access is either through metered public faucets within the community or piped-in water to a cluster of households, each with sub-meters. This approach has benefitted poor communities.

Sources: Manila Water Company, Inc., 2010. Sustainability Report. http://www.manilawater.com.ph/downloada/2010SustainabilityReport.pdf (Accessed on March 25, 2017).

Maynilad Water Services, Inc., 2011. Sustainability Report. Securing our Resources. http://www.mayniladwater.com.ph/downloads/2011MayniladSustainabilityReport.pdf (Accessed on March 25, 2017)

11.3.3.2 Policies and Interventions That Address Social Behavior and Promote Water Conservation

Citizens' behavior relative to water is shaped not only by his knowledge about water as may be obtained through public education programs but also by other factors. By looking at government legislation, one can identify the behaviors that are explicitly promoted.

The 1976 Water Code recognizes the rights of individuals to know who apply for water permits and to protest any issuance of water permits when these infringe on the right to water of other parties. It further confirms the right to water when it states that individuals are free to draw manually water even from private sources for domestic purposes, as long as it does no injury to the owner. The Water Code also has a whole chapter (VI) on conservation and protection of water resources. Those provisions that apply to individual behavior are duties of individuals controlling wells to ensure water is not wasted or contaminated; preference in granting permits in favor of associations (irrigators) rather than individuals when both apply for the same area; and prohibition of individuals from building works or performing acts that may introduce dangerous substances or pollution into the water supply. The rest of the chapter addresses more macro level interventions.

Section 8 of the Water Crisis Act of 1995 identifies a list of 'pilferage' behaviors considered unlawful: destruction of any property or waterworks of any water utility; malicious acts that injure quantity or quality of water or sewage flow; obstruction with works performed on water mains and distribution network; tapping of water connection without appropriate authorization from water utility; tampering with meters resulting in stolen water; use or receipt of illegally obtained water; and possession of stolen or tampered water meters. The Act further stipulates the penalties for pilferage. It is observed that this law only focuses on behaviors that are prohibited but none on advocated actions.

The Clean Water Act of 2004 provides for incentives and rewards. It mandates "due public consultation" (in coming up with a list of industry categories) and inclusion of individuals and civil society among possible recipients of incentives and rewards for outstanding participation in water management initiatives and programs. However, it does not go any further into strategies for achieving these.

Greenpeace (2007) observed that inefficiency in water usage is aggravated by the absence of regulations, economic incentives, and institutional arrangements needed to promote water conservation and rational use of water. Similarly, Gamboa (2011) recommends that the Water Code be amended to integrate the adoption of integrated water resource management (IWRM) as a guiding strategy.

The aforementioned shows that legislation has little to say about individual behavior. This may be because of lack of knowledge from behavioral research that can guide and lead to more comprehensive water policy. There is growing advocacy for non-structural approaches to water management which encompasses scientific research, education, and persuasion to coordinate how humans use water (FAO 1993).

The social psychological literature has looked into the role of intrinsic factors in influencing people's water conservation behavior. Jorgensen et al.'s study (2009) found that interpersonal and institutional trust is a critical attribute in household water consumption. People will not save water if they feel others are not doing their share of reducing their consumption and if they do not trust the water authority. Another study showed that personal normative beliefs related to water had a positive effect on conservation but antisocial behavior inhibited conservation (Corral-Verdugo and Frias-Armenta 2006). Perceptions of inefficacy of water conservation laws produced no effect on water conservation practices. Jones et al. (2011) explored the link between citizens' social capital and perceptions on restrictions imposed from water consumption policy instruments on one hand and their perceived level of effectiveness on the other hand. Social capital consists of social and institutional trust, participation in social networks, and compliance with social norms. The study found that significant connections exist between elements of social capital and perceptions of citizens toward water consumption policies.

Attitudes are usually considered precursors to behavior although decades of attitude-behavior research has not established an absolute relationship. Behavior is better predicted by intentions. The theory of reasoned action and planned behavior proposes that attitudes, subjective norm or the influence of a person's social environment and control over behavior are the predictors of intentions. A study by Syme et al. (1983) among households in Perth, Australia, had mixed results with water consumption being related in some instances to household size and family income and in other instances to attitudinal factors. Kantola et al.'s (1982) study found that influence on intentions to conserve water was explained by subjective normative feelings (perceived peer pressure) and the exogenous variable age. Expressed attitudes did not contribute significantly to predicting intention to consume water when subjective norms were included in the regression model. The results from attitudinal studies suggest pathways to interventions such as providing appropriate information about the behavior, offering information about consequences, and presenting opportunities to compare oneself with others.

The results of a survey on conservation efforts in Melbourne (Syme et al. 2000) showed that, although most community members were confident that their households could succeed in saving water voluntarily, the general view was that voluntary responses were not effective. Research and industry experience suggest that public acceptability of demand management practices depends on the type of proposed measures and the current water supply situation. Sociological methods include appeals, way of living, or legal action. Appeals, through the media or on accounts, rarely last long before consumers forget the urgency. During periods when drought is severe, individual users may be willing to adopt a wide range of short-term measures, including practices that require changes in their normal economic activities and lifestyles.

In the Philippines, water conservation through reuse and recycling has been promoted even in normal times. Recycled water means water which, as a result of treatment of waste, is suitable for a beneficial use that would not otherwise occur. In the urban sector, recycled water can be used in place of potable water in landscape

Box 11.3: Reuse and Recycling of Water in the Cordillera
Domestic water in the upstream (Baguio City) and midstream (La Trinidad) sites is often inadequate and expensive for water users, hence residents resort to recycling and reusing water. Bath water is used to flush the toilet while laundry or dishwashing water is re-used to clean floors or flush the toilet. When suitable, recycled water is used for gardens. Likewise, use of rainwater is not underestimated. Almost all households have storage drums or tanks for rainwater, which is used for cleaning, laundry, flushing the toilet, and even for bathing.

Commercial water, such as that used in water refilling stations, is also recycled in Baguio City. The filtering process results in a large volume ejected as waste water. Business owners use this water to operate a laundry business. It is common to see these two businesses in areas within Baguio City where there is a concentration of dormitories and boarding houses for students.

Even in the downstream town of Sablan where water is abundant for most of the users, some households live too far from water sources. Therefore, the use of rainwater is a necessity and recycling bath water for watering plants is also done.

Source: Focus group discussions in Baguio City, La Trinidad, and Sablan conducted in May 2013.

irrigation and some industrial uses. Government and water agencies should pursue appropriate opportunities for replacing potable water with recycled water. Likewise, the use of grey water has also received a lot of attention as discussed in the earlier section. Box 11.3 demonstrates the reuse and recycling of water in the Cordillera highlands in northern Luzon, Philippines. The area is characterized by differing stages of economic transformation: from a predominantly subsistence-based downstream community to a mixed subsistence-market midstream community and finally, to a predominantly market-based upstream community.

11.4 Recent Initiatives to Improve Access to Water

Water agencies also have a role in areas where the demand for water is not met due to limited coverage of water supply and sanitation services. Plans for satisfying these "unserved" demand are an important component of demand management as are reductions in water use and losses in areas or sectors with "fully served" demand (Dziegielewski 2003). This section discusses initiatives by the private sector to meet the water demand in unserved areas. Although there are two local government initiatives, the *Sagana at Ligtas sa Tubig sa Lahat or Salintubig* (translated as Abundant and Safe Water for All) and Bottom-Up Budgeting – Water, these are at their initial stages and monitoring is still in progress. Both are aimed at attaining significant progress toward the development of providing universal access to clean and safe water supply by 2025.

11.4.1 Private Sector Participation

Within the sphere of publicly financed networks, water systems piped into premises are limited in coverage, and service delivery is irregular at best. LGUs struggle to expand their utilities, leaving both rich and poor residents underserved. A potential solution to accelerate access to piped water services, especially for the poor is public-private partnerships (PPPs). In this arrangement, private sector capital is mobilized for water system improvements and expansion at a scale far larger than that available from public funds. Water services are more reliable as operators face the incentive to match supply with consumer willingness to pay, enabling a sustainable cash flow, and facilitating service coverage expansion. The private sector also brings technical and financial expertise to manage water utilities in a more efficient and sustainable manner.

11.4.1.1 Initiatives of the Metro Manila Water Concessioners

Before the 1997 privatization, the government-owned and -operated Manila Waterworks and Sewerage Services (MWSS) was the water utility mandated to supply water and provide municipal wastewater treatment facilities in Metro Manila. The National Water Crisis Act of 1995 (RA 8041) provided the legal basis for the privatization of MWSS in 1997, a change that brought about some remarkable improvements in water supply coverage and service performance. The PPP arrangement set targets of reaching 24/7 water service within 6 years and universal coverage for water within 11 years, while ensuring drinking water quality according to national standards at a pressure of 16 psi.

Participation was implemented through a concessionaire contract of 25 years where the concessionaires were assigned the task of operating and managing the facilities, whereas MWSS preserved the ownership of the infrastructure. Following the Paris, France water privatization model wherein the utility split its jurisdiction into two to prevent a monopoly, MWSS granted Manila Water Company the East Zone concession and Maynilad Water Services, Inc. the West Zone concession (Northrop 2012). The concession contracts included targets concerning coverage, service quality, and economic efficiency meant to address issues related to supply distribution, finance, and strengthening the government's anti-pilferage efforts.

The story or the Manila water concessions is a remarkable narrative of turnaround in the provision of water and sanitation services over two decades, first in east Manila, and then in west Manila. Both concessions have achieved world-class performance, doubling the number of water connections since the start of the concession period. In the first 5 years of operation, much improvement was noted in service indicators and there was an initial price drop between 1996 and 1997 (Chia et al. 2007; Montemayor 2003). This was due to the low water tariffs offered by the concessionaires. The partnership between MWSS and Maynilad Water Services, Inc. led to 1,129,497 connections in 2013 (MWS, Inc. 2015).

The Manila concessions both feature pro-poor mechanisms that provide differentiated level of services that are more affordable to base-of-the-pyramid customers and contribute significantly to concessionaire revenues. Details of the pro-poor programs of the two concessionaires are discussed in the succeeding section.

11.4.2 PPP Initiatives in the Provinces

This section examines seven water utility PPPs in the provinces using Castro et al.'s (2015) study and the World Bank's Water and Sanitation Program (WSP) study in the Philippines (World Bank 2015). These PPPs varied in size and contract type. Likewise, catalysts that drove both public and private sectors to come together differed across areas, although they shared a commonality in terms of desperate need (Table 11.2).

A key finding in the implementation of PPPs in both Manila and the provinces shows that different arrangements can lead to affordable, reliable, and clean water services, provided there is sufficient market size and willingness to pay. PPPs can thrive in diverse geographies, as long as service is focused on meeting the demand

Table 11.2 PPPs in provinces across the Philippines

		Partners		Connection	Arrangement
Area	Driver of PPP	Public	Private		
Malasique, Pangasinan	Poor service limited to *poblacion*	Municipal government	Inpart Waterworks and Development Corporation	2419	Concession
Tabuk City, Kalinga	Intermittent water supply	City government	Calapan Waterworks Corporation	3600	Lease
Norzagaray, Bulacan	Drying up of wells	Water district	Phil Hydro/ Maynilad	NA	Build-operate-transfer
Laguna	Water-related health issues	Provincial government	Laguna Water Corporation	61,448	Concession (joint venture)
Quezon, Palawan	Complete lack of water supply	Provincial government	Alfonso XII Water Users' Association		Management operation & maintenance contract
Boracay, Aklan	Inadequate water and wastewater infrastructure	Tourism Infrastructure and Enterprises Zone Authority	Boracay Island Water Company	5531	Concession (joint venture)
Sta Cruz, Davao Sur	Intermittent water supply	Municipal government	Sig Construction	3911	Design-build-lease

Source: Castro et al. 2015

for which consumers are willing to pay. Achievements of PPP arrangements are summarized in the following outcomes:

- 24/7 water service
- Water availability ≥100 liters/capita/day
- Water pressure ≥ 7 psi
- Drinking water quality according to Philippine national standards
- Working ratio > 50%, assuring adequate revenue generation to operator
- Collection efficiency >90%
- Non-revenue water ≤20%
- Number of staff per 1000 connections within international benchmarks of 3 to 54

However, achieving universal coverage of households in the Philippines remains a challenge. Llanto (2013) reported that about 15.73 million Filipinos still do not have access to a safe water supply, leaving more room to work with the private sector on innovative financing schemes to increase investments in water supply and sanitation.

11.4.3 Initiatives of Concessionaires and Water Districts to Provide Access to the Poor

Pro-poor approaches are not yet universal, but successful approaches have been implemented in some areas. The two water companies in Manila have made efforts to reach the poor who can ill-afford a private water connection. The issue of illegal settlers had to be addressed inasmuch as private utilities are not allowed to connect illegal settlers to the network. Consequently, community-based innovative solutions, such as the provision of a water system for clustered low-income communities (Manila Water's *Tubig sa Barangay* [Water for Communities] Program) and the installation of a bulk water system or public faucet with a mother meter (Maynilad), have been found to overcome this problem and ensure sustainable solutions that develop self-reliance in some areas of the city (Gazmen 2012). Manila Water's operational initiative, Tubig sa Barangay not only enabled the company to meet its goal of reducing non-revenue water through self-managed water districts among informal communities, but it also earned the company legitimacy and support from the urban poor against opponents of privatization (Hall et al. 2015). Maynilad offered discounted tariffs for consumers with monthly consumption below the minimum bracket of 10 m^3 while also providing a network of tap stands to poor communities.

Water districts carry out programs to reduce utility charges for poor residential customers. These charges are often described by the single category of lifeline rates. The lifeline concept of rate design is frequently proposed as an aid to economically disadvantaged and elderly residential customers who might not be able to pay their bills. Boracay also provide discounted tariffs for consumers with monthly

Box 11.4: Dagupan City Water District Island Water Project
The successful innovation of the Dagupan City Water District (DCWD) is the network of underwater pipelines that extend potable piped water service from the mainland of Dagupan to four island barangays of Dagupan City. The P30 million project was funded from the DCWD funds, with major contribution from the Countryside Development Fund of a congressman and the city government of Dagupan. The scale and novelty of the project is attributed to the support of three succeeding city administrations. This is a perfect example of new mayors lending support to worthy projects of the previous mayor. It is also uniquely an undertaking by local divers who installed the pipes under water. Unlike other underwater pipeline projects that hire professional scuba divers, this project made use of local folks who were also able to earn a living out of the project. Their vast and long experience in diving in these waters allowed them to predict underwater current and other factors that helped in project planning and implementation. The local divers are also relied on for regular maintenance activities of the pipelines.

Source: LWUA (2016).

consumption below the minimum bracket of 10 m^3. Likewise Manila, Laguna, and Boracay provide installment plans for connection charges.

In addition to lifeline and discounted rates, water districts implement special projects that provide access to water in extremely difficult areas. The case of the Dagupan Water District is shown (Box 11.4) to illustrate this point. Pipes have to be installed under water and the collaboration of the recipient community has to be solicited to prevent the pipes from being destroyed by the current and to maintain the facility.

11.5 Summary/Conclusions/Insights/Prospects

Water demand management has, over the years, emerged as both an alternative and complement to the conventional water supply management. Major policy approaches to support and promote WDM can target water users and provide through demand-side and supply-side interventions, respectively. On the consumption side, water demand reduction by avoiding undue consumption can be encouraged by a combination of market incentives and communication and education programs that heavily influence social behavior. On the provider side, promotion of devices and technologies that facilitate less water use and strategies that minimize losses in the system and improve water distribution and operational efficiency could lead to more efficient management and use of available water resources. Exploring alternative sources such as grey water reuse and recycling of wastewater for non-potable uses could also augment existing water supply.

Implementation of the two-pronged approach would require reforms in existing policies on water. On the provider side, policies must provide appropriate and adequate regulatory, institutional, and legal framework for delivery of services; tariff reforms must ensure financial sustainability of operations of the utility; and internal reforms within the providers must aim at improving the operational efficiencies and reducing losses to acceptable levels. Water districts are included in the list of government owned and controlled corporations (GOCCs) and are therefore covered for review under the GOCC Governance Act of 2011. A big challenge for water utilities in the Philippines is to improve efficiency from fuller advantage of economies of scale. Where applicable, policies must be supportive of integrated systems that achieve optimal returns despite relatively lower tariffs. As part of internal reforms, water districts must subject themselves to benchmarking, a tool used to measure performance of a water utility through a set of technical, financial, and social indicators, with the ultimate goal of improving its quality and performance. If undertaken on a regular basis, benchmarking supports utilities in assessing progress and promotes accountability by making information available to the public, decisionmakers, and regulators.

On the consumer side, more policy challenges are faced in terms of long-term changes in ways of life and consumer behavior to reduce water consumption. A "rebound effect" is often observed following a water crisis when customers return to their former patterns of water use. A certain amount of savings is more lasting, partly due to the spread of water-efficient technologies, but also due to lasting behavioral changes. However, little information exists on the effectiveness of non-price conservation programs in reducing water demand. Michelsen et al. (1999) recommended that utilities maintain more detailed and consistent information regarding the implementation of their non-price programs. Turton (1999) emphasized the importance of adaptive social capacity for countries experiencing water scarcity. He agreed with Ohlsson (1999) who suggests that the adaptive capacity of a society should also be regarded as a resource, and therefore societies facing water scarcity will have different development trajectories, depending on their adaptive social capacity.

Our review of national laws related to water showed an obvious bias toward punitive provisions, which cannot be fully satisfied, given the shortcomings in administrative mechanisms for implementation. There is a more promising venue for policies that provide for incentives and rewards for rational water use such as grey water reuse and recycling, and use of water-saving devices and technologies. However, implementation of these policies will remain inadequate and confusing unless we put in place a robust national water resource body with a clear mandate to address water issues both on the supply and demand sides.

References

Abansi, C. L., Doble, M. C. C., Cariño, J. K., & Rola, A. C. (2016). Beyond prices: The cultural economy of water in the Cordillera Highlands of Northern Luzon, Philippines. *Asia Pacific Viewpoint, 57*, 280–293.

Agrawal, A., & Goyal, S. (1999). *Group size and collective action: Third party monitoring in common-pool resources, Leitner Working Paper No. 1999–09*. New Haven: The Leitner Program in International Political Economy, Yale University.

Araral, E., & Wang, Y. (2013). Water demand management: Review of literature and comparison in South-East Asia. *International Journal of Water Resources Development, 29*(3), 434–450.

Bakker, K. (2003). *Good governance in restructuring water supply: A handbook*. Ottawa: Federation of Canadian Municipalities and the Program on Water Issues.

Castro, A., Jagannathan, V., & Navarro, M. (2015). *Beyond one-size-fits-all: Lessons learned from eight water utility public-private partnerships in the Philippines. Water and sanitation program learning note*. Washington, DC: World Bank Group. http://documents.worldbank.org/curated/en/788411468188355625/Beyond-one-size-fits-all-Lessons-learned-from-eight-water-utility-Public-Private-Partnerships-in-the-Philippines. Accessed 3 Jan 2016.

Chia, P. G., Chua, C. K., Kim, F. C., Teo, S., & Toh, K. L. (2007). *Water privatization in Manila, Philippines-should water be privatized? A tale of two water concessionaires in Manila*. Economics and Management in Developing Countries (7 May 2007, pp 21). INSEAD. http://www.circleofblue.org/waternews/wpcontent/uploads/2012/06/Insead_Water_Privatization_Manila_Philippines.pdf. Accessed 15 Mar 2016.

Corral-Verdugo, V., & Frias-Armenta, M. (2006). Personal normative beliefs, anti-social behavior and residential water conservation. *Environment and Behavior, 38*(3), 406–421.

DILG (Department of Interior and Local Government). (2008). *Benchmarking performance of small town water supply*. Quezon City: DILG.

Dziegielewski, B. (2003). Strategies for managing water demand. *Water Resources Update, 126*, 29–39.

Dziegielewski, B., Opitz, E. M., Kiefer, J. C., & Baumann, D. D. (1993). *Evaluating urban water conservation programs: A procedures manual*. Denver: American Water Works Association.

FAO (Food and Agriculture Organization of the United Nations). (1993). *The state of food and agriculture, FAO Agricultural series*. Rome: FAO.

Feitelson, E. (2012). What is water? A normative perspective. *Water Policy, 14*, 52–64.

Fernandez, R. A. (2015). DA promotes water-saving technologies. *The Philippine Star*. Retrieved from http://www.philstar.com/agriculture/2015/11/29/1526946/da-promotes-water-saving-technologies. Accessed 12 Dec 2015.

Freeman, J., & Kolstad, C. D. (2007). *Moving to markets in environmental regulation: Lessons from twenty years of experience*. New York: Oxford University Press.

Gamboa, M. (2011). *Policy brief #1. National College of Public Administration and Governance*. Diliman/Quezon City: University of the Philippines.

Gazmen, P. G. (2012). *Tubig para sa Barangay: Providing clean and affordable water to poor communities using a sustainable model and community participatory approach*. Paper presented at the roundtable discussion on Water Rights and Water Wrongs: Toward Good Water Governance for Development. Social Sciences Division- NAST PHL, January 26, 2012, Hyatt Hotel and Casino Manila, Manila, Philippines.

Gleick, P., Wolff, G., Chalecki, R.R. (2002). The new economy of water: The risks and benefits of globalization and privatization of fresh water." Oakland, California: Pacific Institute for Studies in Development, Environment, and Security.

Glennon, R. (2004). The price of water. *Journal of Land, Resources & Environmental Law, 24*(3), 337–342.

Greenpeace. (2007). *The state of water resources in the Philippines*. Quezon City: Greenpeace Southeast Asia. Greenpeace Southeast Asia. http://www.greenpeace.org/seasia/ph/Global/seasia/report/2007/10/the-state-of-water-in-the-phil.pdf. Accessed 28 Oct 2016.

Hall, R., Lizada, J., Dayo, M. H., Abansi, C., David, M. E., & Rola, A. C. (2015). To the last drop: The political economy of Philippine water policy. *Water Policy, 17*, 946–962.
Hanemann, M. (2005). *The value of water*. Berkeley: University of California.
Herbertson, P. & Tate, E. (2001). *Tools for water use and demand management in South Africa*. Technical Reports in Hydrology and Water Resources, No.73. Geneva: World Meteorological Association. http://www.wmo.int/pages/prog/hwrp/documents/TD73.pdf. Accessed 12 Dec 2016.
Herrfahrdt-Pahle, E. (2010). South African water governance between administrative and hydrological boundaries. *Climate Development, 2*, 111–127.
Inman, D., & Jeffrey, P. (2006). A review of residential water conservation tool performance and influences on implementation effectiveness. *Urban Water Journal, 3*, 127–143.
IPCC (Intergovernmental Panel on Climate Change). (2007). *Climate change 2007: Synthesis report*. Contribution of Working Groups I, II and III to the Fourth Assessment Report of the Intergovernmental Panel on Climate Change [Core Writing Team, Pachauri, R.K and Reisinger, A. (eds.)]. Geneva: IPCC. 104 p.
Jalil, M. A., & Njiru, C. (2006). Water demand management in urban water supplies: Present and future challenges. In M. M. Rahman, M. J. Bin Alam, M. A. Ali, & I. Smout (Eds.), *Environmental sustainability concerns* (pp. 43–57). Dhaka: Environmental Engineering Division (EED), Centre for Environmental and Resource Management (CERM), and International Training Network (ITN) Centre, Bangladesh University of Engineering and Technology.
Jones, N., Evangelinos, K., & Gaganis, P. (2011). Citizens' perceptions on water conservation policies and the role of social capital. *Water Resource Management, 25*, 509–522.
Jorgensen, B., Graymore, M., & O'Toole, K. (2009). Household water use behavior: An integrated model. *Journal of Environmental Management, 91*(1), 227–236.
Kantola, S., Syme, G., & Campbell, N. (1982). Role of individual differences and external variables in a test of sufficiency of Fishbein's model to explain behavioural intentions to conserve water. *Journal of Applied Social Psychology, 12*, 70–83.
Linton, J. (2010). *What is water: The history of a modern abstraction. Vancouver*. Toronto: UBC Press.
Llanto, G. (2013). *Water financing programs: Creating incentives for private sector participation*. Philippine Institute for Development Studies (PIDS) Policy Notes No. 2013–16. Makati City, Philippines: PIDS. http://dirp4.pids.gov.ph/ris/dps/pidsdps1334.pdf. Accessed 14 Mar 2016.
LWUA (Local Water Utilities Administration). (2014a). *A primer on LWUA and the water districts*. http://www.lwua.gov.ph/about_lwua_14/primer.html. Accessed 27 Apr 2016.
LWUA (Local Water Utilities Administration). (2014b). *Manual on water rates: Chapter 3*. http://www.lwua.gov.ph/downloads_10/LWUA_water_rates_manual.pdf. Accessed 29 Apr 2016.
LWUA (Local Water Utilities Administration). (2016). *Water: Water code of the Philippines implementing rules and regulations*. pp. 16–38. http://www.lwua.gov.ph/downloads_14/watercode-phil.pdf. Accessed 27 Apr 2016.
Massarutto, A. (2007). Water pricing and full cost recovery of water services: Economic incentive or instrument of public finance? *Water Policy, 9*(6), 591–613.
Michelsen, J., McGuckin, J. T., & Stumpf, D. (1999). Non-price water conservation programs as a demand management tool. *Journal of American Water Resources Association, 35*, 593–602.
Min, B. (2007). Determinants of water pricing: A regression analysis with cross-country data. *Water Policy, 9*(4), 363–372.
Montemayor, C. (2003). *The Manila Water privatization fiasco and the role of Suez Lyonnaise/Ondeo*. Paper presented at the Summit for Another World, G8 summit in Evian, 2003). http://www.tni.org/sites/www.tni.org/archives/altreg-docs/manila.pdf. Accessed 8 Dec 2015.
MWC (Manila Water Company, Inc.). (2010). *Sustainability report*. http://www.manilawater.com.ph/downloads/2010SustainabilityReport.pdf. Accessed 25 Mar 2017.
MWSI (Maynilad Water Services, Inc.). (2011). *Sustainability report: Securing our resources*. http://www.mayniladwater.com.ph/downloads/2011MayniladSustainabilityReport.pdf. Accessed 25 Mar 2017.

NEDA (National Economic Development Authority). (2010). *Philippine water supply sector roadmap*. Pasig City: NEDA.

Northrop, A. (2012). *Infographic: Water privatization in the Philippines*. Circle of Blue. http://www.circleofblue.org/2012/world/infographic-water-privatization-in-the-philippines/. Accessed 25 Nov 2015.

Ohlsson, L. (1999). *Environment, scarcity and conflict: A study of Malthusian concerns*. Goteborg: Department of Peace and Development Research, Goteborg University.

Olmstead, S., Hanemann, W. M., & Stavins, R. N. (2007). *Water demand under alternative price structures. National Bureau of Economic*. Research working paper series no. 13573. http://www.nber.org/papers/w13573.pdf. Accessed 15 May 2016.

Payero, J. O., Yonts, C. D., Irmak, S., & Tarkalson, D. D. (2005). *Subsurface drip irrigation: Advantages and disadvantages*. Lincoln: University of Nebraska-Lincoln Extension Circular EC776.

PCDSPO (Presidential Communications Development and Strategic Planning Office). (2014). *Infographic: How to conserve water*. Official gazette of the Republic of the Philippines. http://www.gov.ph/2014/05/23/infographic-how-to-conserve-water/. Accessed 7 Nov 2015.

Philippine Water Districts. *Average Production and Consumption Data as of 10/23/2017*. http://122.54.214.222/UnitCons.asp?Date=10/23/2017. Accessed 24 Oct 2017.

Refuerzo, A. (2015). Baguio wastewater treatment facility to ease water pollution. *Northern Philippine Times*. http://northphiltimes.blogspot.com/2015/10/baguio-wastewater-treatment-facility-to.html. Accessed 10 May 2016.

Rola, A. C., Abansi, C. L., Arcala-Hall, R., & Lizada, J. C. (2016). Characterizing local water governance structure in the Philippines: Results of the water managers' 2013 survey. *Water International, 41*(2), 231–250.

Rosellon, E. (2016). *Exploring water-saving technology for the Philippines' national irrigation system*. IRRI. http://news.irri.org/2016/05/philippine-food-security-chief-explores.html. Accessed 26 May 2016.

Syme G. J., Thomas, J. F., & Salerian, S. N. (1983). *Can household attitudes predict water consumption?* Paper presented at the Hydrology and Water Resources Symposium, 8–10 Nov 1983, Hobart, Australia.

Syme, G., Nancarrow, B. E., & Seligman, C. (2000). The evaluation of information campaigns to promote voluntary household water conservation. *Evaluation Review, 24*, 539–578.

Tagar, A., Chandio, F. A., Mari, I. A., & Wagan, B. (2012). *Comparative study of drip and furrow irrigation methods at farmer's field in Umarkot*. http://waset.org/publications/2316/comparative-study-of-drip-and-furrow-irrigation-methods-at-farmer-s-field-in-umarkot. Accessed 26 May 2016.

Turton, A. R. (1999). *Water demand management (WDM): A case study from South Africa*. MEWREW Occasional Paper No. 4. Presented to the Water Issues Study Group, School of Oriental and African Studies.

Van Koeppen, B., Giordano, M., Butterworth, J., & Mapedza, E. (2007). *Community-based water law and water resource management reform in developing countries*. Cambridge, MA: CABI.

World Bank. (2005). *Philippines: Meeting infrastructure challenges*. Washington, DC: The International Bank for Reconstruction and Development.

World Bank. (2015). *Water and Sanitation Program end of year report*. http://www.wsp.org/sites/wsp.org/files/publications/WSP-End-Year-Report-FY15.pdf. Accessed 27 May 2016.

Yonts, C. D. (2008). *Soil physical properties and irrigation basics [CD-ROM]*. Paper presented at the Irrigation and Energy Workshop for Corn Growers, 11 Feb 2008, Clay Center. https://masters.agron.iastate.edu/files/reeschristopher-finalccproj.pdf. Accessed 26 May 2016.

Young, O. (2002). *The institutional dimensions of environmental change*. Cambridge: MIT Press.

Dr. Corazon L. Abansi is Professor of the Institute of Management, University of the Philippines Baguio and holds a Ph.D. in Agricultural Economics major in Resource Economics. She is an ICM practitioner of UNDP-IMO facility "Environmental Management of the Seas of East Asia". She did research and development work on economics of water pollution, resource valuation and community mobilization while serving as Officer and resource economist of UNDP-IMO Program for the Prevention of Marine Pollution in East Asian Seas. Dr. Abansi is actively engaged in research in the Cordilleras focusing on the economics of hydropower generation and ecosystem services and resource management of indigenous communities. As site coordinator and resource economist of the project, "Towards Good Water Governance for Development" she is working towards developing an adaptive-collaborative water governance mechanism for the Cordilleras and has co-authored three papers published in the Water Policy, International Journal of Water Resources Development and Water International.

Dr. Rosalie Arcala Hall is a Full Professor at the University of the Philippines Visayas (UPV). She completed her Master's degree in Political Science and PhD in International and Public Affairs (2002) at Northeastern University, Boston, Massachusetts. She completed research projects with The Asia Foundation, The Nippon Foundation and Toyota Foundation concerning policy frameworks, their localization, resistance and competing narratives behind contestations in the areas of hard and soft security. She has contributed chapters to the books Rosalina Palanca-Tan ed. Nature and Culture: Environmental Issues in Asia (Ateneo de Manila Press, 2014), and Justine Vaz and Narumol Aphineves eds. Living Landscapes Connected Communities: Culture, Environment and Change Across Asia (Areca Books, 2014). She is co-author to two articles in Water Policy and International Journal of Water Resources Development. Currently, she is a member of the Philippine Commission on Higher Education Technical Committee on Political Science and the Philippine Political Science Association Board.

Dr. Ida M.L. Siason is Professor Emeritus of the University of the Philippines, and served as faculty and Chancellor of the UP Visayas. She obtained a Masters in Psychology degree from the Ateneo de Manila University and a PhD Social Psychology from the Pennsylvania State University, USA. She has been involved in the 'Water Governance for Development Project' of the University of the Philippines. Dr. Siason's research and publications have been in social analysis, women's studies, social-psychological characterization of fishing communities, gender issues, enforcement and compliance, and community-based coastal resource management. She has presented papers in national and international conferences in psychology, fisheries, and more recently on water governance. She has spearheaded activities to promote fisheries social science in the Philippines and was the charter president of the National Network on Women in Fisheries in the Philippines.

Chapter 12
Towards a More Responsive Water Policy and Practice: Challenges and Prospects

Agnes C. Rola, Rosalie Arcala Hall, and Juan M. Pulhin

Abstract This final chapter discusses key findings of the book; one significant finding, among others, is that the Philippines will suffer from an impending water crisis if institutions were not strengthened. Sectoral issues are also summarized. Challenges and ways forward are discussed.

Keywords Water sustainability • Water planning • Science- based water management • Water pricing • Integrated Watershed Management

12.1 Introduction

Distilling from the key messages given and the challenges described in previous chapters, this concluding chapter synthesizes the context, key issues, and sectoral performance of water policy and practice in the Philippines with focus on domestic, agricultural, aquaculture/marine, integrated water resource utilization, and environmental protection to ensure the sustainability of water resources. The ways forward are discussed in the last section.

A.C. Rola (✉)
Institute for Governance and Rural Development (IGRD), College of Public Affairs and Development, University of the Philippines Los Baños, College, Los Baños, Laguna, Philippines
e-mail: acrola@up.edu.ph

R.A. Hall
Division of Social Sciences, College of Arts and Sciences, University of the Philippines Visayas, Miagao, Iloilo, Philippines

J.M. Pulhin
Department of Social Forestry and Forest Governance, College of Forestry and Natural Re-sources (CFNR), University of the Philippines Los Baños, Los Baños, Laguna, Philippines

A.C. Rola et al. (eds.), *Water Policy in the Philippines*, Global Issues in Water Policy 8, https://doi.org/10.1007/978-3-319-70969-7_12

12.2 Key Findings

This section summarizes the key findings.

12.2.1 An Impending Water Crisis in the Philippines Is Real

Available data show that water demand is increasing and water supply is dwindling. Some water resource regions are already experiencing water shortage, especially during dry months. Majority of the major river basins are undergoing massive land conversions to other uses, threatening the sustainability of water supply and recharge from this precious resource. Projections indicate that the situation is likely to worsen in the future in the absence of appropriate intervention.

The 1998 JICA-NWRB study projected that, by the year 2025, there will be water availability deficit in major river basins in at least six of the 12 water resource regions, even under the low economic growth scenario and not even considering the likely adverse impacts of climate change (JICA/NWRB 1998).

Many of the major rivers and lakes are heavily polluted. Domestic wastewater is the main contributor of bacterial contamination to groundwater supply. Water-borne diseases such as diarrhea, cholera, dysentery, hepatitis A, and others can be caused by the presence of coliform bacteria in drinking water. An estimated 2.2 million tons of organic pollution is produced annually by the domestic (48%), agricultural (37%), and industrial (15%) sectors. Groundwater has been contaminated with coliform in 58% of the areas sampled. The adverse impact of water pollution costs the Philippine economy an estimated PhP 67 billion (more than US$1.3 billion) annually *(World Bank* 2003). Saline water intrusion has emerged as a problem in some areas, reducing the availability of groundwater supply. This is caused by overexploitation or excessive withdrawal of groundwater. The rapid decline of the forest cover in the Philippines is also a cause of concern as this directly affects surface water supply.

Urbanizing, highly populated, and high-economic-growth areas were observed to have negative surface water balance. While these areas are mostly served by formal water providers, non-revenue water remains high, especially in the cities. Climate change also threatens long-term water supply as longer dry spells would induce further stress on the agricultural, domestic, and energy sectors. In addition, the available data only capture users attached to formal water providers and do not include informal users or those who engage in self-provisioning. Most domestic water users are informal.

12.2.2 Institutional Mechanisms Remain Weak and Fragmented

Formal authority over water matters is dispersed among many government agencies and between the national and local levels of government. The national government agencies have overlapping or competitive mandates. Regulatory power is given to three different agencies, each with its own capability and personnel deficits. Agencies collect their own data in keeping with their mandates, be it water quality, resource management, or supply. Interagency coordination platforms exist for flood control, water crisis, and sector planning, but these are weak and focus only on these singular dimensions. Water apex bodies are institutionally weak and are structured as subline agency bodies that are intermittently transferred, depending on the political mood of the executive branch. The decentralization of decision making on water is limited, with local governments not having the ability to supervise water organizations operating within their administrative jurisdiction and being highly dependent on financial transfers from the national government for water development projects.

Formalization of the water sector through the state apparatus is incomplete. Formal institutional arrangements are more prevalent in urban areas and town centers serviced by water districts. In remote locations, informal water governance schemes based on indigenous and customary rules prevail rather than water laws set by the government. In these places, the state's exclusive claim on water (legal rights) is contested or not observed. The state's regulatory reach is weak; the more numerous private water providers, rural and barangay (village) water and sanitation associations, and local-government-run water utilities are operating without permit and not observing standards on tariff setting.

There is poor compliance with water rules, including permit-based access and withdrawal, as well as pollution. Sanitation is neglected by the government in favor of public investments toward water provisioning (irrigation and domestic).

There remains a serious gap in mechanisms for transfers between sectors and across administrative jurisdictions. The lack of clarity regarding these mechanisms generates conflicts among actors and, subsequently, political mobilization by losing groups for favorable outcomes. Poor legal grounding has also produced conflicts involving local government units (LGUs) over water transfers reaching the courts. Social agreements among multiple users of a common water source provide a temporary fix to the tensions arising from incoherent water rights claims.

Water pricing policy is vague in the Philippines. Seemingly, water is not treated as a commodity. There are several schemes for subsidies, which reveal that Philippine policy treats water as a social good and should be accessible to all as a human right. The pricing policy covers payment for the delivery and other maintenance services but does not include a definite bulk water price. From this, it is inferred that, since water is an abundant resource, then it should be free. Increasingly, water transfers, competition for multiple uses, and payments for environmental services are emerging issues as water scarcity becomes real.

12.2.3 Sectoral Issues: Key Findings

12.2.3.1 Domestic Water

Low domestic water service coverage and high tariffs are observed among water districts. Water districts with level III piped-in water connections only service less than 50% of the households. The remaining households either self-provide or are served by the informal sector such as community-based systems or LGUs that are neither registered nor regulated by state authorities. Water district tariffs are high because all their capital expenditure requirements are funded from loans.

LGU-operated systems have the worst performance among all the utilities benchmarked. Water provision is simply politically motivated, thus, no emphasis is made on skills development, professional buildup, or financial sustainability. Most LGU systems are not ring-fenced; hence, revenue is not linked to expenses. Most LGU-run systems have operations and maintenance costs heavily funded by the LGU internal revenue allotment. Level II or level III water system infrastructure is also mostly provided under the national water agencies' water programs or through Congressional funds. Dole-out mentality practices still exist. LGUs have also shown little interest in pursuing water supply projects due to leadership uncertainties brought about by the 3-year electoral term.

Water cooperatives need a lot of technical and financial support as the Cooperative Development Agency provides only administrative support. Community-based-organizations do not have access to commercial funds for expansion. They also need a lot of technical and financial support as no national agency effectively assists them, even if, on paper, they are technically under the supervision of LWUA.

Financing packages tend to cater only to credit-worthy utilities with no concessional financing for the non-credit worthy or those that are still in the process of becoming credit worthy. Subsidy policies for LGU systems are not strictly implemented.

12.2.3.2 Industry Sector

The industry sector wastewater is not being strictly monitored. Based on case studies presented in Chap. 5, aside from silver, lead, and nitric acid contaminants, water samples in public areas around one area studied showed contaminations of copper and chromium, while plant samples showed the presence of manganese, lead, and copper. In samples taken from pilot sites, cadmium, nickel, zinc, and manganese were present and exceeded the limits set by the Philippine drinking water standards. Heavy metals were found in fish and shellfish samples, exceeding the limits set by the Bureau of Fisheries and Aquatic Resources.

There is no ambient water quality data collected from agencies that monitor industry sector water quality. It was observed that data are being gathered regularly, but these have not been encoded and analyzed for regulatory purposes.

Policy instruments also exist to mitigate the behavior of firms not to pollute. As cited earlier, an "environmental user's fee" is imposed on establishments that discharge their wastewater into Laguna Lake. In spite of this, the Laguna lake ecosystem, in general, is in a sorry state.

On a bright note, on the other hand, there are available technologies in the country that convert wastes into beneficial assets. One of these is the agro-recycling of biodigester effluent through land application. Aside from nitrogen, phosphorus, and potassium, the bio digester effluent contains a wide array of nutrients such as calcium, magnesium, sodium, chloride, copper, iron, manganese, and zinc, which are essential for plant growth and development.

12.2.3.3 Agriculture Water Management

In spite of the huge irrigation investments, data showed that the rate of irrigation system deterioration is faster than the establishment of new systems. The deterioration of the systems, coupled with reduced surface water for big dams due to both high sedimentation and watershed degradation, lead to water scarcity, resulting in a decline in irrigation water supply and a reduction in irrigation service area. This water scarcity can also be attributed to several factors beyond the control of the National Irrigation Administration (NIA), such as sectoral water allocation (priority for domestic use over irrigation in cases of water crisis) and climate change. On the latter, reduced precipitation or overly long dry spells have resulted in serious deficit in surface water, on which many irrigation systems depend.

The decreasing dependable flow of surface water sources and the inability to control sedimentation are exacerbated by the prevalence of damaged or dilapidated dams, headworks, and control structures. Some of these are caused by typhoons, but others are simply due to faulty design and poor operation and maintenance.

Irrigation water pricing does not aim to recover the full cost of building the dam but to be able to provide for operations and maintenance (O&M). Like pricing for domestic water, irrigation fees also do not reflect the bulk price of water, only conveyance. The rate of payment per hectare cultivated, rather than actual use, is enshrined in a national law that is difficult to amend. Intermittently, the Philippine president declares waivers of irrigation fees for farmers (as did former President Estrada and current President Duterte), thus making the national government absorb all O&M costs. The impact of this policy decisions on farmer productivity and income remains to be seen.

Discharge of both domestic and agricultural wastewater and agricultural run-off has caused extensive pollution of receiving water bodies such as rivers and lakes. Domestic wastewater contains raw sewage and detergents; agricultural wastewater has fertilizers and chemicals; and industrial wastewater can have heavy metals and oils.

12.2.3.4 Aquaculture

Just like in the industry sector, worsening water quality is a serious concern for aquaculture. In Metro Manila alone, nine river sub-basins are used as dump sites. Almost 90% of wastewater being discharged in water bodies (rivers, lakes, and the sea) have little or no primary treatment. The loss in fishery production due to pollution has been estimated to be PhP 17 billion (World Bank 2003).

In spite of the presence of legal frameworks and institutions, there is evidence that points to the operation of informal systems and networks for land and water access/use. Typically, households or communities closer to water bodies with secure rights to use the water and land resources have better chances of setting up cages or ponds and generating income from aquaculture than households who do not have such rights. Several communities in the coastal areas have traditional rights to use coastal waters, which could be formalized to favor the entry of the poor into aquaculture. Traditional rights to use water were upheld by virtue of their profession and/or residence that was closer to the water bodies or by formal permits from the local government. The local government regulates the number of fish pens and cages that can maintain the carrying capacities of the lake and other bodies of water. But the number of fish pens/cages and feeds/feeding practices in lakes, rivers, and coastal waters has increased more than what is provided for in the law. The "carrying capacity" of such water bodies has been exploited, resulting in "fish kills."

12.2.3.5 Multiple/Integrated Water Resource Utilization

The primary issue in multiple/integrated water resource system is the absence of rules for water allocation and concrete policies for water use and water quality standards. In the country, very limited systems have agreed upon water allocation rules. This is not yet available in most multi-use systems, such as natural lakes or reservoirs. This is observed to cause conflicts. In addition, serious gaps in understanding the interactions of hydrology, geomorphology, and ecology in watersheds, rivers, wetlands, and estuaries inhibit effective water planning of multi-use systems.

12.2.3.6 Watersheds and Water

In a watershed from where freshwater supply commonly emanates, soil, climate, forest cover, land use, and land use practices largely influence the attributes of water. Environmental protection as provided for in various policies and programs is bound to positively impact water resources owing to the inherent connection of water with other natural resources and the environment in general. There are three major reasons for the limited positive impacts of environmental policies on water: asynchrony and disunity of policies, ineffective policy implementation, and inadequate infusion of science in policy formulation.

The perennial lack of interconnection of various environmental policies and programs is attributable primarily to the absence of institutional mechanisms to compel linkaging between policies and programs of various government agencies, including those agencies with closely related mandates and functions. Sector-based and commodity-based planning and budgeting system does not facilitate inter-sectoral and interagency policies and programs and perpetuates the silo mentality of various government agencies.

A common constraining factor is the limited and inefficient use of financial resources that are available for the proper implementation of environmental policies and programs. Associated with this constraint is the unavailability of adequate human resources and facilities that are necessary to enforce regulations and to execute program of action.

Many environmental policies in the country are at best founded on knowledge and information that are too general and are generated in areas under different circumstances. What is worse is that some of these policies are grounded on institutions shaped by partial knowledge or truths, myths, and personal prejudices that give birth to flawed assumptions and unrealistic expectations. These are consequences of lack of access to correct information and knowledge that could be due to either absence of locally specific knowledge and information or absence of mechanisms to bring available knowledge and information to policymakers.

12.3 Challenges

12.3.1 Need for Data for Water Planning

Lack of more updated and reliable data/information on water supply and demand is one of the major constraints that limit the effectiveness and efficiency of water management in the Philippines. Such information is crucial for policy and decision making both at the national and local levels to develop more appropriate and strategic policies and programs to advance water resource management and sustainability. Planning at all levels is hampered by lack of reliable data and the absence of a systematic and regular monitoring of sector activities at the LGU level. Unless the country can overcome this lack of a water data base, future generations will inherit a legacy of declining and degraded water resources that threaten their livelihoods and well-being, particularly the poorer segments of society.

12.3.2 Institutional Reforms and Political Capital to Carry Them Out

There is no centralized regulatory agency for the water and sanitation sector. The existing national regulator, the NWRB, does not have the authority and resources to do economic regulation on water and sewerage on all water service providers.

Some agencies have several functions that should not be housed in one agency—i.e., service provider and regulation, financing, supervision, and regulation. There are also no consequences for government utilities that perform poorly.

LGUs have no regulatory capacity, except on business permitting. LGU-run utilities are not required by any agency to submit regular reports. DILG, which exercises authority over LGU-run utilities, is unable to monitor the performance of LGU-run water utilities mainly because of lack of resources.

No one is clearly accountable for implementing water reforms–there is a multitude of agencies involved in the water sector. Each agency has its respective role in the sector and because the reform process cuts across the mandates of all agencies, they must all be involved. This means that all decisions are made by a committee, and the responsibility for implementation is often diluted. This lack of accountability for implementing reforms has affected most reform initiatives in the sector. For example, regarding reform proposals on sector financing, no single agency has had the responsibility of ensuring that financing is available to the sector and the mandate to make sure that policies are in place to make this a reality. The institutional arrangements and mechanisms to promote synergy among the various water-related agencies and overall efficiency in the water sector remains a challenge.

To effect further reforms on institutional arrangements would require enormous investment in political capital to bring not only agencies with entrenched bureaucracies that may not want their mandates challenged but also water service providers and local communities on board. Legitimacy or widespread acceptance of policy changes is a good starting point.

12.3.3 Sectoral Challenges

12.3.3.1 Domestic Water

The domestic water sector suffers from persistent low level III coverage, especially in rural areas, and very low investment toward sanitation systems. This is in large part due to the policy assigning the provision of water and sanitation systems to LGUs that have neither the funds nor the fiscal incentive to finance such an undertaking. Given the 3-year terms of local executives, local governments cannot be relied upon to provide satisfactory and sustainable service levels for water and sanitation as the project cycle of water and sanitation infrastructure usually takes longer than 3 years.

Tariff setting for LGU-run utilities tends also to be politically sensitive. As such, LGUs should be tasked with providing the services, but they should not be the service providers themselves, unless it is a last recourse. Using IRA funds for subsidizing water and sanitation operations does not motivate LGUs to improve service delivery and collect the proper tariffs. Most of the IRA funds should instead be used for capital expenditures.

There is a need to form a planning and monitoring body at the provincial level to oversee the water and sanitation sector within their boundaries. Having plans made at this local-government level allows for nuancing, taking into account the specificities of hydrological conditions. It also brings greater accountability to the water service provider performance and subsequent decisions on public investments or financing.

12.3.3.2 Industrial Water

The challenges related to water use policy for the industrial sector in the Philippines reveal further the need for improved water governance. The backyard smelting and other polluting industries around river systems demonstrate the necessity of establishing reliable baseline data on which subsequent monitoring of water pollution can be based. The mining companies' case illustrates the challenge of imposing legal and political remedies for violations of industrial standards to minimize water pollution and associated environmental destruction. On the other hand, agro-recycling of industrial wastewater presents the importance of tapping the support of a research institution in finding technical solutions to the wastewater problem. It also highlights the need to incentivize the reuse of water in the industrial sector and to scale up initiatives using economic instruments.

12.3.3.3 Agricultural Water Management

The very low dry-season irrigation intensity is due to design shortcomings at the headworks, including underestimation of flood flows and sediment loads, inadequate provisions for sediment control, and underestimation of reservoir inflow and outflow hydrographs. These problems are more evident in the case of the communal irrigation system, where most dams are already old, with exposed rock cores, damaged spillways, and silted storage area. Possibly due to limited funds, there is apparent neglect in the estimation of dependable flow and sediment discharge, relying on old design criteria or adopting design parameters from other systems.

NIA can also make irrigator associations (IAs) accountable for the rehabilitation of existing systems. A progressive transfer of rehabilitation responsibilities can be facilitated through an agreed-upon schedule of reduction in the share of costs.

With the expansion of IA roles and downloading of responsibilities from NIA to IAs, NIA can focus on the higher level role of supervising the devolution, managing the headwork (reservoirs, dams, and main canals), implementing volumetric charges to IAs at the head gate, and providing technical support to IAs. The IAs will take care of transferred assets, collect water fees to cover their O&M costs, and manage water efficiently and equitably. NIA has to provide financial support for asset rehabilitation, if not done before the transfer; and technical support for O&M. As a transition arrangement, the IAs may need to hire professional support. Sediment management control measures must be implemented to maintain at the minimum the sediment inflow scenarios.

12.3.3.4 Aquaculture

Unclear water rights for aquaculture result in extensive farming, even unto areas where no fish pond licenses and leases have been awarded. The number of fish pens/cages and feeds/feeding practices in lakes, rivers, and coastal waters should be regulated within the "carrying capacity" of such water bodies to prevent "fish kill." This would entail intergovernmental coordination and multi stakeholder efforts to come up with evidence-based indicators for lake/river/coastal health, as well as a monitoring system and enforcement mechanism that is efficacious.

There is a need to review the fees and basis for awarding fishpond leases and permits to consider the traditional rights of fishers and fishing communities and within shorter time periods that will allow for a more rational way of accessing water resources. Strengthening institutional mechanisms and sustaining multi sectoral participation in water quality management are essential to promote water quality management. Although the responsibility for monitoring may be delegated to provincial or local government agencies, support must be shared for maximum results.

Good aquaculture practices such as the use of settling ponds, recirculating water systems, and probiotics for preventing shrimp diseases and "self-pollution" of brackish water ponds should be applied. Integrated multi trophic aquaculture is an ecosystem approach in integrated marine aquaculture that can be explored.

There is an urgent need to address climate change concerns and their impacts on aquaculture, water resources, and the diverse communities that rely on them.

12.3.3.5 Multiple/Integrated Water Resource Utilization

For this multisector or integrated water resource utilization, the policy challenge is to develop a framework to be able to continuously monitor and periodically review and assess strategies to plan and manage major water resources. It should be capable of describing and capturing the hydrologic, geomorphologic, and ecologic interactions of these systems. The understanding is that, while a science-based, computerized DSS is needed when dealing with complex, large-scale, and dynamic water systems, this tool will not replace humans (i.e., stakeholders and actors) in the ultimate planning and management decision making.

For large-scale, complex, and dynamic multi-use water resource systems, managing these systems require an understanding of the physical system (natural processes, climate, weather), social system (societal, political, economic), and human system (cultural, behavioral, lifestyles). Since there are definitely various stakeholders as well as private and government institutions involved in the governance of such multi-use water systems, there is the need to strengthen the synergy and coordination of these various stakeholders and institutions to efficiently and effectively manage such water resource systems.

12.4 Prospects and Ways Forward

12.4.1 Science-Based Water Management

Water is a fluctuating resource, making it difficult to measure in time and in space. This means that coordinating and harmonizing data collection on both spatial and temporal scales is critical. As information needs may differ at local, regional, and global levels, indicators developed for one spatial scale may not be applicable to another (UNESCO 2006).

The mandate for monitoring requires collection of standardized data for surface water quality and safety, especially for drinking water. These data are currently lodged with the DENR and DOH, respectively. There are also data collected on erosion, land movement (slides), and sedimentation in rivers currently undertaken by the DENR Mining and Geosciences Bureau in select localities, but they have not been analyzed in conjunction with their effects on water flows. Spatial-scale data collection (i.e., river basins, watersheds, lakes) has also been noted in some areas. Water-related concerns require long-term horizons. Meeting future water demand in light of climate change supply projections and pollution scenarios necessitates longitudinal data.

To this end, there is a need for an inventory, assessment, consolidation, and standardization of monitoring systems across all government agencies with built-in long-term horizons (e.g., 50–100 years), complemented by an integrated knowledge management system accessible to researchers, planners, and decision makers. Landform, aquatic, geologic, socioeconomic, and political databases (most especially local government ordinances) at applicable local spatial scale must be brought together to inform research and as bases for crafting complementary water resource plans.

Science-based data must also inform the valuation of water rights in the form of licenses and permits. To the extent that the rights to develop, extract, and distribute are determined by these permits and licenses, they must accurately reflect the current or future projected state of the resource. Much of the value of current permits/licenses do not reflect environmental services; penalties for violations of effluent standards are ridiculously outdated and do not include costs for cleanup or health impact. For example, extensive fish farming from underpriced long-term fishpond licenses and leases have resulted in fish kills; industrial and commercial discharge of untreated water into river systems have rendered many biologically dead; and uncontrolled putting up of fish cages obstructs fish migration. Without sufficient science-based data, water rights could not be an effective carrot-and-stick tool to direct behavior.

12.4.2 Governance Mechanisms

The framing of water challenges in terms of governance has allowed a broadening of the water agenda. The scrutiny of corruption, democratization processes, and power imbalances between both rich and poor countries and between rich and poor people is increasingly being accepted. Indeed, governance and politics are increasingly viewed as a part of the problem and therefore an essential part of any solution to water crises. There is enough water for everyone. The problem we face today is largely one of governance: equitably sharing this water while ensuring the sustainability of natural ecosystems. At this point in time, the Philippines has not yet achieved this balance (UNESCO 2009).

Philippine water policy suffers from lack of coherence between environmental, development, and social equity goals. The fragmentation of mandates on water among various national agencies with distinct goals (environmental preservation, agricultural productivity, population health, among others); and between levels of government (national and local) makes governance particularly challenging. A key governance reform needed is the creation of a national agenda for water and sanitation that dovetails all three goals and commits the state, regardless of changes in political-partisan leadership, to a package of policies and institutional changes. Central to these reforms is the reconstitution of the NWRB with stronger regulatory reach over all water organizations (water districts, local government-run utilities, and community-based organizations). While improvements in water district performance have been realized through the use of benchmarking tools tied to financial incentives, there is a need to extend this regulatory ambit to other types of water organizations. The rebooted NWRB must also serve as a permanent planning platform for public expenditures and investments and alternative financing on water projects. Beyond the current interagency framework chaired by NEDA, there is a need to open dialogues and decision making to those in private and non-profit sectors that have convergent interests.

The reconstituted National Water Resources Agency should have regional presence for monitoring, to be complemented by provincial-government-level monitoring of municipal water systems. In domestic water, LGUs should be discouraged as service provider; their role can be directed toward improving satisfactory and sustainable domestic water by other providers by way of supervision and policy guidance. In the same vein, NIA's role should also be streamlined toward supervising and managing headwork, setting volumetric charges at head gate and providing technical support to IAs. The move toward participatory and inclusive governance must proceed with government actors providing broad direction and steering and away from sector capture and dependency. Alternative provisioning by government in domestic, irrigation, or multi-use systems should be explored. In irrigation, public-private partnership (PPP) contractual arrangements covering management or financing or both is encouraged. Along these lines, the development of large-scale water resource systems (dams, lakes, reservoirs, river works) could be treated as public works with multi-century horizons, but which actual distribution could admit more participation by local stakeholders with productivity, food sustainability, and social policy as entry points. To bridge national policy and local preferences, decisions on water must be a locally devolved process, giving IAs and community-based domestic water organizations (e.g., barangay and rural waterworks) the ability to adjust tariffs in line with their poverty situation.

Regulation and monitoring could only be realized with permanent science and technology support. These tasks could be given to regional water resource centers housed within local academic institutions with built-in expertise. These water resource centers can serve as data collectors and repositories. It should also have the capacity to acquire real-time data through satellite imagery and cellular network.

Reliance on the punitive provisions of the law, in the light of serious administrative shortcomings, is not expected to generate desirable effects on water demand. Rather, a combination of market incentives (i.e., promoting devices and technologies to decrease water use and manage losses; grey water reuse and recycling) and non-price programs for inculcating "water-friendly" behavior is needed. To this end, information, education, and communication materials intended for water conservation and recycling should be based on the adaptive capacity of society as a resource.

Given these shortcomings, reforms in water policy should be directed in these areas: (1) strengthened NWRB in the form of a Water Department with robust regulatory, coordinative, oversight, conflict resolution, and overall planning capabilities for all watershed-based use and allocation decisions to meet broad national goals; (2) polycentric governance arrangements that admit informal actors, indigenous and customary rules, and conflict-resolution mechanisms, alongside those prescribed by law; (3) clarity and local grounding of state property rights to water, but with accommodation of communal and private rights claims; (4) institutional arrangements for water governance based on the physical connectivity of surface water resource (upstream, midstream, downstream) and between water, land, and forest resources; (5) substantive decentralization that puts meaningful decision making on water concerns at the local level, through inclusive and participatory platforms with government, the private sector, and civil society; and (6) science and empirics-based decision making through the institutionalization of research centers that generate accurate water data.

Water pricing as a policy instrument for water allocation in the Philippines needs a lot of improvement. Within the domestic sector, water price varies from block pricing practiced by the water districts to free water delivered by community-based water service providers. Across sectors, agriculture is the highest consumer of water. Irrigation service fee is currently proposed to be scrapped, in effect making this a free commodity. Water pricing is a complex issue because of the perception that water is not only an economic good but also a public good. Access to water is a human right, therefore governments need to ensure that this is so. Water is also an integral part of the ecosystem as a natural resource. Given these, the Philippines should be able to design a water pricing policy across sectors that will entice more efficient use by consumers and make it a self-sustaining enterprise for the water supply industry.

Furthermore, payment of environmental services as a policy instrument to pay the upland stakeholders for the conservation measures they need to conduct so there will be sustainable water supply in the lowlands was piloted in some areas of the country. Its replication to other areas will be determined by the availability of institutions that will pursue this and the availability of data that will be used in the valuation process, as resource valuation is contingent upon the community's own perception of values. This means that planning for water use should seriously consider the watershed as a unit of planning.

12.4.3 Overcoming and Adapting to Driving Forces

It is crucial to overcome or adapt to the different driving forces such as demographic factors, urbanization, land use change, and climate change for the future generations not to inherit a legacy of declining and poor quality water resources. Incidentally, many of these factors are outside the water resource sector and hence are not being taken into account by existing water policies and governance mechanisms. Meeting the water sustainability challenge requires an understanding of the complex and interrelated drivers of unsustainability and an acceptance of high-quality water as a human right (Schnoor 2015). It also requires a holistic and integrated approach that balances the economic, social, and environmental goals in the pursuit of sustainable water resources.

Reversing the tide of declining and degraded water resources to serve domestic and industrial sectors, agriculture, and aquaculture means implementing policies and the corresponding policy instruments in an integrated manner instead of the usual piecemeal approach. It also requires putting investments into human capacities in terms of regular monitoring of water bodies and being strict with sanctions, especially for polluters. In water planning, there is some sense in using the watershed as the unit of analysis.

12.4.4 Integrated Watershed Management

Forest cover loss in many watersheds in the country has been severe (Cruz et al. 2010). The ratio of forest cover to irrigated and irrigable lands is generally quite low, and this could have serious implications on soil erosion and the availability and quality of water for irrigation, domestic, and power requirements. The fight toward ensuring water sustainability in the country is therefore largely hinged on the sustainable management of watersheds, which currently are under intense anthropogenic pressure.

It is crucial to adopt an integrated approach to watershed management at the local level to address multiple demands from the watershed by the different stakeholders. This requires collective efforts from different stakeholders who should be engaged in an iterative learning process of integrated situation analysis, planning, implementation, and monitoring and evaluation, guided by an agreed upon watershed vision. It also requires a new mode of collaboration that allows breaking of disciplinary and administrative boundaries, political affiliations, and institutional mandates and promotes sharing of responsibilities, resources, and accountabilities toward the common goal of pursuing water sustainability. Realizing this requires a local champion who could catalyze the process either from the LGUs, civil society, or the national government like the DENR.

References

Cruz, R. V. O., Pulhin, J. M., & Mendoza, M. D. (2010). *Reinventing CFNR: Leading the way in integrated tropical forest and natural resource management education, research and governance (2011–2035).* Centennial Professorial Lecture, University of the Philippines Los Baños, January to December 2010.

JICA (Japan International Cooperation Agency)/NWRB (National Water Resources Board). (1998). *Master plan study on water resources management in the Republic of the Philippines.* Tokyo: Nippon Koei Ltd./Nippon Jogesuido Sekkei, Ltd.. 527 p.

Schnoor, J. L. (2015). Water unsustainability. *Dædalus, Journal of the American Academy of Arts & Sciences, 144*(3), 1–11.

UNESCO (United Nations Educational, Scientific and Cultural Organization) (2006). UN world water development report 2: Water: A shared responsibility. http://www.unesco.org/water/wwap/wwdr/wwdr2. Accessed 20 Apr 2017.

UNESCO (United Nations Educational, Scientific and Cultural Organization) (2009). UN world water development report 3: Water in a changing world and Facing the challenges. http://www.unesco.org/new/en/natural-sciences/environment/water/wwap/wwdr/wwdr3-2009/downloads-wwdr3. Accessed 20 April 2017.

World Bank (2003). *Philippines – Environment monitor 2003.* Washington, DC: World Bank. http://documents.worldbank.org/curated/en/144581468776089600/Philippines-Environment-monitor-2003, Accessed 24 May 2017.

Dr. Agnes C. Rola is full Professor at the University of the Philippines Los Baños (UPLB), former Dean of the College of Public Affairs and Development, UPLB, and member of the National Academy of Science and Technology- Philippines. She has degrees in Statistics (BS) and Agricultural Economics (MS) from the UP; and PhD in Agricultural Economics (Major in Natural Resource Economics) from the University of Wisconsin Madison, USA. She attended the Summer Certificate on Environmental Leadership Program at the University of California-Berkeley and has more than 20 years' research experience in sustainable agriculture at the watershed level with a research focus on water governance. With colleagues, she has written and edited an award winning book, "Winning the water wars: watersheds, water policies and water institutions" (2004), whose recommendations were adopted in the Philippines' Clean Water Act. For the past four years, she led two major research programs on water in the Philippines, namely, water governance for development and water security under climate risks.

Dr. Rosalie Arcala Hall is a Full Professor at the University of the Philippines Visayas (UPV). She completed her Master's degree in Political Science and PhD in International and Public Affairs (2002) at Northeastern University, Boston, Massachusetts. She completed research projects with The Asia Foundation, The Nippon Foundation and Toyota Foundation concerning policy frameworks, their localization, resistance and competing narratives behind contestations in the areas of hard and soft security. She has contributed chapters to the books Rosalina Palanca-Tan ed. Nature and Culture: Environmental Issues in Asia (Ateneo de Manila Press, 2014), and Justine Vaz and Narumol Aphineves eds. Living Landscapes Connected Communities: Culture, Environment and Change Across Asia (Areca Books, 2014). She is co-author to two articles in Water Policy and International Journal of Water Resources Development. Currently, she is a member of the Philippine Commission on Higher Education Technical Committee on Political Science and the Philippine Political Science Association Board.

Dr. Juan M. Pulhin is full Professor and former Dean of the College of Forestry and Natural Re-sources, University of the Philippines Los Baños (UPLB). He earned his Bachelor of Science and Master of Science in Forestry degrees in UPLB and Ph.D. degree in Geographical Sciences from The Australian National University. He was a Visiting Professor at The University of Tokyo for four times and has more than 30 years of experience in natural resources education, research and development. He has authored more than 100 technical publications on various aspects of natural resources management and climate change. He was a Coordinating Lead Author of the Intergovernmental Panel on Climate Change (IPCC) Fifth Assessment Report and a Lead Author of the Fourth Assessment Report. He has been involved in an interdisciplinary research project on water governance for development of the University of the Philippines and in numerous watershed development planning projects since 1998.

Index

A.C. Rola et al. (eds.), *Water Policy in the Philippines*, Global Issues in Water Policy 8, https://doi.org/10.1007/978-3-319-70969-7

Printed by Printforce, the Netherlands